Lecture Notes in Physics

For information about Vols. 1–90, please contact your bookseller or Springer-Verlag.

Vol. 91: Computing Methods in Applied Sciences and Engineering, 1977, II. Proceedings, 1977. Edited by R. Glowinski and J. L. Lions. VI, 359 pages. 1979.

Vol. 92: Nuclear Interactions. Proceedings, 1978. Edited by B. A. Robson. XXIV, 507 pages. 1979.

Vol. 93: Stochastic Behavior in Classical and Quantum Hamiltonian Systems. Proceedings, 1977. Edited by G. Casati and J. Ford. VI, 375 pages. 1979.

Vol. 94: Group Theoretical Methods in Physics. Proceedings, 1978. Edited by W. Beiglböck, A. Böhm and E. Takasugi. XIII, 540 pages. 1979.

Vol. 95: Quasi One-Dimensional Conductors I. Proceedings, 1978. Edited by S. Barišić, A. Bjeliš, J. R. Cooper and B. Leontič. X, 371 pages. 1979.

Vol. 96: Quasi One-Dimensional Conductors II. Proceedings 1978. Edited by S. Barišić, A. Bjeliš, J. R. Cooper and B. Leontić. XII, 461 pages. 1979.

Vol. 97: Hughston, Twistors and Particles. VIII, 153 pages. 1979.

Vol. 98: Nonlinear Problems in Theoretical Physics. Proceedings, 1978. Edited by A. F. Rañada. X, 216 pages. 1979.

Vol. 99: M. Drieschner, Voraussage – Wahrscheinlichkeit – Objekt. XI, 308 Seiten. 1979.

Vol. 100: Einstein Symposion Berlin. Proceedings 1979. Edited by H. Nelkowski et al. VIII, 550 pages. 1979.

Vol. 101: A. Martin-Löf, Statistical Mechanics and the Foundations of Thermodynamics. V, 120 pages. 1979.

Vol. 102: H. Hora, Nonlinear Plasma Dynamics at Laser Irradiation. VIII, 242 pages. 1979.

Vol. 103: P. A. Martin, Modèles en Mécanique Statistique des Processus Irréversibles. IV, 134 pages. 1979.

Vol. 104: Dynamical Critical Phenomena and Related Topics. Proceedings, 1979. Edited by Ch. P. Enz. XII, 390 pages. 1979.

Vol. 105: Dynamics and Instability of Fluid Interfaces. Proceedings, 1978. Edited by T. S. Sørensen. V, 315 pages. 1979.

Vol. 106: Feynman Path Integrals, Proceedings, 1978. Edited by S. Albeverio et al. XI, 451 pages. 1979.

Vol. 107: J. Kijowski, W. M. Tulczyjew, A Symplectic Framework for Field Theories. IV, 257 pages. 1979.

Vol. 108: Nuclear Physics with Electromagnetic Interactions. Proceedings, 1979. Edited by H. Arenhövel and D. Drechsel. IX, 509 pages. 1979.

Vol. 109: Physics of the Expanding Universe. Proceedings, 1978. Edited by M. Demiański. V, 210 pages. 1979.

Vol. 110: D. A. Park, Classical Dynamics and Its Quantum Analogues. VIII, 339 pages. 1979.

Vol. 111: H.-J. Schmidt, Axiomatic Characterization of Physical Geometry. V, 163 pages. 1979.

Vol. 112: Imaging Processes and Coherence in Physics. Proceedings, 1979. Edited by M. Schlenker et al. XIX, 577 pages. 1980.

Vol. 113: Recent Advances in the Quantum Theory of Polymers. Proceedings 1979. Edited by J.-M. André et al. V, 306 pages. 1980.

Vol. 114: Stellar Turbulence. Proceedings, 1979. Edited by D. F. Gray and J. L. Linsky. IX, 308 pages. 1980.

Vol. 115: Modern Trends in the Theory of Condensed Matter. Proceedings, 1979. Edited by A. Pekalski and J. A. Przystawa. IX, 597 pages. 1980.

Vol. 116: Mathematical Problems in Theoretical Physics. Proceedings, 1979. Edited by K. Osterwalder. VIII, 412 pages. 1980.

Vol. 117: Deep-Inelastic and Fusion Reactions with Heavy Ions. Proceedings, 1979. Edited by W. von Oertzen. XIII, 394 pages. 1980.

Vol. 118: Quantum Chromodynamics. Proceedings, 1979. Edited by J. L. Alonso and R. Tarrach. IX, 424 pages. 1980.

Vol. 119: Nuclear Spectroscopy. Proceedings, 1979. Edited by G. F. Bertsch and D. Kurath. VII, 250 pages. 1980.

Vol. 120: Nonlinear Evolution Equations and Dynamical Systems. Proceedings, 1979. Edited by M. Boiti, F. Pempinelli and G. Soliani. VI, 368 pages. 1980.

Vol. 121: F. W. Wiegel, Fluid Flow Through Porous Macromolecular Systems. V, 102 pages. 1980.

Vol. 122: New Developments in Semiconductor Physics. Proceedings, 1979. Edited by F. Beleznay et al. V, 276 pages. 1980.

Vol. 123: D. H. Mayer, The Ruelle-Araki Transfer Operator in Classical Statistical Mechanics. VIII, 154 pages. 1980.

Vol. 124: Gravitational Radiation, Collapsed Objects and Exact Solutions. Proceedings, 1979. Edited by C. Edwards. VI, 487 pages. 1980.

Vol. 125: Nonradial and Nonlinear Stellar Pulsation. Proceedings, 1980. Edited by H. A. Hill and W. A. Dziembowski. VIII, 497 pages. 1980.

Vol. 126: Complex Analysis, Microlocal Calculus and Relativistic Quantum Theory. Proceedings, 1979. Edited by D. Iagolnitzer. VIII, 502 pages. 1980.

Vol. 127: E. Sanchez-Palencia, Non-Homogeneous Media and Vibration Theory. IX, 398 pages. 1980.

Vol. 128: Neutron Spin Echo. Proceedings, 1979. Edited by F. Mezei. VI, 253 pages. 1980.

Vol. 129: Geometrical and Topological Methods in Gauge Theories. Proceedings, 1979. Edited by J. Harnad and S. Shnider. VIII, 155 pages. 1980.

Vol. 130: Mathematical Methods and Applications of Scattering Theory. Proceedings, 1979. Edited by J. A. DeSanto, A. W. Sáenz and W. W. Zachary. XIII, 331 pages. 1980.

Vol. 131: H. C. Fogedby, Theoretical Aspects of Mainly Low Dimensional Magnetic Systems. XI, 163 pages. 1980.

Vol. 132: Systems Far from Equilibrium. Proceedings, 1980. Edited by L. Garrido. XV, 403 pages. 1980.

Vol. 133: Narrow Gap Semiconductors Physics and Applications. Proceedings, 1979. Edited by W. Zawadzki. X, 572 pages. 1980.

Vol. 134: $\gamma\gamma$ Collisions. Proceedings, 1980. Edited by G. Cochard and P. Kessler. XIII, 400 pages. 1980.

Vol. 135: Group Theoretical Methods in Physics. Proceedings, 1980. Edited by K. B. Wolf. XXVI, 629 pages. 1980.

Lecture Notes in Physics

Edited by H. Araki, Kyoto, J. Ehlers, München, K. Hepp, Zürich
R. Kippenhahn, München, H. A. Weidenmüller, Heidelberg
and J. Zittartz, Köln

175

Defect Complexes in Semiconductor Structures

Proceedings of the International School
Held in Mátrafüred, Hungary
September 13–17, 1982

Edited by J. Giber, F. Beleznay, I. C. Szép, and J. László

Springer-Verlag
Berlin Heidelberg GmbH 1983

Editors

J. Giber
J. László
Physical Institute of the Technical University
H-1111 Budapest, Hungary

F. Beleznay
I.C. Szép
Research Institute for Technical Physics
of the Hungarian Academy of Sciences
H-1325 Budapest, Hungary

Sponsored by: Hungarian Academy of Sciences
Roland Eötvös Physical Society

Organized by: Research Institute for Technical Physics of the Hungarian Academy of Sciences

Supported by: Microelectronics Company
United Incandescent Lamp
and Electrical Co., Ltd., Tungsram

ISBN 978-3-540-11986-9 ISBN 978-3-540-39456-3 (eBook)
DOI 10.1007/978-3-540-39456-3

Originally published by Springer-Verlag Berlin Heidelberg New York in 1983

2153/3140-543210

PREFACE

This book is a compilation of the lectures delivered at the International School on Defect Complexes in Semiconductor Structures held in Mátrafüred, Hungary, from September 13 to 17, 1982.

The school was attended by 105 experts from 15 countries and thus offered ample proof of the actuality and importance of the topic. In addition to the 16 invited lectures, 17 papers were presented and 30 posters placed on exhibit. These proceedings contain the most interesting of the contributed papers, some graphic material adapted from the posters, and, of course, the invited papers. The novelty value of the papers presented was the principal criterion for selection.

The propagation of semiconductor devices in microelectronics has yielded a development emphasized in the material of this school. The review papers give the reader an overview of recent results from investigations of the defects occurring in semiconductor structures, as well as of applied methods and problems which remain to be solved. The contributed papers provide the reader with some details about specific problems.

For the most part the volume is organized along the same lines as the sessions of the school. It begins with defect diagnostics and the results of theoretical examinations, and then covers the concrete defects which develop in silicon and compound semiconductors. Investigations of the Si/SiO_2 interface are also discussed. The last part deals with features of semiconductor devices which are changed by implantation, annealing and irradiation. The book concludes with a summary prepared on the basis of a round-table discussion held at the close of the school. The questions and answers which emerged offered new insights and pointed ahead to challenging problems.

The material included in these proceedings was selected and reviewed by the editorial board. The editors are grateful to Dr. L. Zsoldos and Dr. G. Ferenczi for their assistance, and to Mrs. I. Bársony, Mr. Z. Soha, and Mrs. Z. Pákh for their tireless efforts in the technical editing of the manuscript.

The editors hope that this volume will help those unable to attend the school to become familiar with the subjects of this branch of science. The references should also serve as a learning aid for beginners in the field.

Budapest, Hungary J. Giber, F. Beleznay, I.C. Szép, J. László
October 1982

CONTENTS

A TECHNOLOGIST'S VIEW ON DEFECTS

Iván C. Szép

Research Institute for Technical Physics
of the Hungarian Academy of Sciences
H-1325 Budapest, Újpest 1, P.O.Box 76, Hungary

Organizing Committe Chairman

This introductory talk is meant to be rather a headmaster's opening address at the beginning of the new school term. Because this convention was declared by the organizers to be a *school*, an introduction of this kind may seem appropriate.

It is customary that headmasters - in their opening speech dwell longer on past events and glory than on future prospects, applying the teaching of the past to the tasks of the future. If you do not mind I shall follow their example.

Having spent half of my life in the realms of semiconductor technology I have gathered a lot of practical experience. The wisdom collected during these years had boiled down to the following maxim:

> "Life is full of defects.
> And so are semiconductors!"
>
> The Ancient Technologist

Looking back to the milestones of semiconductors technology from the heights of our present knowledge one really admires *that* sort of persistence by which we have been after semiconductors with ever better structural and compositional perfection, higher degree of homogeneity on one side and the demand for single or multiple sharp, step-like heterogeneity on the other side. Having reached such goals as for instance practically dislocation-free single crystals, almost ideally abrupt p-n junctions produced by ion-implantation or atomically clean surfaces it seems now that we have to a great extent misunderstood the behaviour of solid matter by imposing such rules on Nature which she apparently is unwilling to obey. We have to comply with the situtation that *Nature abhors perfection* because perpetual motion can exist only due to local irregularities and imperfections. The heroic strife of scientists and technologists to produce perfect semiconducting materials and structures by now has left us with the lesson that:

1.) Defects are a constant attribute of solid matter;
2.) There is a steady increase in time in the variety of defects;
3.) The number and state of defects is governed by the laws of thermo-

dynamics including the law of mass action;

4.) Hence suppressing or removing one type of a defect may bring to light or strengthen the role of an other type of defect;

5.) Semiconductors should be treated therefore as dynamic systems of defects, created and influenced by various stages of technological processing.

It is a practical experience that the removal of *primary structural* defects (grain boundaries, misfits, dislocations, etc.) leads to reinforced activity of *compositional* (generally: point) *defects* in semiconductors. In elemental semiconductors substitutional impurities may control the level of interstitials, again their removal by purification may strengthen the interaction between residual impurities (segregation of precipitates), leading to associations and defect complexes. Striations, swirls, microprecipitates are examples of such behaviour. More will follow in the lectures.

Compound semiconductors composed of two, three or more constituents represent a formidable assembly of complex defects. One meaning of the word "complex", quoted from Webster's Dictionary: a group of obviously related units of which the degree and nature of the relationship is imperfectly known. This enlightens fairly well the degree of our understanding complex defects today. The growing importance of compound semiconductors underlines the necessity to investigate and identify the properties and the ways to control these defects.

Talking about defects in semiconductors let me stress my point that *these* are the primary constituents responsible for device properties; the ordered or even disordered atomic framework of the material itself, due to the presence of atoms in excess by six-seven orders of magnitude can be regarded as a continuum.

Any such a continuum with periodic or statistically homogeneous distribution of atoms is completely uninteresting from a device point of view, unless we introduce a defect, e.g. carve out a finite region and surround it with a *surface*.

I am sure technologists will agree with me that a *semiconductor surface* is one of the most wicked kind of defect accumulation, which needed many thousand hours of strenous work to get it under control. Of course, without this no MOS or MIS devices would exist. Unfortunately, to get something under control does not mean that we have a detailed and undisputed explanation for all properties connected to the surface.

In addition to this so called *external surface* technologists are frequently confronted with *internal surfaces*.Grain boundaries, slip, twin and dislocation planes, p-n or hetero-junction boundaries are a two-dimensional interruption in the continuum, i.e. a defect complex of specific properties. The last two of them nobody would take for harmful.

Another flourishing group of defects is registered under the heading of *point defects*. Vacancies, interstitials, foreign atoms- single or compounded- belong to this group. These are natural constituents of every solid system, because *Nature abhors not only perfection but also purity*(Remember entropy of mixing!). It seems that at present this group of defects is of attracting growing interest. New and ingenious methods are devised and applied in many research institutions to gain insight into the intricate behaviour of imperfections such as site exchange in compounds, differently charged ionic states of impurities, especially of transition metal-ions, bound states of hydrogen, oxygen and other gases, etc. Again a number of examples will follow in the program. Because of the more complex electronic structure of compound semiconductors the interaction with defects manifests itself in new phenomena as controlled light emission of different wavelength, optical waveguiding etc. It is certain that by controlled introduction or development of specific defect complexes new effects will be discovered based on interaction of different forms of energy, such as light, acoustical, magnetic, etc. This may lead to new device structures and broader selection of semiconductor devices. We should bear in mind that 98 % of the volume of semiconductor devices, which are today produced all over the world, are based on p-n junction properties - a very limited utilization of semiconductor features. Bulk and also size effects have been insufficiently exploited. Work on super-lattices and submicron sized heterostructures is pointing to such possibilities.

The necessity of highly perfect single crystals is diminishing as we arrive to scaled-down VLSI structures, where even the smallest diffusion length of minority carriers or a 1/3 value of bulk mobility is acceptable, due to submicron base-width of bipolar or analogous gate length in MIS devices, respectively. Deposited semiconductor layers of rather poor crystalline perfection, with remnants of ion-implantation damage, are functioning satisfactorily. Layers of hydrogenated amorphous silicon, deposited by plasma discharge or wafers of polysilicon cast in moulds after proper processing show sufficient photovoltage to be used for cost-effective solar cells.

These new developments underline once again the importance to understand and control the properties of defects in semiconductor device structures, using the expression "defect" in its broadest sense of meaning. Economic pressure may bring us to the point, where we shall have to think it over whether it is sound to use expensive single crystal wafers, in which only 1 % of the thickness is performing electronic functions, the rest is mechanical support. It may turn out that it is more profitable to use glass or ceramic support carrier sheets with deposited thin recrystallized silicon or other semiconductor. The success of such technology will depend on the degree by which the residual defects in play are under control.

I have to point out that these trends are by no means absolutely new, they rather indicate a return to old practise, of course with a higher level of understanding and elaborate technique. Just to demonstrate how all this has started, I shall show you some historic pictures.

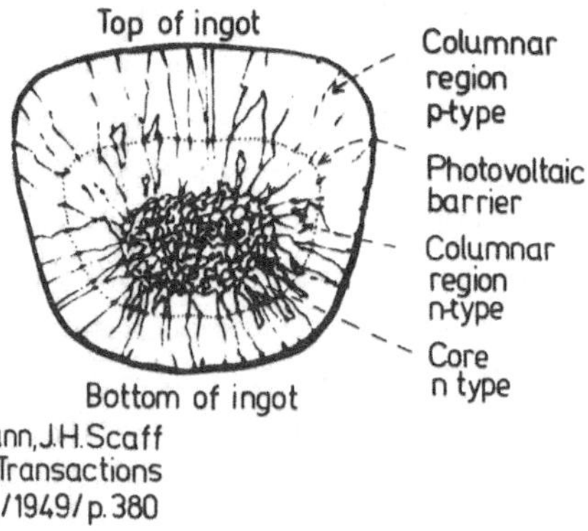

Fig. 1.

Figure 1 is the polished and etched cross-section of a silicon ingot, solidified from a melt containing boron and phosphorus. This drawing was published by Pfann and Scaff in 1949 [1], taken for clarity from the photo to indicate the site of the photovoltaic barrier. This was the first time a p-n junction was deliberately produced. The apparent polycrystallinity of the ingot seems to have little influence on the shape of the p-n junction which develops due to difference in segregation coefficients of the acceptor and donor impurity. For the correct functioning of injection-type p-n junction devices, as invented by Shockley, a flat boundary was a must, in junction transistors even two close, parallel p-n boundaries were needed, one for the emitter-base and one for the base-collector junction. To achieve a high current amplification the lifetime of injected carriers had to be sufficiently high, so anything promoting their early recombination was undesirable. So began the single crystal era, with all the cutting, slicing, lapping, polishing and etching operations, resulting in 50 to 60 % loss of the starting weight of an ingot.

Of course, this was not just as simple. On Fig. 2 Part A may look like a snail, but really it is the outline of a germanium single crystal, grown by Gordon Teal in Bell Labs at the end of 1949, as it appears in an early 1950 patent specification [2]. In the claims it is called to be a single crystal with one or two slip boundaries at the tapering end. Part B is a photograph of an etched section of a

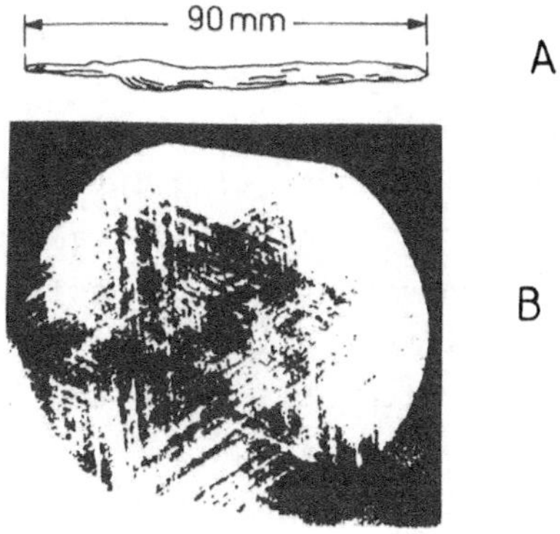

Fig. 2.

Czochralski grown silicon crystal of mere 50 g weight, which was badly deformed during cooling after growth. The density of etch pits indicates the density of dislocations intersecting the surface of the section. This picture appeared in the paper of Bennett and Sawyer in the 1956 volume of Bell System Technical Journal [3].

Since then we have learnt to grow silicon single crystal weighting 30-40 kg's, almost to completely remove the dislocations, discovering instead the presence of vacancy clusters, swirls (which persist because of the absence of dislocations), struggle with compositional inhomogeneities called striations, which arise due to kinetics of the crystallization process, wrangle with dissolved gases, which react with silicon to yield precipitates of oxides, carbides, nitrides, etc. You may see impressive pictures about all this in the following lectures.

Thermal treatment of single crystal wafers they undergo during various steps of processing gives rise to changes and transformations, leading to appearence of stacking faults, vacancy clusters, segregated phases, etc.

Deposited layers of silicon, produced by molecular beam epitaxy or by plasma decomposition of gaseous silicon compounds are much less liable to such contamination, by appropriate treatments (e.g. laser annealing) their crystalline perfection can be raised almost to the level of pulled single crystals. In early 50's few people were acquainted how to grow silicon single crystals. Hydrogen reduction of $SiCl_4$ was sometimes used to deposit elemental silicon on graphite studs (early attempts for epitaxy!). However, because of impure materials these layers were of inferior quality and could be used only for the purpose of point-contact detectors. Today the deposition technique has reached a high level and in a number of fields gradually will replace the classical technology based on bulk single crystals.

With computer-aided-manufacturing methods this thin-film technology will afford the construction of complete sub-systems and systems on a common carrier substrate, utilizing the broadest variety of solid-state properties, not just those of semiconductors. Now, to end this report, let me explain why the Organizing Committee has decided to begin the curriculum by the presentation of a technologist's view on defects.

In our electronic age the trend of continuous break-up of scientific disciplines keeps on. Narrow and highly specialized sub-divisions appear with frontiers guarded to prevent infiltration of unfamiliar thoughts and alien problems. To that purpose usually a special lingo is employed mostly unintelligible to outsiders. Without going into details such is the well known situation in the field of the so called *basic sciences*. Needless to say this is very convenient for the scientific individual because he is able for himself to follow and understand developments in his particular discipline without looking to others.

Just the opposite seems to be true in the field of *engineering sciences*. Although subdivision into smaller and smaller chapters follows in the same way as with basic sciences, however for a successful mastering of any engineering technology a vast knowledge from *various* scientific and engineering disciplines is required. Semiconductor technologist is the best example for this case. Now it is clear that anybody whom we call "technologist", is in a desperate position when he tries to solve his problems alone. Being the last person in the specialists' queue he becomes a sort of scapegoat for the faults committed mostly by others. Crying for help he may seldom find assistance from the camps of aforementioned narrow-subdivided specialists. Barricades, unfamiliar language and sometimes the fear not to get soiled by becoming involved in technology prevent prompt and effective cooperation of the respective parties.

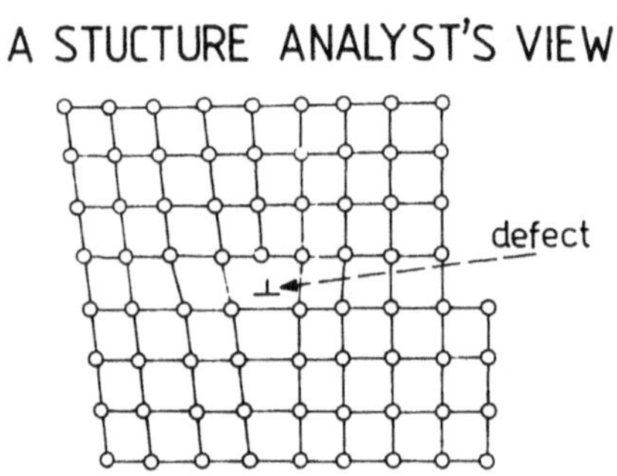

Fig. 3.

Well, take for instance the expression "defect" once again. Speaking to a *structure analyst* about defects, a lot of concepts may emerge in this brain, but mostly in a way as represented on Fig. 3.

If you approach a *solid-state physicist* with your defect problem he will rush to the blackboard to show you his view on defects (Fig. 4).

Now, the Organizing Committee sticked to the idea that these and also other views can be reconciled to mutual benefit of all parties involved, with the result that victimized technolgists will be able to better cope with the problem of defects, acquiring the necessary and sufficient amount of intelligible knowledge from other disciplines. I am sure we shall have a

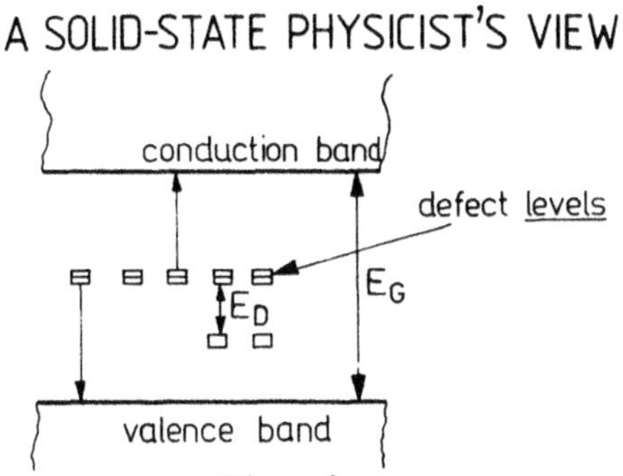

Fig. 4.

Fig. 5.

favourable atmosphere during our school to do this, the views of technologists on defects will change and possibly we will save them from such nightmares as shown on Fig. 5.

References

1. W.G. Pfann, J.H. Scaff, Metals Transactions 185, 389 in J. Metals 1 (1949 June)
2. Brit. Patent Specification No. 706849
3. D.C. Bennett, B. Sawyer, Bell System Techn. J. 35, 637 (1956)

PART I

DEFECT DIAGNOSTICS

CHARACTERIZATION OF IMPURITIES AND DEFECTS BY ELECTRON PARAMAGNETIC RESONANCE AND RELATED TECHNIQUES*

James W. Corbett, Richard L. Kleinhenz and You Zhi-pu+

Institute for the Study of Defects in Solids
Physics Department, SUNY/Albany
Albany, New York, 12222 USA

+Department of Physics, Sichuan University,
Chengdu, China

ABSTRACT

The use of electron paramagnetic resonance (EPR and ENDOR) in the study of defects in semiconductors is briefly reviewed, including group IV (C-diamond, Si, Ge, SiC), III-V (AlSb, GaAs, GaSb, GaP, InAs, InP, InSb), II-VI (BaO, BaS, BeO, CaO, CaS, CaSe, CdO, CdS, CdSe, CdTe, MgO, SrO, SrS, ZnC, ZnS, ZnSe, ZnTe) and miscellaneous systems. The identification of defects via EPR is described as is the exploitation of that identification as a tool in future studies. Particular attention is paid to Si. In the category of defect identification in silicon there is emerging an integrated panorama of identified defects ranging from point defects to their complexes and aggregates to so-called intermediate defect configurations on to dislocations and stacking faults. In addition impurities and their interactions are discussed, including mention of a possible origin of the so-called oxygen shallow donor and of the understanding of hydrogen in silicon. As examples of EPR studies as a tool in defect studies the concentration dependence of the phosphorus resonance is described from the regime of isolated defects to the metallic regime. The ENDOR results on defects in silicon are discussed at some length, and it is emphasized that even for the shallow donors there is far from a satisfactory theoretical treatment of the data, and that the situation is even worse for other defects in silicon.

I. Introduction

Electron paramagnetic resonance (EPR) is just one of the many forms of spectroscopy being used to study defects in semiconductors. All spectroscopies contribute to defect studies in two ways: 1) by aiding in the identification of the defect; 2) by studying the properties of the defect. We will survey here EPR studies in semiconductors and see examples of how they have contributed to defect identification and the investigation of defect properties. In the limited space of such an article we cannot give a comprehensive account of the detailed work that EPR studies have yielded. We will try, however, to give an accurate perspective of the past and potential contributions of such studies.

Central to the question of defect identification is the establishment of the defect symmetry. This symmetry may be probed by a spectral

*Supported in part by the Office of Naval Research Contract No. N000-14-75-C0919

variation with respect to crystal orientation; EPR is a form of Zeeman spectroscopy and hence the spectral variation is studied vs magnetic field as it is oriented with respect to the crystal.

For an isolated electron in free space the interaction energy (E) of the electron's magnetic moment ($\underline{\mu}$)

$$\underline{\mu}_e = - g \beta \underline{s} \quad (1)$$

with

$$\beta = eh/(2mc), \\ = 9.2732\text{x}10^{-21} \text{ erg/gauss}, \quad (2)$$

$\underline{s}$ the electron spin and g(=2.0023) a scalar determined by the Dirac theory of the electron, with a magnetic field ($\underline{H}$) is given by

$$E_m = g\beta H\ m \quad (3)$$

with m the electron spin quantum number ($m = \pm 1/2$). The selection rule for a magnetic dipole transition is $\Delta m = \pm 1$. For a photon frequency ν the electron paramagnetic (or spin) resonance (EPR or ESR) condition is given by

$$h\nu = g\beta H, \quad (4)$$

i.e., for g = 2 and H = 3300 gauss, ν = 10,000 MHz - the X-band microwave range. An electron on an atom can experience an interaction with the nucleus of that atom. For example, for a hydrogen atoms, the proton has a nuclear spin ($\underline{I}$) of 1/2 and a magnetic moment ($\underline{\mu}_N$) given by

$$\mu_N = g_N \beta_N \underline{I} \quad (5)$$

where the nuclear magneton (β_N)is given by

$$\beta_N = |e|h/(2M_p c) \quad (6)$$

with M_p the mass of the proton,

$$\beta_N = \frac{m}{M_p} \beta \cong \frac{1}{1836} \beta \quad (7)$$

$$= 5.0505 \times 10^{-24} \text{ erg/gauss},$$

i.e., three orders of magnitude smaller than the Bohr magneton. In the high field limit one can treat the nuclear interaction as a perturbation on the widely split electron energy levels with the perturbation described by a hyperfine constant A in the coupling of $\underline{I}$ to $\underline{s}$, i.e., $A\underline{I}\cdot\underline{s}$; now the energy levels are given by

$$E(m,M) = g\beta Hm + AMm \tag{8}$$

with M the nuclear spin quantum number. Nuclear magnetic resonance (NMR) transitions occur for $\Delta M = \pm 1/2$; for hydrogen A = 506 gauss and the transitions are in the radio frequency range. In the presence of the nuclear interaction the electron transitions are split into two transitions ($\Delta m = \pm 1/2$, $\Delta M = 0$) separated by A. Feher [1] introduced the powerful technique of Electron-Nuclear-Double Resonance (ENDOR) in which one monitors the perturbation in the EPR as one achieves the NMR transition; this technique permits the sensitive and unambigous identification of nuclei associated with the electron transition being observed and has given rise to a number of multiple transition techniques (see, for example, ref. [2]), including the detection of magnetic resonance by monitoring an optical transition (ODMR), recently reviewed by Cavenett [3]. For a simple paramagnetic ion similarly experiencing a magnetic field in free space the energy levels (E_M) are given by

$$E_M = g\beta HM \tag{9}$$

with M the projection of the total angular momentum $\underline{J}$ vector along $\underline{H}$. Now the g factor is given by the Landé expression:

$$g = 1 + [J(J+1)+S(S+1)-L(L+1)]/[2J(J+1)] \tag{10}$$

with $\underline{S}$ and $\underline{L}$ the electron spin and orbital angular momentum vectors. A more complex ion can have additional energy contributions due to the nuclear spin (I) and quadrupole interactions.

The free electron, the hydrogen atom and a paramagnetic ion in free space all have their analogs in semiconductors which were observed at the very beginning of EPR in semiconductors; the first report of EPR in a semiconductor was that of a paramagnetic manganese ion in powdered ZnS by Schneider and England [4], followed soon by the observation of conduction electron spin resonance (CESR) in silicon by Portis et al. [5] and of electrons bound at the hydrogen-like, shallow donors in silicon by Fletcher et al. [6]. The resonances in solids are potentially richer in information and concomitantly more complex than those in free space in two primary aspects; first, the resonant electron may be delocalized, partially residing on many atoms, and, second, the symmetry experienced by the resonant electron will be lower than that of free space. The word "potentially" was chosen advisedly for these, and other, complications in solids may make the resonance unobservable! Like other forms of spectroscopy, EPR provides a limited "window" with which to view defects; if the defects are not visible through that window the spectroscopy is no help, but if it is visible, EPR may provide extensive information about the atomic and electronic configuration of the defect.

The resonant electron experiences a variety of interactions in the solid and these are described by tensors, e.g., a g-tensor or A-tensor, rather than scalars in foregoing equations. For many resonances, the g-tensor describes the positions of the resonance lines with respect to the orientation of the magnetic field with respect to the crystal axes, i.e., commonly the g-tensor describes the overall symmetry of the resonant electron. If that symmetry is cubic (O_h = m3m) (the point group symmetry of the conduction electron in the diamond lattice which has the space group label O_h^7 = Fd3m) or tetrahedral ((T_d = $\bar{4}$3m), the highest point group symmetry in the diamond, or the zincblende (T_d^2 = F$\bar{4}$3m)) structure, then the g-tensor becomes a scalar and the resonance is isotropic, i.e., independent of crystal orientation. If the symmetry is lower than cubic in these lattices then there will be more than one line and more than one equivalent defect in the lattice. Consider a defect with an axial g-tensor along a <100> axis. For that defect the resonance line will be given by $g_{\parallel}$ when the magnetic field is along *its* <100>, say [100], and by $g_{\perp}$ when the field is perpendicular to that axis. But when that defect has the magnetic field aligned along the [100] direction, defects of the same type but with g = [010] and [001] will give rise to a resonance line twice as large as the one at g (= [100]) line at the $g_{\perp}$ position. The equivalence of these defects is referred to as orientational degeneracy and that degeneracy may be "lifted" by the magnetic field as just described.

Similarly the hyperfine interaction (A) can be an anisotropic tensor. The Hamiltonian for the dipolar interaction, H_d, between $\underline{\mu}_e$ and μ_N separated by the vector $\underline{r}$ is given [7] by

$$H_d = \left\{ \frac{\underline{\mu}_e \cdot \mu_N}{r^3} - 3\,\frac{(\underline{\mu}_e \cdot r)(\underline{\mu}_N \cdot r)}{r^5} \right\} - 4\pi\delta(r)\underline{\mu}_e \cdot \underline{\mu}_N \qquad (11)$$

with $\delta(r)$ the Dirac delta function. The hyperfine energy due to the last term in Eq.(11) is due solely to the s-state component of the electron wave function at the nucleus in question; it contributes an isotropic contribution to the hyperfine tensor and in the case of the ground state of the hydrogen atom is all of A. The bracketed terms in Eq.(11) give non-zero values for p-, d-, etc., functions and give rise to an anisotropic contribution to A.

As mentioned earlier the hyperfine and additional interactions can be a severe disadvantage since the resultant lines may be so broadened as to be effectively unobservable. The overall situation may be seen in Table I where we list the isotopes making up the several semiconductors that we consider, the isotopic abundances, their spins, magnetic moments and quadrupole moments. For silicon we see that the isotopes Si^{28} and Si^{30} have spin zero and so contribute nothing in the way of spectral splittings; the remaining isotope (Si^{29}) has spin 1/2 so that it creates the simplest of splittings and its abundance (4.7%) is just about ideal to provide a dilute label on lattice sites. In the III-V compounds on the other hand we find the lattice constituents often have several magnetically active isotopes with large spins which split the lines into many sub-lines with strong (and overlapping) anisotropies; little wonder that the lines (when observed) in III-V compounds tend to be very broad. As we have indicated earlier, however, when the resonances are observable, these interactions can provide a luxury of information; for example, as we will discuss more fully below, ENDOR studies in silicon have permitted mapping the electron wave-function over thiry *shells* of atoms surrounding the shallow donors in silicon, an abundance of information not fully digested by theory.

An abundance of EPR data is available in semiconductors - too much to present in detail here. We will present a survey of this data in the next section and in the following section will present (briefly still!) more details on the work in silicon to give a better perspective of the potential depth of the contribution of EPR studies.

Table I Relevant properties of elements [8]

Isotope	Natural abundance [%]	Spin	Magnetic moment [μN]	Electric quadrupole moment [$10^{-24}cm^2$]
Al^{27}	100	5/2	+3.6414	+0.15
As^{75}	100	3/2	+1.439	+0.29
B^{10}	19.78	3	+1.8007	±0.08
B^{11}	80.22	3/2	+2.6885	+0.04
Ba^{132}	0.097	-	-	-
Ba^{134}	2.42	0	0	0
Ba^{135}	2.59	3/2	+0.8365	+0.18
Ba^{136}	7.81	-	-	-
Ba^{137}	11.32	3/2	+0.9357	+0.28
Ba^{138}	71.66	0	0	0
Be^{9}	100	3/2	-1.1776	+0.05
C^{12}	98.89	0	0	0
C^{13}	1.11	1/2	+0.7024	-
Ca^{40}	96.97	0	0	0
Ca^{42}	0.64	-	-	-
Ca^{43}	0.145	7/2	-1.317	-
Ca^{44}	2.06	-	-	-
Ca^{46}	0.0033	-	-	-
Ca^{48}	0.18	-	-	-
Cd^{106}	1.22	-	-	-
Cd^{108}	0.88	0	0	0
Cd^{110}	12.39	0	0	0
Cd^{111}	12.75	1/2	-0.5943	-
Cd^{112}	24.07	0	0	0
Cd^{113}	12.26	1/2	-0.6217	-
Cd^{114}	28.86	0	0	0
Cd^{116}	7.58	0	0	0
Ga^{69}	60.4	3/2	+2.016	+0.19
Ga^{71}	39.6	3/2	+2.562	+0.12
Ge^{70}	20.53	0	0	0
Ge^{72}	27.43	0	0	0
Ge^{73}	7.76	9/2	-0.8792	-0.22
Ge^{74}	36.54	0	0	0
Ge^{76}	7.76	0	0	0
In^{113}	4.28	9/2	+5.523	+0.82
In^{115}	95.72	9/2	+5.534	+0.83
Mg^{24}	78.70	0	0	0
Mg^{25}	10.13	5/2	-0.8553	+0.22
Mg^{26}	11.17	0	0	0
O^{16}	99.759	0	0	0
O^{17}	0.037	5/2	-1.8937	-0.026
O^{18}	0.204	0	0	0
P^{31}	100	1/2	+1.1317	-
S^{32}	95.0	0	0	0
S^{33}	0.76	3/2	+0.6433	-0.055

Table I (Continued)

Isotope	Natural abundance [%]	Spin	Magnetic moment [μN]	Electric quadrupole moment $[10^{-24}cm^2]$
S^{34}	4.22	0	0	0
S^{36}	0.014	0	0	0
Sb^{121}	57.25	5/2	+3.359	-0.29
Sb^{123}	42.75	7/2	+2.547	-0.37
Se^{74}	0.87	0	0	0
Se^{76}	9.02	0	0	0
Se^{77}	7.58	1/2	+0.534	-
Se^{78}	23.52	0	0	0
Se^{80}	49.82	0	0	0
Si^{28}	92.21	0	-	-
Si^{29}	4.70	1/2	-0.5533	-
Si^{30}	3.09	0	-	-
Sr^{84}	0.56	-	-	-
Sr^{86}	9.86	0	0	0
Sr^{87}	7.02	9/2	-1.093	+0.36
Sr^{88}	82.56	0	0	0
Te^{120}	0.089	-	-	-
Te^{122}	2.46	-	-	-
Te^{123}	0.87	1/2	-0.7359	-
Te^{124}	4.61	-	-	-
Te^{125}	6.99	2/2	-0.8871	-
Te^{126}	18.71	0	0	0
Te^{128}	31.79	0	0	0
Te^{130}	34.48	0	0	0
Zn^{64}	48.89	0	0	0
Zn^{66}	27.81	0	0	0
Zn^{67}	4.11	5/2	+0.8755	+0.17
Zn^{68}	18.57	0	0	0
Zn^{70}	0.62	-	-	-

II. Survey of EPR Studies in Semiconductors

In this section we present lists of defects for which an at least tentative identification exists in the EPR literature. Certainly there are defects which have been well identified without the help of EPR, many of which are inaccessible to EPR because they do not involve an unpaired spin. For example, the oxygen interstitial in silicon has a well-established configuration, arrived at by studies of the vibrational absorption lines [9-13]. As a further example, one should realize that the defects which we list represent a specific charge state of the defect; many of the defects have several charge states in the forbidden gap and typically a charge state change converts a paramagnetic state to a non-paramagnetic state (or vice-versa) and the latter is inaccessible to EPR.

In Tables II, III and IV we list the defects for the group II-V, III-V and IV semiconductors [14]. In addition to the information in Tables II-IV, there is some information in the literature on EPR studies in other semiconductor system: As, As_2S_3, As_2Se_3, B, BN, $Cd_xHg_{1-x}Se$, $Cd_xHg_{1-x}Te$, $CdIn_2S_4$, $Cd_xMg_{1-x}Te$, $CdS_{1-x}Se_x$, $CdSiP_2$, GaSe, a-(for amorphous)Ge-Al, a-Ge-Cu, a-Ge-S-Ag, a-Ge-S, a-Ge-Te-Si, HgSe, HgTe, In_2Te_3, PbTe, $Pb_xSn_{1-x}Te$, Se, SiTe, SnTe, Te, a-Te-As-Ge-S, $Zn_xHg_{1-x}Se$, $Zn_xHg_{1-x}Te$, $ZnSiAs_2$, $ZnSiP_2$.

What is included as a "semiconductor" is somewhat arbitrary. We have included work on diamond, with its ~5.5 eV forbidden gap, because it can under some circumstances be a semiconductor; we include $Pb_xSn_{1-x}Te$, which may have a zero-energy band gap, because it is termed a "small-band gap semiconductor". We include purely covalent systems, highly ionic systems and systems with a wide variety of lattice structures (and in the amorphous cases, no "lattice" at all!). We include them because the technique unifies the systems, and we expect that the physics of defects in the different systems may have common features (as do semiconductor defects with defects in purely ionic systems and even metals [463]).

Table II Defects with identified EPR spectra in II-VI compounds in the literature [14]. In each system there may be additional spectra due to unidentified defects. V_X denotes a vacancy at an X site; X_{INT} - the element X in an interstitial site; X - the element (X) in a substitutional site; (X.Y) - an association of X and Y; CESR - conduction electron spin resonance; VHSR - valence hole spin resonance; a-X, Surface - a resonance attributed to the amorphous X and to the surface, resp.; etc.

BaO

V_O
Gd, Mn

BaS

V_S

BeO

V_{Be}, V_O
Al, B, Cu, F, Li
B_{INT}

CaO

V_{Ca}, V_O; $(V_O \cdot V_O)$; $(V_O \cdot V_{Ca} \cdot V_O)$
O_{INT}
Ag, Ce, Co, Cr, Cu, Eu, F, Fe, Gd, Ir, K, Li, Mg, Mn, Mo
Na, Nd, Ni, Pb, Ti, Tm, V, Yb
(Li•H), Li•Li), (OH), $(V_O \cdot Mo)$, (Mn•Mn)
Surface

CaS

Eu, Mn, Sn, Yb
Surface

CaSe

Pb, Sn

CdO

Cu, Ni
$(V_{Ca} \cdot V_O$ + OH)
CESR

CdS V_{Cd}, V_S
Ag, Br, Cl, Co, Cr, Cu, Dy, Er, Eu, Fe, Ga, Gd, Ge, I
Mn, Ni, Nd, Pb, Sn, Ti, Tm, V, Yb
(Fe•Ag), (Fe•Cu), (Fe•Li), (Gd•?), (N•N)
CESR, Surface

Table II (Continued)

CdSe

V_{Cd}
Co, Eu, Fe, Gd, Mn, Nd, Sn, Ti, V
CESR, Surface

CdTe

V_{Cd}
Ag, Cr, Co, Cu, Er, Eu, Fe, Gd, Li, Mn, Nd,
Ti, Tl, Tm, Yb
$(Fe \cdot F_6)$ $(Yb \cdot Ag_3)$ $(Yb \cdot Ag_4)$ $(Yb \cdot Au_3)$ $(Yb \cdot Au_4)$ $(Yb \cdot Li)$
CESR

MgO

V_{Mg}, V_O; $(V_{Mg}\ V_O)$; $(V_O\ V_{Mg}\ V_O)$
O_{INT}
Ag, Al, Co, Cr, Cu, Dy, Er, Eu, F, Fe, Gd, H, Ir,
Li, Mn, Na, Ni, Pd, Rh, Ru, Sn, V, Yb
$(V_{Mg} \cdot Al)$, $(V_{Mg} \cdot Cr)$, $(V_{Mg} \cdot F)$, $(V_{Mg} \cdot Fe)$, $(V_{Mg} \cdot Li)$, $(V_{Mg} \cdot Mn)$,
$(V_{Mg} \cdot Na)$ $(V_{Mg} \cdot OH)$, (OH), (Cr•Cr), (Mn•Mn), (V•V)
Surface

SrO

V_{Sr}, V_O
Ag, Ce, Co, Cr, Cr, Cu, Er, Eu, F, Fe, Gd, K, Li, Mn, Na, Ni, V
$(V_{Sr} \cdot OH)$, (OH)
Surface

SrS

Mn

ZnO

V_{Zn}, V_O; $(V_{Zn}\ V_{Zn})$
Zn_{INT}, O_{INT}
Al, Cl, Co, Cr, Cu, Fe, Ga, Gd, In, Li, Mn, Nb
Ni, Pb, Sn, V, Yb
(Cu•H), (Cu•OH), (Fe•Ag), (Fe•Cu), (Fe•Li), (OH)
CESR, Surface

ZnS

V_{Zn}, V_S
Al, Br, Co, Cr, Cu, Dy, Er, Eu, F, Fe, Ga, Gd, Ge,
I, In, Mn, Nd, Ni, Pb, Sc, Si, Sn, Tl, Ti, V, Yb
$(Y_{Zn} \cdot Al)$, $(V_{Zn} \cdot Br)$, $(V_{Zn} \cdot Cl)$, $(V_{Zn} \cdot Cu)$, $(V_{Zn} \cdot Fe)$
$(V_{Zn} \cdot Ga)$, $(V_{Zn} \cdot In)$, (Ag•Fe), (Ag•Ga), Cu•Fe), (Cu•Ga),
(Fe•Li)

CESR, Surface

Table II (Continued)

<u>ZnSe</u>

V_{Zn}; $(V_{Zn}\cdot Zn_{INT})$
As, Co, Cr, Dy, Er, Eu, Fe, Ge, Mn, Ni, P,
Pb, Si, Sn, V, Yb
$(V_{Zn}\cdot Te)$; (Ag•Fe), (As•Fe), (Cu•Fe), (Fe•Li), (P_4)
Surface

<u>ZnTe</u>

Co, Cr, Er, Fe, Ge, Mn, Pb, V, Yb
(Ag•Fe), (Cu•Fe), (Fe•Li)

<u>Table III</u> Defects with identified EPR spectra in III-V compounds in the literature [14]. (See Table II for explanation of symbols.)

<u>AlSb</u>

Surface

<u>GaAs</u>

Cd, Co, Cd, Cr, Cu, Fe, Mn, Ni, S, Si, Te, Ti, V, Zn
(Mn•O), As_{Ga}
CESR, Surface, a-GaAs

<u>GaSb</u>

CESR, Surface

<u>GaP</u>

V_{Ga}
P_{Ga}
C, Cd, Co, Cr, Fe, Ge, Mn, N, Ni, O, S, Se, Si, Sn, Te, Zn
(Cd•O), (Li_3) (Li•Mn), (N_{INT}) (Mn•S)
Surface

<u>InAs</u>

Fe
CESR, Surface

<u>InP</u>

Co, Cr, Fe, Ni
$(Fe\cdot In_{INT})$
CESR

<u>InSb</u>

CESR, Surface

Table IV Defects with identified EPR spectra in IV systems in the literature [14]. (See Table II for explanation of symbols.)

C (Diamond)

V

C_{INT}

B, Co, N, Ni

(V•N), $(V{\cdot}N_2)$, $(V{\cdot}N_3)$, $(V_2{\cdot}N_4)$, (N_2), (N_3), (N•Al), (N•Si)

Surface, a-C

Silicon	References [ ]
V	90,98,244,341,392,417,420,423,424,426,427,434
V_2	23,24,32,56,57,59,60,79,84,89,94,102,135,137,152,249,257, 274,288,315,320,366,371,372,373,378,411,415,419,420,423, 424,434,440
V_3	247,249,252
V_4	38,49,59,60,63,64,65,88,89,124,135,151,152,178,247,249,252, 288,290,315,467
V_5	38,49,59,63,88,248,249,252,287,290,315,329,317,414
V•O	17,32,55,56,63-66,89,130,133,178,233,234,254,261,274,287, 320,406,407,412,415-417,423,427,440
$V{\cdot}O_2$	412
$V{\cdot}O_3$	412
$V_2{\cdot}O$	255,257,261
$V_2{\cdot}O_2$	63,255,257,261
$V_3{\cdot}O$	49,63,255,257,261
$V_3{\cdot}O_2$	49,63,255,257,261
$V_3{\cdot}O_3$	255
V•Al	420,412,423,424
V•As	423
V•B	432
V•Ge	392,424,427
V•Li	147
V•P	17,40,64,88-90,178,247,274,370,415-418,420,423,427
V•Sb	423
V•Sn	392,428,429
V•C•O	21,260-262,371,422
$V{\cdot}P_2$	370
$V_2{\cdot}P$	247
Al	44,45,59,128,212,274,332,361,453
As	22,47,48,61,75,76,85,106,116-118,120,126,161-166,203,206, 207,209,211,213,274,294,295,337,338,361,382,383,387,456, 458,459,465-468

Table IV (Continued)

Silicon (Cont.)

Au	191,224,263,374
B	36,44,45,59,118,119,121,128,212,266,274,321,328,451,453,462
Bi	75,76,274
Co	42,169,254,259,274-276
Cr	22,72,92,121,126,167-170,273,274,380,443,445
Cu	169,273,274,445
Fe	22,41,43,91,118,121,125,126,167-170,199,259,268,269,272-276, 293,296,380,404,405,436,437,443
Ga	44,45,274,361,453
In	44,274,361,453
Li	22,47,118,121,147,189,190,192,274,320,327,384,425,439
Mg	37
Mn	22,77,121,126,167-170,270,273-276,443,445
N	70,71
Ni	126,273,274
O	118,121
P	1,6,22,26,27,29,30,33,36,47,64,65,75,76,80,81,85,86,89,102, 106,108,115,117,118,120,121,122,126,141-143,146,155,161-166, 181,182,183,195,196,197,200,203,204,206,207,209,211,213,215, 219,226-233,234,236,241,265-267,274,278-282,283-275,288,294, 295,301-306,309,316-318,321,322,333,334,337,338,340,344,348, 349,352,356-358,363,364,369,375,377,381-383,385,387,389, 393-399,402,403,409-411,433,441,444,453,456-461,465-468
Pd	121,274,380,446
Pt	121,184,274,380,446
S	117,277,363,370,380
Sb	6,22,47,48,61,75,76,114,117,120,121,161-166,203,206,207,209, 213,274,287,294,316,331,338,387,465-468
Te	505
V	22,121,126,169,266,267,273,274,443
Zn	145
$(Al)_{INT}$	62,424
$(B)_{INT}$	430
$(C)_{INT}$	67,258,431
$(C_2)_{INT}$	67
$(P)_{INT}$	247
$(Si)_{INT}$	434
$(Si_2)_{INT}$	68,246,250,256
(Al•Al)	424
(Al•Cr)	274
(Al•Mn)	274

Table IV (Continued)

Silicon (Cont.)	
(Au•Cr)	274
(Au•Fe)	191,224,263,374
(Au•Mn)	274
(B•Cr)	274
(B•Fe)	274
(B•Mn)	274
(Cr•Ga)	273,274
(Fe•Ga)	274
(Fe•In)	273,274
(Ga•Ga)	274
(Li•O)	174,274,327
(Li•O•Si_{INT})	148,149
(Li•Si_{INT})	147,148,149
(Mn_4)	274
(Mn•Pt)	274
(Mn•Zn)	274
(P•P)	370
CESR	5,108,109,127,138,151,152,158,160,174,175,177,201, 202,218,242,287,289,316,317,351,377,450
VHSR	46,127,218
Surface	16,17,73,74,78,83,118,132,137,171-173,178,185,186, 188,214-216,235,239,240,264,284,285,286,291,292, 297,298,330,331,347,367,368,386,412,413
Dislocation	18-30,35,36,58,156-159,189,190,215,282,296,353,388, 435,442,447
a-Si	15,28,31,32,38,51-54,56,59,64,65,82,97-101,105,107, 108,110-113,123,124,131,134-137,139,140,152,153, 154,173,174,175,176,178-181,187,215,220,222,232, 233,237,238,241,246,251,253,287-290,292,300,301, 307,308,310-313,315,316,318,323-325,335,336,355, 360,365-367,376,378,390,391,406-408,411,454,455, 467
Ge	As, Bi, Mn, Ni, P, Sb
	P_{IBT}
	(V•O)
	CESR, Surface, a-Ge
SiC	V_C, (V_C•V_C)
	Al, B, N
	(V_{Si})•B)
	Surface

It will be seen in Tables II, III and IV that there is a wide variation in the amount of information available, Si having the most, followed by MgO, etc. The disparities reflect a combination of factors (e.g., Is the system widely studied? Are high quality, suitably-doped, single crystals available?) besides the questions related to the ease of EPR studies discussed above. It will be seen that information on some impurity centers are available, less on intrinsic defects, such as vacancies and interstitials, and varying amounts on defect complexes. The tables give a perspective view of the past effort and probably of balance of effort and impact of EPR studies in the future. But we are optimistic about the opportunities for EPR work in III-V compounds, and fully persuaded about the importance of such work, though Table III clearly reflects the difficulty of EPR in these compounds, particularly in contrast to Tables II and IV.

We should include a warning about the immutability of the identifications listed in these tables, for the benefit of the non-expert (or perhaps even more as a reminder for the expert, for we all may be snared by feelings of priority, paternity and pride!). Any identification is like a "house of cards"; it is made up of a lot of bits and pieces of information which assemble into a (more or less) consistent whole, but which may collapse under the impact of "one-too-many" experiments. Some identifications are so buttressed that they might as well be cast in concrete, but not all parts of each edifice are as well established as others. Any assessment of the validity of a model would be subjective. Instead we simply say *caveat canem.*

It should also be noted that there are many EPR spectra in the literature for which even a tentative defect identification has not been advanced. No attempt was made to include such information.

The full perspective concerning the utility of EPR studies in semiconductors cannot be seen from the listings in these tables. We will try to amplify that perspective by considering in more depth the defects in silicon in the next section.

SILICON

At this point we will assume that the identifications shown are established and try to give a perspective of the uses of these identified defects (via EPR measurements) as tools in related studies. (In practive the separation of identification and use is rarely clean, one aiding the other.) There are many examples which could be given, however, space limits our treatment.

The first topic exploits the special nature of EPR spectra. It is the study [1,86,87,121,122,182,183,209,211,222,226-231,236,278-281, 387] of the EPR of shallow donors (say, phosphorus) in silicon as a function of phosphorus concentration. The phosphorus concentration ranges from ca. $10^{15}/cm^3$, in which case the donors are isolated in a semiconductor, up to ca. $10^{19}/cm^3$ in which regime the donors are strongly interacting in a metallic system. In the dilute regime the electron resides on the phosphorus atom which is 100% P^{31} (see Table I) with a nuclear spin 1/2 and consequently the electron resonance is split into two lines by this hyperfine interaction. As the phosphorus concentration is increased a new line [375] occurs mid-way between these lines which is due to the anti-ferromagnetic [209] exchange coupling between pairs of lines. In more concentrated samples there are additional lines due to transitions in clusters of three and four nearby donors [464]. In more concentrated samples these lines broaden and shift in ways which can be correlated [222,306,349] with the formation of an impurity band by the hopping of electrons from phosphorus to phosphorus, which process can be followed with increasing phosphorus concentration through the metal-insulator (or semiconductor) transition [470,471] (which occurs [349] at a donor concentration of $\sim 4 \times 10^{18}/cm^3$) on into the metallic regime. In fact the spectral behaviour is a complicated function of donor concentration, sample temperature, microwave measuring power, etc., and although the general features are established Si:P remains a simple model system in which the percolation of charge carriers among disordered centers may still fruitfully be studied.

The EPR, or more particularly, the ENDOR, of the shallow donors provides another aspect of EPR studies of semiconductors, namely a detailed mapping of the resonant-electron wave-function and a presisting challenge to the theory of defects in solids. There have been a number of papers [1,115-117,161-166,206,207,294,338,465-468] on the measurement of the ENDOR of the shallow substitutional donors (As, P and Sb) in silicon and on the interpretation of those measurements, since the first measurements by Feher [1,115-117]. He observed five distinct shells of atoms as distinguished by differing hyperfine interactions and he suggested an identification of five sets of lattice sites with these shells, his results based on the isotropic term in Eq.(11), and the variation in results vs. donor type. Hale and Mieher [161-163, 468] carried out even more comprehensive and improved measurements. They observed a total of twenty-two shells of atoms (ca. 150 lattice sites), with perhaps an additional twenty shells left potentially resolvable but untabulated [206]. They argued that one of Feher's identifications was erroneous but could provide only one additional match of lattice shell to ENDOR line. Hale and Mieher [165] then calculated the anisotropic term in Eq.(11) (the dipole-dipole interaction), but could provide no additional matches. Hale and Castner [164] studied the change in the hyperfine values of the various ENDOR lines in response to various uniaxial stress, defining piezohyperfine constants. They compared theory to their measurements and succeeded in matching an additional shell and suggesting several other matches.

The difficulty in matching more shells can be seen by considering the theory of the shallow donors, which is still essentially the effective mass theory (EMT) as given in detail by Kohn and Luttinger, although there are many papers [166,465,466,472-488] making substantial refinements in that theory. The attempts at matching shells by Feher, by Hale and Mieher, and by Hale and Castner, used as a wave-function for the resonant electron at $\underline{r}$

$$\psi(\underline{r}) = \left(\frac{1}{6}\right)^{1/2} \sum_{j=1}^{6} F_j(\underline{r})\psi_o(\underline{k}_{oj},\underline{r}) \qquad (12)$$

with the $\psi_o(\underline{k}_{oj},\underline{r})$ Bloch functions associated at the jth conduction band minima at $\underline{k}_{oj}$ along the <100>, or Δ, axes of the Brillouin zone, and the $F_j(\underline{r})$ are hydrogenic-like envelope functions derived from the EMT Hamiltonian. These solutions correspond to a specific value of the ionization energy (E) of the ground state of the donor and on the value of $\underline{k}_{oj}$. The hyperfine values ascribed to each shell in the lattice also depend on E and $\underline{k}_{oj}$ and the experimental uncertainty in E and in $\underline{k}_{oj}$ hampered the identification. Castner [465,466] attempted to improve the matching of shells to hyperfine data by admixing into the wave-functions components from the L, K and U points in the Brillouin zone, i.e., carrying out configuration mixing for the system, as well as including the lack of inversion symmetry inherent in T_d. Ning and Sah [481,482] included the valley-orbit interaction in the treatment of the problem calculating the energies of the ground state and some excited states and the isotropic hyperfine parameters. Neither of these works succeeded in assigning any new ENDOR shells.

A major step forward was made by Ivey and Mieher [206,207,468]. They used a model potential to represent the impurity potential, a wave-vector-dependent dielectric function to screen that potential and pseudo-potential, complex-valued Bloch functions to calculate the hyperfine and piezohyperfine constants. In solving for E they included not just the conduction band minima, but rather the four valence bands and the lowest four conduction bands and carried their numerical integration over the whole Brillouin zone, i.e., they extended the treatment a long way from the conceptionally simple EMT theory! They considered how the isotropic hyperfine constants varied versus the values of E and $\underline{k}_{oj}$. By this boot-strap procedure and, as have the others, by considering the variation vs. the type of donor atom, they matched twenty-two shells of atoms with isotropic hyperfine and the piezohyperfine values; they supported these identifications by calculating [468] the anisotropic hyperfine constants as well. Onffroy [338] used generalized Wannier functions to treat the shallow donors with a strong central-cell perturbation. She calculated the isotropic hyperfine value. She disagreed with some of the Ivey-Mieher identifications, and suggested several additonal identifications. She concluded that (with her approach) a wavefunction consisting of Bloch functions from the vicinity of the conduction band minima would suffice. Mieher [294] replied to the Onffroy paper disagreeing on a number of points. For our purposes it sufficies to emphasize (*as all would probably agree*) the comparison for the shallow, substitutional donors in silicon of theory to experiment (albeit the very penetrating and revealing ENDOR experiment) is simply unsatisfactory! And this for the (theoretically) simplest defects in semiconductors.

We can make our point as follows. The envelope-function given in Eq.(11) falls off smoothly versus distance from the donor but the wave-function does not, due to the modulation implied in $\underline{k}_{oj}$. This can be seen in Fig. 1 (after Sieverts [373]) where ψ^2 and the observed experimental values are simply arrayed versus $\underline{r}$ and in Fig. 2 (from Ivey-Mieher [207]) in which ψ^2 as a linear combination of pseudopotential Bloch functions for the six conduction band minima (<u>not</u> multiplied by an envelope function) is presented as a spatial contour map on which various lattice sites are identified. The strongest hyperfine value is agreed by all to correspond to the (0,0,4) site while the nearest neighbour to the donor (the (1,1,1) site) is much lower in the rankings (15th for As and 22nd for P and Sb). Ivey and Mieher, and Onffroy both present "figures of merit" for comparing their theoretical treatments to experiment. In Fig. 3 we show the ratio of the theoretical value of the isotropic hyperfine constant to the experimental values for the case of P and for EMT (as given by Ivey-Mieher (IM)), for Ivey-Mieher and for Onffroy (O); the average for each is also shown. In some cases the agreement is within 10%, but in many cases wildly off! And the same conclusion obtains for As and Sb. *One can simply not gloss over these problems in treating the theoretically unsimplest of defects in silicon.*

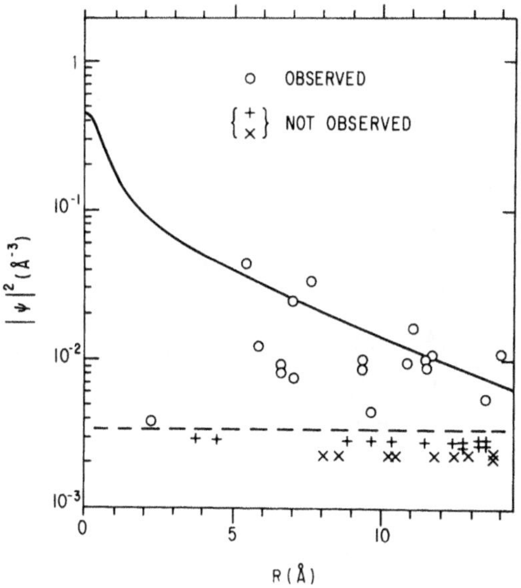

Fig. 1.
The resonant electron density versus distance from the phosphorus atom in silicon. The solid curve shows the envelope function (i.e., without interference between Bloch waves). (After E.G. Sieverts, ref.[373].)

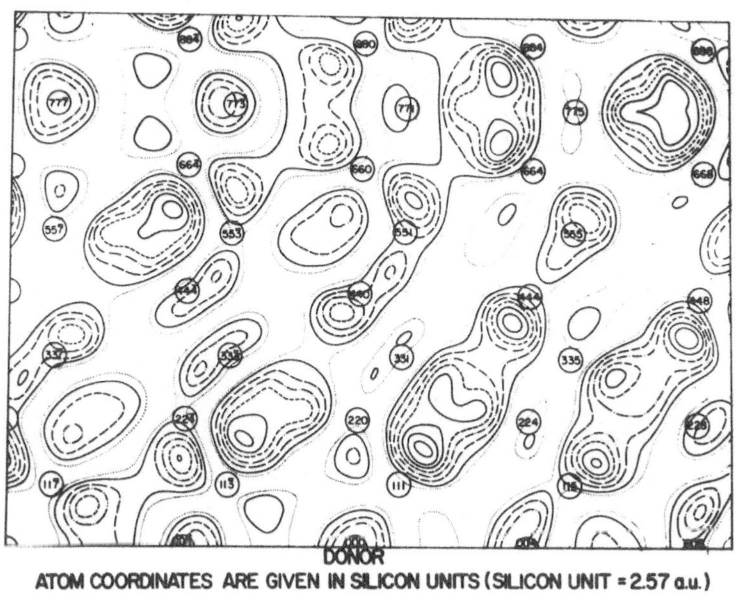

Fig. 2. Contour plot of the electron density for the linear combination of the six conduction-band minima Bloch functions (i.e., without the envelope function of effective mass theory). The dotted lines are the lowest contours. (From Ivey and Mieher [207].)

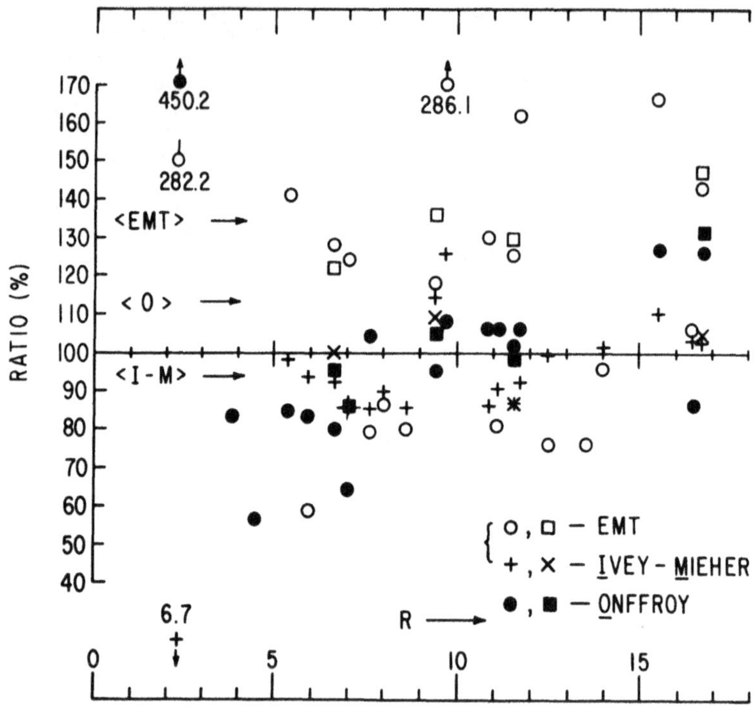

Fig. 3. Ratio (including average values in <>) of the theoretical to experimental values of the isotropic component of the hyperfine value for various sites at the distance (R) from phosphorus in silicon for three theories: effective mass theory (EMT), Ivey and Mieher [207], and Onffroy [338]

Simple as the problem is supposed to be, it is still very complicated, and the problem must be solved completely. For example, as can be seen in Fig. 2 a small displacement in the atom at (1,1,1) makes a dramatic change in the hyperfine constant, but lattice strains (surely different for the different donors) have not been included. And at the outset Feher [1,486] recognized that the anisotropy in the hyperfine interaction at the (0,0,4) site implied that d-band character had to be included in the treatment, i.e., more bands than Ivey-Mieher and Onffroy considered.

So much for the "simplest problem", what about other problems?

There is another shallow donor which might have proven simple, namely interstitial lithium. In fact lithium has an ionization energy (E_c-0.033 eV) closest to the simplest EMT value, but, as studied in detail by Watkins and Ham [425], the resonance properties are quite complicated. For the substitutional shallow donors the ground state is an A_1 state with the E and T_2 states higher; for interstitial lithium the A_1 state is ca. 1.8 meV higher than the E and T_2 states [490]. As a consequence the states of the system, and hence the resonance properties, are sensitive to internal and external strains. Moreover apparently the equilibrium site for the lithium atom is not the T_d interstitial position. Watkins and Ham studied the anisotropic resonance properties of the lithium, as well as observed the ENDOR of the lithium nuclei and eight shells of silicon atoms (they did not study the anisotropy of the latter). They were able to fit their results by permitting the Li to distort in a <100> direction from the T_d interstitial site, although they did not claim this distortion to be uniquely determined. Such a distortion is, however, in keeping with the configuration of the Si:H center in Ge as obtained by Haller et al. [491], and with the configuration that was calculated for interstitial hydrogen in silicon [492] using a MNDO (modified neglect of differential overlap) quantum theory (although we should note that a less sophisticated theory (extended Hückel theory) gave the result [493] that the interstitial hydrogen atom is stable at the T_d site).

There are other shallow donors, namely those which arise as so-called heat-treatment defects. A number of anisotropic resonance centers have been associated with these heat-treatment defects by Muller et al. [314], but they advanced no models for the defects. The defects occur as oxygen precipitates, and although Kaiser et al. [495] showed that rate of donor formation was proportional to the fourth power of the oxygen concentration suggesting that "the" oxygen shallow donor contains four oxygens, they realized that there were in fact several heat treatment donors, a result emphasized in subsequent studies. Recently Synder and Corbett [494] have argued that the shallow donor might arise from the interaction of two oxygens. The oxygens each bond with two silicon lattice atoms leaving no dangling bonds. However, the "lone pair" electrons (i.e., the remaining four electrons on each oxygen) will from a bonding and an antibonding state each filled with two electrons; the remaining four electrons are distributed two each in non-interacting states on the oxygens. Thus far there is still no associated electrical activity. If, however, the filled antibonding state is substantially above the conduction band it can lose its two electrons yielding a doubly positively charged center which can form Rydberg states that correspond to the shallow double donor. *Ab initio* calculations on small clusters suggest that this mechanism can give rise to donor states [494], but more extensive work is needed. In any event these shallow donor states are not simple either.

The substitutional acceptors could in principle be a simple oppositely charged analog of the shallow donors, since they have comparable energy levels (with the ionization energies ranging from 45,69,73 and 156 meV for B, Al, Ga and In, respectively, versus the theoretical value of 34 meV) and the top of the valence band is at $\underline{k}$=o. The six-fold degeneracy (from p-like, L=1 states) at the top of the valence band is lifted by spin-orbit coupling between the electron (s=1/2) yielding an upper, four-fold degenerate (J=3/2) state and a two-fold (J=1/2) state ca. 0.035 eV lower. Thus the treatment of the acceptor ground state must include all six states, and an anisotropic mass besides [474,475, 496-502]. The high degeneracy of the ground state makes it very sensitive to internal and external strains, and, as recognized by Kohn [475], Feher et al. [119] found it necessary to apply an external stress to observe the acceptor resonances: B, Al, Ga and In. Subsequently higher purity and more perfect crystals permitted Neubrand [328] to observe the boron resonance without an external stress. Ludwig and Woodbury [274] observed the resonance of a deeper acceptor, Zn, which has levels at 310 and 650 meV. No ENDOR has been done on any of these resonances.

Ludwig [276] has also observed the resonance of a deep, double donor, S, and has observed eight shells of silicon neighbours by ENDOR. He attempted to use EMT to explain his ENDOR data but found the agreement poor. He identified only one shell, namely the nearest neighbour silicon (1,1,1), which has the largest isotropic hyperfine value in contrast to the shallow donors where this atom has almost uniformly the lowest observed isotropic hyperfine value. Others have also treated the problem [466,503]. Castner [466] attempted to apply his treatment to S^+ which includes the subsidiary conduction band "minima" at the L, K. and U points. He obtained and "improved qualitative agreement between theory and experiment" [504], including getting the (1,1,1) site to have the largest isotropic hyperfine value. He was not content with the final result, however, for a variety of reasons not the least of which is the fact that a "comparable qualitative agreement can also be obtained for S^+ in the tetrahedral interstitial site" [505]. This same ambiguity between interstitial and substitutional sites was found by Spaeth [506] in the ENDOR measurements on tellurium in silicon, but these measurements were not complete at our last knowledge.

As Table IV shows there have been EPR measurements on a number of other impurities. The iron-series transition elements form a particularly interesting system, but, since Prof. Ammerlaan is writing for this proceedings on this topic, we will omit it. In any event there have unfortunately been no ENDOR on these elements.

There have been other ENDOR measurements, and that brings us to our final subject, namely, intrinsic defects and their complexes. The silicon vacancy has been identified by Watkins [417,426,427]. There have been a number of calculations concerning the vacancy in semiconductors using a variety of approaches [507-520], notably among these is the one-electron linear-combination-of-atomic-orbitals treatment [426] which has also proven so useful in treating other defects. The vacancy in silicon has been found [518,519,521] to be a negative-U system [522,523] so that the change in bonding energy in going from V^{++} to V^{+} to V° more than compensates the electron-electron repulsion energy so that V^{+} is a meta-stable state. ENDOR measurements on the vacancy have been under taken by the Amsterdam group and we look forward to their results.

Probably the most thoroughly studied intrinsic defect in silicon and the one for which the model is best established in the divacancy [23,57,84,102-104,257,260,371-373,410,419,524-526], including ENDOR results revealing 19 shells (or 62 lattice sites) of neighbours of V_2^{+} [104] and 33 shells (or 106 lattice sites) for V_2^{-} [372]. There has been substantial theoretical activity on the divacancy in silicon [372,373,526-532], but only the nearest neighbours to the divacancy have been associated with ENDOR shells. In view of our discussion of the earlier theoretical treatment of ENDOR results, this difficulty is not surprising. *Still we must conclude that the ENDOR results on shallow- and deep level defects in silicon are* *the* *testing ground for defect theories in semiconductors and they still remain an unmet challenge.*

In addition to the vacancy and the divacancy, the trivacancy [252], the planar- [63,210,252,329] and a non-planar tetravacancy [252], and a non-planar pentavacancy [248,252,239] have been identified and some of the properties of these defects have been established. Moreover it has been recognized [533-535] that this hierarchy of multivacancy defects leads naturally to larger vacancy aggregates such as have been observed in the electron microscope, and that this hierarchy is also consistent [533] with the nature of amorphization by heavy-ion bombardment.

As shown in Table IV the vacancy has also been observed in complexes with substitutional impurities. In all cases but Sn, the impurity remains on a substitutional lattice site. The V·Sn complex [428] has the Sn atom at a bond-centered site very much like what is thought to be the saddle-point configuration for vacancy migration (or the split-vacancy pair [536]). This vacancy configuration is also reminiscent of the partial dissociation observed for the divacancy [526] and the chemically driven partial-dissociation of vacancy-hydrogen complexes [533,537]. These complexes arise from the bonding of hydrogen with dangling silicon bonds. There has been considerable activity [533, 537-555] associated with the infrared vibrations of the Si-H bonds in crystalline- and in amorphous-silicon, including work which explains these discrete vibration frequencies in terms of the electronegativity of the other neighbours of the silicon. Eelctronegativity is simply another way of saying that the strength (or bond order [554]) of neighbouring bonds influences the nature of the Si-H bond; it is a concept that we expect to be of great help in explaining the spectra of other systems, including EPR. In spite of a great deal of effort on our own part and on the part of the workers at Sandia [556], EPR spectra with the characteristic hydrogen hyperfine signature have not been found; our efforts continue on this problem.

Table IV also shows that a number of vacancy-oxygen complexes have been identified, and some of their properties established. The oxygen related centers tend to yield triplet states, presumably due to the interaction with the oxygen, but this point needs further exploration.

Much less is known about interstitials. To be sure some impurities (H and H_2, Li, O, the iron-group transition elements, Au, etc.) are known to occur or thought to occur as interstitials in silicon. (We show H_2 as a possible interstitial defect because it has been found [557] theoretically that H_2 can occur and that this result can explain the channeling results found [542,558] for deuterium in silicon). The intrinsic silicon interstitial has not been positively identified, although theory has argued [559-561] that it should be a split-<100> interstitial, a conclusion supported by identification of the carbon split-<100> interstitial [481]. It has been argued [256] that the P6 center [210] is due to a diinterstitial but there is no firm model for a number of other EPR spectra thought [533] to be due to interstitials. It has been noted [562] that interstitials may aggregate in <110> chains and Tan [533,534] has pointed out that these chains can grow to form interstitial defects observed in the electron microscope.

In summary there is an emerging panorama of defects covering the range from point defects to macroscopic defect configurations which are building blocks of amorphous silicon.

SUMMARY

EPR studies have been widely applied in semiconductors. In some cases these studies have penetrated to considerable depth, in other cases the progress has been slow. Looking at Table II, III and IV may lead the reader to believe that essentially everything has been done. In this text and elsewhere [535], however, we have argued that much remains to do but that the opportunity for great progress still remains before us.

References

1. G. Feher, Phys. Rev. 103, 834 (1956)
2. "Multiple Electron Resonance Spectroscopy", eds. M.M. Dorio and J.H. Freed (Plenum Press, NY, 1979)
3. B.C. Cavenett, Adv. in Phys. 30, 475 (1981)
4. E.E. Schneider and T.S. England, Physica 17, 221 (1951)
5. A.M. Portis, A.F. Kip, C. Kittel and W.H. Brattain, Phys. Rev. 90, 988 (1953)
6. R.C. Fletcher, W.A. Yager, G.L. Pearson, A.N. Holden, W.T. Read and F.R. Merritt, Phys. Rev. 94, 1392 (1954)
7. For a careful discussion of Eqn.(11) see Appendix B of "Introduction to Electron Paramagnetic Resonance" by M. Bersohn and J.C. Baird (W.A. Benjamin, NY, 1966) pp. 208-220
8. "Handbook of Chemistry and Physics" (Chem. Rubber Co., Cleveland, 1971) 52nd edition, p. B245 et seq.
9. W. Kaiser, P.H. Keck and C.F. Lange, Phys. Rev. 101, 1264 (1956)
10. H.J. Hrostowski and B.J. Alder, J. Chem. Phys. 33, 980 (1960)
11. J.W. Corbett, R.S. McDonald and G.W. Watkins, J. Phys.Chem. Solids 25, 873 (1964)
12. D.R. Bosomworth, W. Hayes, A.R.L. Spray and G.D. Watkins, Proc. Roy. Soc. Lond. A 317, 133 (1970)
13. G.D. Watkins, J.W. Corbett and R.S. McDonald, to be published
14. There are over one thousand references which should be cited for the information in Tables II, III and IV. Space does not permit our including these references. We have shown the references for work on silicon (refs. 1,5,6,15-469) to illustrate this point. Additions or comments on the information in these tables would be very welcome. The references were obtained by searching Phys. Abs. through 1980 and all of Current Contents.
15. D.Adler and E.J. Yoffa, Phys. Rev. Lett 36, 1197 (1976)
16. L.N. Aleksandrov and L.V. Ovchinnikova, Sov. Phys.-Solid State 15, 1708 (1974)
17. L.N. Aleksandrov and L.V. Ovchinnikova, Sov. Phys.-Solid State 18, 1209 (1976)
18. H. Alexander, R. Labusch and W. Sander, Solid State Comm. 3, 357 (1965)
19. H. Alexander, M. Kenn, B. Nordhofen, E. Weber and W. Sander, in "Lattice Defects in Semiconductors, 1974" (The Institute of Physics, London 1975) pp. 433-438
20. H. Alexander, B. Nordhofen and E. Weber in "Magnetic Resonance and Related Phenomena", Vol. 2 (North-Holland Publishing Co., Amsterdam 1975), eds. P.S. Allen, E.R. Andrew and C.A. Bates, pp. 453-454
21. N. Almeleh and B. Goldstein, Phys. Rev. 149, 687 (1966)
22. C.A.J. Ammerlaan, Atomic Energy Review 12, 743 (1971)

23. C.A.J. Ammerlaan and G.D. Watkins, Phys. Rev. B 5, 3988 (1972)
24. C.A.J. Ammerlaan and J.C. Wolfrat, Phys. Stat. Sol. B 89, 85 (1978)
25. E. Antoncik, Hyperfine Interactions 8, 161 (1980)
26. A.Kh. Antonenko, N.N. Gerasimenko, A.V. Dvurechenskii, L.S. Smirnov and G.M. Tseitlin, Sov. Phys.-Semicond. 10, 81 (1976)
27. A.Kh. Antonenkor, N.N. Gerasimenko and V.V. Dvurechenskii, Sov. Phys.-Semicond. 11, 322 (1977)
28. R. Bachus, B. Movaghar, L. Schweitzer and U. Voget-Grote, Phil. Mag. B 39, 27 (1979)
29. N.T. Bagraev, L.S. Vlasenko and R.A. Zhitnikov, Jetp. Lett. 24, 366 (1976)
30. N.T. Bagraev, L.S. Vlasenko and R.A. Zhitnikov, Sov. Phys.-Solid State 19, 1467 (1977)
31. A.I. Baranov, A.V. Dvurechensky, N.N. Gerasimenko and L.S. Smirnov, in "Radiation Effects in Semiconductors, 1976" (Institute of Physics, London, 1977) p. 505
32. A.I. Baranov, N.N. Gerasimenko, A.V. Dvurechenskii and L.S. Smirnov, Sov. Phys.-Semicond. 11, 53 (1977)
33. P.I. Baranskii, A.A. Bugai, V.M. Maksimenko, V.V. Savyak and V.P. Shapovalov, Soviet Physics-Semicond. 14, 854 (1980)
34. P.I. Baranskii, V.M. Maksimenko and V.V. Savyak, Sov. Phys.-Semicond. 14, 856 (1980)
35. L. Bartelsen and H. Alexander, Preprint: 1977
36. L. Bartelson, Phys. Stat. Sol. B 81, 471 (1977)
37. J.E. Baxter and G. Ascarelli, Phys. Rev. B 7, 2630 (1973)
38. W. Beezhold and K.L. Brower, IEEE Trans. Nucl. Sci. NS-20, 209 (1973)
39. Yu.M. Belousov, V.N. Gorelkin and V.P. Smilga, Sov. Phys.-Jetp. 47, 331 (1978)
40. G. Bemski, B. Szymanski and K. Wright, J. Phys. Chem. Solids 24, 1 (1963)
41. N.T. Bendik, V.S. Garnyk and L.S. Milevskii, Sov. Phys.-Solid State 12, 1340 (1970)
42. W. Bergholz, J. Phys. D: Applied Physics 14, 1099 (1981)
43. M. Berke, E. Weber, H. Alexander, H. Luft and B. Elschner, Sol. Stat. Comm. 20, 881 (1976)
44. G.L. Bir, E.I. Butikov and G.E. Pikus, J. Phys. Chem. Solids 24, 1467 (1963)
45. G.L. Bir, E.I. Butikov and G.E. Pikus, J. Phys. Chem. Solids 24, 1475 (1963)
46. G.L. Bir, E.I. Butikov and G.E. Pikus, Sov. Phys.-Solid State 9, 835 (1967)
47. L.D. Bogomolova, V.N. Lazukin and I.V. Chepeleva, Sov. Phys.-USP. 7, 508 (1965)
48. J.A. Borders and K.L. Brower, Rad. Effects 6, 135 (1970)
49. V.A. Botvin, Yu.V. Gorelkinskii, V.O. Sigle and M.A. Chubisov, Sov. Phys.-Semicond. 6, 1453 (1973)
50. V.A. Botvin, Yu.V. Gorelkinskii, V.A. Kudryashev and V.O. Sigle, Sov. Phys.-Semicond. 8, 1049 (1975)
51. M.H. Brodsky and R.S. Title, Phys. Rev. Lett. 23, 581 (1969)
52. M.H. Brodsky, R.S. Title, K. Weiser and G.D. Pettit, Phys. Rev. B 1, 2632 (1970)
53. M.H. Brodsky, D.M. Kaplan and J.F. Ziegler, in "Proc. 11th Int'l. Conf. on Physics of Semiconductors, 1972" Vol. 1, (Elsevier Publishing Co., Amsterdam, 1972) pp. 529-535
54. M.H. Brodsky and R.S. Title, in "Structure and Excitations of Amorphous Solids" (American Institute of Physics, New York, 1976) pp. 97-101
55. P.R. Brosious, Appl. Phys. Lett. 29, 265 (1976)
56. P.R. Brosious, in "Ion Implantation in Semiconductors 1976" (Plenum Press, New York, 1977) pp. 417-425

57. P.R. Brosious, in "Defects and Radiation Effects in Semiconductors, 1978", ed. J.H. Albany (Inst. of Phys., Bristol-London, 1979) pp. 248-257
58. S.V. Broude, V.A. Grazhulis, V.V. Kveder and Yu.A. Osipyan, Sov. Phys.-JETP. 39, 721 (1974)
59. K.L. Brower, F.L. Vook and J.A. Borders, Appl. Phys. Lett. 15, 208 (1969)
60. K.L. Brower, F.L. Vook and J.A. Borders, Appl. Phys. Lett. 16, 108 (1970)
61. K.L. Brower and J.A. Borders, Appl. Phys. Lett. 16, 169 (1970)
62. K.L. Brower, Phys. Rev. B 1, 1908 (1970)
63. K.L. Brower, Rad. Effects 8, 213 (1971)
64. K.L. Brower and W. Beezhold, in "Ion Implantation in Semiconductors" (Springer-Verlag, Berlin, 1971), eds. I. Ruge and J. Graul, p. 7
65. K.L. Brower and W. Beezhold, J. Appl. Phys. 43, 3499 (1972)
66. K.L. Brower, Phys. Rev. B 5, 4274 (1972)
67. K.L. Brower, Phys. Rev. B 9, 2607 (1974)
68. K.L. Brower, Phys. Rev. B 14, 872 (1976)
69. K.L. Brower, in "Ion Implantation in Semiconductors 1976" (Plenum Press, New York, 1977) pp. 427-434
70. K.L. Brower, in "Nuclear and Electron Resonance Spectroscopies Applied to Material Science", eds. E.N. Kaufmann and G.K. Shenoy (Norht-Holland, New York, 1981) pp. 71-84
71. K.L. Brower, Phys. Rev. Lett. 44, 1627 (1980)
72. A.A. Bugai, V.S. Vikhnin, V.E. Kustov, V.M. Maksimenko and B.K. Krulikovskii, Sov. Phys.-JETP 47, 1170 (1978)
73. T.T. Bykova and I.V. Vinokurov, Sov. Phys.-Solid State 7, 2103 (1966)
74. P.J. Caplan, J.N. Helbert, B.E. Wagner and E.H. Poindexter, Surf. Sci. 54, 33 (1976)
75. T.G. Castner, Phys. Rev. 130, 58 (1963)
76. T.G. Castner, Phys. Rev. 155, 816 (1967)
77. T.G. Castner and H.S. Tan, Sol. Stat. Comm. 26, 389 (1978)
78. P. Chan and A. Steinemann, Surface Science 5, 267 (1966)
79. L.J. Cheng, J.C. Corelli, J.W. Corbett and G.D. Watkins, Phys. Rev. 152, 761 (1966)
80. M. Chiba and A. Hirai, J. Phys. Soc. Japan 27, 1700 (1969)
81. M. Chiba and A. Hirai, J. Phys. Soc. Japan 33, 730 (1972)
82. K.C. Chu, W.R. Hurren, E. Hale and J. Reigle, J. Appl. Phys. 44, 4243 (1973)
83. M.F. Chung, J. Phys. Chem. Solids 32, 475 (1971)
84. J.W. Corbett and G.D. Watkins, Phys. Rev. 138, A555 (1965)
85. B.L. Crowder, R.S. Title, M.H. Brodsky and G.D. Pettit, Appl. Phys. Lett. 16, 205 (1970)
86. P.R. Cullis and J.R. Marko, Phys. Rev. B 1, 632 (1970)
87. P.R. Cullis and J.R. Marko, Phys. Rev. B 11, 4184 (1975)
88. D.F. Daly and K.A. Pickar, Appl. Phys. Lett. 15, 267 (1969)
89. D.F. Daly and H.E. Noffke, in "Radiation Effects in Semiconductors" (Gordon and Breach, New York, 1971) pp. 179-187
90. D.F. Daly, J. Appl. Phys. 42, 864 (1971)
91. M.F. Deigen, V.Ya. Bratus, B.E. Vugmeister, I.M. Zeritskii, A.A. Zolotukhin, A.A. Konchits and L.S. Milevskii, Sov. Phys.-JETP 42, 1073 (1975)
92. E.S. Demidov and V.A. Uskov, Sov. Phys.-Solid State 16, 624 (1974)
93. E.S. Demidov and V.A. Uskov, Sov. Phys.-Semicond. 8, 919 (1975)
94. E.S. Demidov, Sov. Phys.-Solid State 19, 100 (1977)
95. E.S. Demidov, A.A. Ezhevskii and A.F. Khokhlov, Sov. Phys.-Semicond. 11, 459 (1977)
96. G.B. Demidovich and V.F. Kiselev, Phys. Stat. Sol. B 50, K33 (1972)
97. J.R. Dennis and E.B. Hale, Rad. Effects 19, 67 (1973)

98. J.R. Dennis, G.K. Woodward and E.B. Hale, in "Lattice Defects in Semiconductors, 1974" (The Institute of Physics, London, 1975) pp.467-473
99. J.R. Dennis and E.B. Hale, Rad. Effects 30, 219 (1976)
100. J.R. Dennis and E.B. Hale, Appl. Phys. Lett. 29, 523 (1976)
101. J.R. Dennis and E.B. Hale, J. Appl. Phys. 49, 1119 (1978)
102. J.G. DeWit and C.A.J. Ammerlaan, in "Ion Implantation in Semiconductors", eds. I.Ruge and J.Graul (Springer-Verlag, Berlin, 1971) pp. 39-46
103. J.G. DeWit, C.A.J. Ammerlaan and E.G. Sieverts in "Lattice Defects in Semiconductors, 1974" (The Institute of Physics, London, 1975) pp. 178-184
104. J.G. DeWit, E.G. Sieverts and C.A.J. Ammerlaan, Phys. Rev. B 14, 3494 (1976)
105. Z.Z. Ditina, L.P. Strakhov and H.H. Helms, Sov. Phys.-Semicond. 2, 1006 (1969)
106. D.E. Dugdale, S.D. Lacey and G. Lancaster, J. Phys. C: Solid State Phys. 4, 654 (1971)
107. J. Dumas, Phys. Stat. Sol. B 86, K75 (1978)
108. A.V. Dvurechenskii, N.N. Gerasimenko and V.B. Glazman, in "Applications of Ion Beams to Materials, 1975", eds. G.Carter, J.S.Colligon and W.A.Grant (Institute of Physics, London, 1976) pp. 18-23
109. A.V. Dvurechenskii, N.N. Gerasimenko and V.B. Glazman, Rad. Effects 31, 37 (1976)
110. A.V. Dvurechenskii, N.N. Gerasimenko, S.I. Romanov and L.S. Smirnov, Rad. Effects 30, 67 (1976)
111. A.V. Dvurechenskii and L.S. Smirnov, Rad. Effects 37, 173 (1978)
112. A.V. Dvurenskii, I.A. Rjazantsev and L.S. Smirnov, in "International Conference on Ion Beam Modification of Materials, Budapest, Hungary, 1978", eds. J.Gyulai, T.Lohner and E.Pasztor (Central Research Institute for Physics, Budapest, 1979) pp. 1049-1056
113. A.V. Dvurechenskii and I.A. Ryazantsev, Sov. Phys.-Semicond. 14, 51 (1980)
114. J. Eisinger and G. Feher, Phys. Rev. 109, 1172 (1958)
115. G. Feher, C.S. Fuller and E.A. Gere, Phys. Rev. 107, 1462 (1957)
116. G. Feher, J. Phys. Chem. Solids 2, 486 (1958)
117. G. Feher, Phys. Rev. 114, 1219 (1959)
118. G. Feher and E.A. Gere, Phys. Rev. 114, 1245 (1959)
119. F. Feher, J.C. Hensel and E.A. Gere, Phys. Rev. Lett. 5, 309 (1960)
120. G. Feher, in "Semiconductor Physics" (Academic Press, New York, 1961) pp. 579-586
121. G. Feher, in "Paramagnetic Resonance" Vol. 2, ed. W. Low (Academic Press, New York, 1963) pp. 715-738
122. R.C. Fletcher, W.A. Yager, G.L. Pearson, A.N. Holden, W.T. Read and F.R. Merritt, Phys. Rev. 94, 1392 (1954)
123. K. Gamo, A. Doi, K. Masuda and S. Namba, Journal of the Japan Society of Applied Physics 39, 78 (1970)
124. D.V.N. Gashtol', N.N. Gerasimenko, A.V. Dvurechenski and L.S. Smirnov, Sov. Phys.-Semicond. 9, 551 (1975)
125. W. Gehloff and K.H. Segsa, Phys. Stat. Sol. A 41, K21 (1977)
126. D. Geist, Phys. Stat. Sol. 5, 217 (1964)
127. D. Geist, Phys. Stat. Sol. B 46, 283 (1971)
128. B.L. Gel'Mont and M.I. D'Yakonov, Sov. Phys.-Semicond. 7, 1345 (1974)
129. N.N. Gersimenko, A.V. Dvurechenskii, L.S. Smirnov and E.G. Shirokov, Crystal Lattice Defects 2, 125 (1971)
130. N.N. Gerasimenko, A.V. Dvurechenskii, V.I. Panov and L.S. Smirnov, Sov. Phys.-Semicond. 5, 1439 (1972)
131. N.N. Gerasimenko, A.V. Dvurechenskii and L.S. Smirnov, Sov. Phys.-Semicond. 5, 1487 (1972)

132. N.N. Gerasimenko, A.V. Dvurechenskii and L.S. Smirnov, Sov. Phys.-Semicond. 6, 862 (1972)
133. N.N. Gerasimenko, A.V. Dvurechenskii, V.I. Panov and L.S. Smirnov, Sov. Phys.-Semicond. 6, 867 (1972)
134. N.N. Gerasimenko, A.V. Dvurechenskii and L.S. Smirnov, Sov. Phys.-Semicond. 6, 965 (1972)
135. N.N. Gerasimenko, A.V. Dvurechenskii and G.P. Lebedev, Sov. Phys.-Semicond. 7, 1530 (1973)
136. N.N. Gerasimenko, A.V. Dvurechenskii, S.I. Romanov and L.S. Smirnov, in "Radiation Damage and Defects in Semiconductors", ed. J.E. Whitehouse (Institute of Physics, London, 1973) pp. 72-80
137. N.N. Gerasimenko and A.V. Dvurechenskii, in "Elementary Physico-Chemical Processes at the Surface of Single Crystal Semiconductors", ed. A.V. Rzhanov (Nauka Press, Novosibirs, 1975) pp. 73-82
138. N.N. Gerasimenko, V.B. Glazman and A.V. Dvurechenskii, Sov. Phys.-Semicond. 9, 1139 (1976)
139. N.N. Gerasimenko, G.A. Gudaev, A.V. Dvurechenskii, L.S. Smirnov and S.A. Sokolov, Sov. Phys.-Semicond. 10, 735 (1976)
140. N.N. Gerasimenko, A.V. Dvurechenskii, A.I. Mashin and A.F. Khokhlov, Sov. Phys.-Semicond. 11, 108 (1977)
141. E.M. Gershenzon, N.M. Pevin and M.S. Fogelson, Phys. Stat. Sol. B 49, 287 (1972)
142. E.M. Gershenzon, I.T. Semenov, M.S. Fogelson and D.P. Chambootkin, Phys. Stat. Sol. B 69, 679 (1975)
143. V.B. Ginodman, P.S. Gladkov, B.G. Zhurkin and N.A. Penin, Sov. Phys.-Semicond. 4, 264 (1970)
144. V.B. Ginodman, P.S. Gladkov, B.G. Zhurkin, V.G. Mikhalevich and N.A. Nenin, Sov. Phys.-Semicond. 4, 1489 (1971)
145. V.B. Ginodman, P.G. Gladkov, B.G. Zhurkin and B.V. Kornilov, Sov. Phys.-Semicond. 5, 1930 (1972)
146. V.G. Ginodman, B.G. Zhurkin, A.N. Lykov and L.A. Shul'man, Sov. Phys.-Semicond. 7, 1135 (1974)
147. B. Goldstein, Phys. Rev. Lett. 17, 428 (1966)
148. B. Goldstein, in "Radiation Effects in Semiconductors" (Gordon and Breach, New Yir, 1971) pp. 223-231
149. B. Goldstein, Rad. Effects 8, 229 (1971)
150. V.A. Gordeev, S.P. Kruglov, V.I. Kudinov, L.A. Kuz'min, S.M. Mikirtych'yants, E.V. Minaichev, Yu.V. Obukhov, G.I. Savel'ev, V.G. Firsov and G.V. Shcherbakov, JETP Lett. 27, 394 (1978)
151. Yu.V. Gorelkinskii, V.O. Sigle and S.Zh. Takibaev, Phys. Stat. Sol. A 22, K55 (1974)
152. Yu.V. Gorelkinskii, V.O. Single and V.A. Botvin, Sov. Phys.-Semicond. 10, 1339 (1976)
153. G. Götz, W. Karthe, B. Schnabel and N. Sobolev, Phys. Stat. Sol. A 50, K209 (1978)
154. J.C. Gourdon, P. Fretier and J. Pescia, J. Physiquie - Lettres 42, L21 (1981)
155. I. Grancher and W. Czaja, J. Phys. Chem. Solids 28, 231 (1967)
156. V.A. Grazhulis and Yu.A. Osip'yan, Sov. Phys.-JETP 31, 677 (1970)
157. V.A. Grazhulis and Yu.A. Osip'yan, Sov. Phys.-JETP 33, 623 (1971)
158. V.A. Grazhulis, V.V. Kveder and Yu.A. Osip'yan, JEPT Lett. 21, 335 (1975)
159. V.A. Grazhulis, V.V. Kveder, Yu.A. Osip'yan, Y.H. Lee, R.L. Kleinhenz, C.P. Scholes and J.W. Corbett, Phys. Lett. 66A, 398 (1978)
160. I.G. Gverdtsiteli, L.N. Aleksandrov and V.V. Boiko, Sov. Phys.-Solid State 9, 1328 (1967)
161. E.B. Hale and R.L. Mieher, Physics of Semiconductors II, 1134 (1968)
162. E.B. Hale and R.L. Mieher, Phys. Rev. 184, 839 (1969)
163. E.B. Hale and R.L. Mieher, Phys. Rev. 184, 751 (1969)

164. E.B. Hale and T.G. Castner, Phys. Rev. B 1, 4763 (1970)
165. E.B. Hale and R.L. Mieher, Phys. Rev. B 3, 1955 (1971)
166. E.B. Hale, J. Phys. Chem. Solids 34, 621 (1973)
167. F.S. Ham, Phys. Rev. Lett. 7, 242 (1961)
168. F.S. Ham, J. Phys. Chem. Solids 24, 1165 (1963)
169. F.S. Ham and G.W. Ludwig, in "Paramagnetic Resonance", Vol. 1 (Academic Press, New York, 1963) pp. 131-143
170. F.S. Ham, Phys. Rev. 138, A1727 (1965)
171. D. Haneman, Phys. Rev. 170, 705 (1968)
172. D. Haneman, M.F. Chung and A. Taloni, Phys. Rev. 170, 719 (1968)
173. D. Haneman, Japanese J. Appl. Phys. 2, 371 (1974)
174. S. Hasegawa, R. Kontani and T. Shimizu, Jap. J. Appl. Phys. 10, 1641 (1971)
175. S. Hasegawa, R. Kontani and T. Shimizu, Jap. J. Appl. Phys. 10, 655 (1971)
176. S. Hasegawa, K. Ichida and T. Shimizu, Jap. J. Appl. Phys. 12, 1181 (1973)
177. S. Hasegawa, H. Karimoto and T. Shimizu, Jap. J. Appl. Phys. 12, 1190 (1973)
178. S. Hasegawa and T. Shimizu, J. Phys. Doc. Japan 38, 766 (1975)
179. S. Hasegawa and S. Yazaki, Solid State Comm. 23, 41 (1977)
180. S. Hasegawa, S. Yazaki and T. Shimizu, Solid State Comm. 23, 901 (1977)
181. S. Hasegawa, S. Yazaki and T. Shimizu, J. Non-Cryst. Solids 27, 215 (1978)
182. F.T. Hedgcock and T.W. Raudorf, Can. J. Phys. 49, 2930 (1970)
183. F.T. Hedgcock, F. Heininger and A. Paoli, J. Magnetism & Magnetic Materials 1, 39 (1975)
184. J.C.M. Henning and E.C.J. Egelmeers, Sol. State Comm. 38, 1037 (1981)
185. J. Higinbotham and D. Haneman, Surface Science 32, 466 (1972)
186. J. Higinbotham and D. Haneman, Phys. Can. 29, 10 (1973)
187. M. Hirose, M. Taniguchi, T. Nakashita, Y. Osaka, T. Suzuki, S. Hasegawa and T. Shimizu, Journal of Non-Crystalline Solids 35 & 36, 297 (1980)
188. G. Hochstrasser and J.C. Courvoisier, Helv. Phys. Acta 39, 189 (1966)
189. M. Höhne, Phys. Stat. Sol. B 86, 119 (1978)
190. M. Höhne, Phys. Stat. Sol. B 85, 525 (1978)
191. M. Höhne, Phys. Stat. Sol. (b) 99, 651 (1980)
192. A. Honig and A.F. Kp, Phys. Rev. 95, 1686 (1954)
193. A. Honig, Phys. Rev. 96, 234 (1954)
194. A. Honig and J. Combrisson, Phys. Rev. 102, 917 (1956)
195. A. Honig and R. Levitt, Phys. Rev. Lett. 5, 93 (1960)
196. A. Honig and E. Stupp, Phys. Rev. 117, 69 (1960)
197. A. Honig, in "Semiconductor Physics (Academic Press, New York, 1961) pp. 610-612
198. H. Horiye and E.G. Wikner, J. Appl. Phys. 40, 3879 (1969)
199. T. Igo, J. Phys. Soc. Japan 16, 1472 (1961)
200. T. Igo, J. Phys. Soc. Japan 18, 28 (1963)
201. T. Igo, J. Phys. Soc. Japan 19, 1086 (1964)
202. T. Igo, Review of the Electrical Communication Laboratory 12, 488 (1964)
203. T. Igo, J. Phys. Soc. Japan 21, 1013 (1966)
204. T. Igo, J. Phys. Soc. Japan 21, 874 (1966)
205. I.G. Ivanter, B.A. Nikol'skii, A.N. Ponomarev, V.I. Selivanov and V.A. Suetin, Sov. Phys.-JETP 48, 189 (1978)
206. J.L. Ivey and R.L. Mieher, Phys. Rev. Lett. 29, 176 (1972)
207. J.L. Ivey and R.L. Mieher, Phys. Rev. B 11, 822 (1975)

208. T. Izumi, M. Satoh, M. Kawamura and T. Matsumori, "International Conf. on Ion Beam Modification of Materials, Budapest, Hungary, 1978", eds. J. Gyulai, T. Lohner and E. Pasztor (Central Research Institute for Physics, Budapest, 1979) pp. 83-92
209. D. Jerome and J.M. Winter, Phys. Rev. 134, A1001 (1964)
210. W. Jung and G.S. Newell, Phys. Rev. 132, 648 (1963)
211. H. Kamimura and N.F. Mott, J. Phys. Soc. Japan 40, 1351 (1976)
212. E.O. Kane, Phys. Rev. 119, 40 (1960)
213. S.T. Kang, Progress of Theoretical Physics 38, 525 (1967)
214. D. Kaplan, D. Lepine, Y. Petroff and P. Thirry, Phys. Rev. Lett. 35, 1376 (1975)
215. D. Kaplan, I. Solomon and N.F. Mott, Journal de Physique Lettres 39, L51 (1978)
216. D. Kaplan and M. Pepper, Solid State Comm. 34, 803 (1980)
217. L. Katz and E.B. Hale, in "Neutron Transmission Doping in Semiconductors" (Plenum Press, New York, 1979) pp. 307-316
218. T. Kawabata, K. Muro and S. Narita, Solid State Comm. 23, 267 (1977)
219. T.A. Kennedy and J.H. Pifer, Phys. Rev. B 11, 2017 (1975)
220. A.F. Khokhlov and P.V. Pavlov, JEPT Lett. 24, 211 (1976)
221. A.F. Khokhlov, A.I. Mashin and S.M. Polyakov, Sov. Phys.-Semicond. 12, 549 (1978)
222. N. Kishimoto, K. Morigaki, K. Murakami, A. Shimizu and A. Hiraki, Phys. Stat. Sol. B 80, K113 (1977)
223. N. Kishimoto and K. Morigaki, J. Phys. Soc. Japan 42, 137 (1977)
224. R.L. Kleinhenz, Y.H. Lee, J.W. Corbett, E.G. Sieverts, S.H. Muller and C.A.J. Ammerlaan, Phys. Stat. Sol. (b) 108, 363 (1981)
225. J.C. Knights, D.K. Biegelsen and I. Solomon, Solid State Comm. 22, 133 (1977)
226. H. Kodera, J. Phys. Soc. Japan 19, 915 (1964)
227. H. Kodera, J. Phys. Soc. Japan 21, 1040 (1966)
228. H. Kodera, J. Phys. Soc. Japan 21, 578 (1966)
229. H. Kodera, J. Phys. Soc. Japan 27, 1197 (1969)
230. H. Kodera, J. Phys. Soc. Japan 26, 377 (1969)
231. H. Kodera, J. Phys. Soc. Japan 28, 89 (1970)
232. O.G. Koshelev, Sov. Phys.-Solid State 8, 473 (1966)
233. O.G. Koshelev, Yu.P. Koval' and Ya.G.Klyava, Sov. Phys.-Semicond. 1, 1057 (1968)
234. O.G. Koshelev, Yu.P. Kival' and Ya.G. Klyava, Sov. Phys.-Semicond. 1, 1435 (1968)
235. D. Kropman, A. Sugis and M. Vinnal, Phys. Stat. Sol. A 44, K1 (1977)
236. N. Kumar and K.P. Sinha, Z. Physik 197, 26 (1966)
237. M. Kumeda, Y. Jinno and T. Shimizu, Phys. Stat. Sol. B 81, K71 (1977)
238. M. Kumeda, Y. Jinno, I. Watanabe and T. Shimizu, Solid State Comm. 23, 833 (1977)
239. V.V. Kurylev and S.N. Karyagin, Phys. Stat. Sol. A 21, K127 (1964)
240. H. Kusomoto and M. Shoji, J. Phys. Soc. Japan 17, 1678 (1962)
241. S.D. Lacey and G. Lancaster, Phys. Lett. 22, 386 (1966)
242. G. Lancaster and E.E. Schneider, in "Semiconductor Physics" (Academic Press, New York, 1961) pp. 589-591
243. G. Lancaster, J.A. Van Wyk and E.E. Schneider, Prof. Phys. Soc. 84, 19 (1964)
244. M. Lannoo, Annales de Physique 3, 391 (1968)
245. G.L. Latshaw, P.B. Russell and S.S. Hanna, Hype-fine Interactions 8, 105 (1980)
246. P.G. LeComber, R.J. Loveland, W.E. Spear and R.A. Vaughn, in "Amorphous and Liquid Semiconductors", Vol. 1 (Taylor and Francis, London, 1974) pp. 245-250

247. Y.H. Lee, Y.M. Kim and J.W. Corbett, Rad. Effects 15, 77 (1972)
248. Y.H. Lee and J.W. Corbett, Phys. Rev. B 8 2810 (1973)
249. Y.H. Lee, P.R. Brosious and J.W. Corbett, Rad. Effects 22, 169 (1974)
250. Y.H. Lee and J.W. Corbett, Sol. State Comm. 15, 1781 (1974)
251. Y.H. Lee, P.R. Brosious and J.W. Corbett, Phys. Lett. 49A, 425 (1974)
252. Y.H. Lee and J.W. Corbett, Phys. Rev. B 9, 4351 (1974)
253. Y.H. Lee, P.R. Brosious, L.J. Cheng and J.W. Corbett, in "Ion Implantation in Semiconductors", ed. Susumu Namba (Plenum Press, New York, 1975) pp. 519-524
254. Y.H. Lee, J.C. Corelli and J.W. Corbett, Physics Letters A 59, 238 (1976)
255. Y.H. Lee and J.W. Corbett, Phys. Rev. B 13, 2653 (1976)
256. Y.H. Lee, N.N. Gerasimenko and J.W. Corbett, Phys. Rev. B 14, 4506 (1976)
257. Y.H. Lee, T.D. Bilash and J.W. Corbett, Rad. Effects 29, 7 (1976)
258. Y.H. Lee, L.J. Cheng, J.D. Gerson, P.M. Mooney and J.W. Corbett, Solid State Comm. 21, 109 (1977)
259. Y.H. Lee, R.L. Kleinhenz and J.W. Corbett, Appl. Phys. Lett. 31, 142 (1977)
260. Y.H. Lee, L.J. Cheng, P.M. Mooney and J.W. Corbett, in "Solar Cell High Efficiency and Radiation Damage" (National Aeonautics and Space Administration Scientific and Technical Information Office, Washington, D.C., 1977) pp. 179-186
261. Y.H. Lee, J.C. Corelli and J.W. Corbett, Physics Letters 60A, 55 (1977)
262. Y.H. Lee, J.W. Corbett and K.L. Brower, Phys. Stat. Sol. A 41, 637 (1977)
263. Y.H. Lee, R.L. Kleinhenz and J.W. Corbett, in "Defects and Radiation Effects in Semiconductors, 1978" ed. J.H. Albany (Institute of Physics, Bristol-London, 1979) pp. 521-527
264. B.P. Lemke and D. Haneman, Phys. Rev. Lett. 35, 1379 (1975)
265. D.J. Lepine, Phys. Rev. B 2, 2429 (1970)
266. R.A. Levy, Phys. Rev. Lett. 5, 425 (1960)
267. R.T. Longo, R. Magno and J.H. Pifer, J. Phys. Soc. Japan 29, 237 (1970)
268. W. Low and M. Weger, Phys. Rev. 118, 1119 (1960)
269. W. Low and M. Weger, Phys. Rev. 118, 1130 (1960)
270. G.W. Ludwig, H.H. Woodbury and R.O. Carlson, J. Phys. Chem. Solids 2, 490 (1958)
271. G.W. Ludwig and H.H. Woodbury, Phys. Rev. 117, 1286 (1960)
272. G.W. Ludwig and H.H. Woodbury, Phys. Rev. Lett. 7, 240 (1961)
273. G.W. Ludwig and H.H. Woodbury, in "Semiconductor Physics" (Academic Press, New York, 1961) pp. 596-598
274. G.W. Ludwig and H.H. Woodbury, in "Solid State Physics", eds. F. Seitz and D. Turnbull (Academic Press, New York, 1962) Vol. 13, pp. 223-304
275. G.W. Ludwig and F.S. Ham, Phys. Rev. Lett. 8, 210 (1962)
276. G.W. Ludwig and F.S. Ham, in "Paramagnetic Resonance", Vol. 1, ed. W. Low (Academic Press, New York, 1963) pp. 620-627
277. G.W. Ludwig, Phys. Rev. 137, A1520 (1965)
278. S. Maekawa and N. Kinoshita, J. Phys. Soc. Japan 19, 2240 (1964)
279. S. Maekawa and N. Kinishita, J. Phys. Soc. Japan 20, 1447 (1965)
280. S. Maekawa, J. Phys. Soc. Japan 21, 1221 (1966)
281. S. Maekawa, J. Phys. Soc. Japan 21, 574 (1966)
282. A. Makosa and T. Figielski, Phys. Stat. Sol. B 91, K65 (1969)
283. J.R. Marko, Phys. Lett. 27A, 119 (1968)
284. J.R. Marko and A. Honig, Phys. Rev. B 1, 718 (1970)
285. J.R. Marko, Can. J. Phys. 48, 834 (1970)
286. W.A. Martin, Phys. Stat. Sol. A 44, K159 (1977)

287. K. Masuda, S. Namba, K. Gamo and K. Murakami, in "US-Japan Seminar on Ion Implantation in Semiconductors" ed. S. Namba (Japan Society for the Promotion of Science, 1972) pp. 19-26
288. K. Masuda, in "Ion Implantation in Semiconductors", ed. S. Namba (Plenum Press, New York, 1975) pp. 473-483
289. K. Masuda, in "Radiation Effects in Semiconductors, 1976" (Institute of Physics, London, 1977) pp. 174-185
290. T. Matsumori, T. Kobayashi, H. Maekawa and T. Izumi in "Ion Implantation in Semiconductors and Other Materials", ed. B.L. Crowder (Plenum Press, New York, 1973) p. 32
291. G. Mendz and D. Haneman, J. Phys. C 11, L197 (1978)
292. G. Mendz and D. Haneman, Journal of Physics 13, 6737 (1980)
293. M. Merke, E. Weber, H. Alexander, H. Luft and B. Elschner, Solid State Comm. 20, 881 (1976)
294. R.L. Mieher, Phys. Rev. B 17, 2069 (1978)
295. N. Mikoshiba, K. Suzuki and T. Ishiguro, J. Phys. Soc. Japan 24, 337 (1968)
296. L.S. Milevskii, I.G. Varvarina and N.T. Bendik, Inorganic Materials 12, 1143 (1976)
297. J.J. Miller and D. Haneman, Surf. Sci. 19, 45 (1970)
298. D.J. Miller and D. Haneman, Surf. Sci. 33, 477 (1972)
299. D.J. Miller and D. Haneman, Solid State Comm. 27, 91 (1978)
300. F.F. Morehead, B.L. Crowder and R.S. Title, J. Appl. Phys. 43, 1112 (1972)
301. K. Morigaki and J.M. Winter, J. Phys. Soc. Japan 21, 570 (1966)
302. K. Morigaki and S. Toyotomi, J. Phys. Soc. Japan 31, 1091 (1971)
303. K. Morigaki and S. Maekawa, J. Phys. Soc. Japan 32, 462 (1972)
304. K. Morigaki and M. Onda, J. Phys. Soc. Jap. 36, 1049 (1974)
305. K. Morigaki, N. Kishimoto and D.J. Lepine, Solid State Comm. 17, 1017 (1975)
306. K. Morigaki and M. Rosso, Journal de Physique 36, 1131 (1975)
307. K. Morigaki, D.J. Dunstan, B.C. Cavenett, P. Dawson, J.E. Nicholls, S. Nitta and K. Shimakawa, Solid State Comm. 26, 981 (1978)
308. K. Morigaki, P. Dawson, B.C. Cavenett, D.J. Dunstan, S. Nitta and K. Shimakawa, in "Physics of Semiconductors, 1978", ed. B.L.H. Wilson (Institute of Physics, London, 1979) pp. 1163-1166
309. K. Morigaki, in "Physics of Semiconductors, 1978", ed. B.L.H. Wilson (Institute of Physics, London, 1979) pp. 943-952
310. B. Movaghar and L. Schweitzer, Phys. Stat. Sol. B 80, 491 (1977)
311. B. Movaghar, L. Schweitzer and H. Overhof, Phil. Mag. B 37, 683 (1978)
312. M. Movaghar, B. Ries and L. Schweitzer, Phil. Mag. B 41, 141 (1980)
313. B. Movaghar, B. Ries and L. Schweitzer, Phil. Mag. B 41, 159 (1980)
314. S.H. Muller, M. Sprenger, E.G. Sieverts and C.A.J. Ammerlaan, Solid State Comm. 25 , 987 (1978)
315. K. Murakami, K. Masuda, K. Gamo and S. Namba, Jap. J. Appl. Phys. 12, 1307 (1973)
316. K. Murakami, K. Masuda and S. Namba, Solid State Comm. 18, 663 (1976)
317. K. Murakami, K. Masuda, K. Gamo and S. Namba, Appl. Phys. Lett. 30, 300 (1977)
318. K. Murakami, S. Namba, N. Kishimoto, K. Masuda and K. Gamo, J. Appl. Phys. 49, 2401 (1978)
319. K. Murakami, E. Ikawa, K. Gamo, S. Namba, Y. Akasaka and Y. Masuda, Applied Phys. Lett. 35, 413 (1979)
320. J.A. Naber, H. Horiye and B.C. Passenheim, Rad. Effects 8, 239 (1971)
321. H. Nagashima, Y. Ochiai and E. Matsuura, Solid State Comm. 11, 733 (1972)
322. H. Nagashima and H. Yamazaki, J. Phys. Soc. Japan 41, 711 (1976)

323. H. Nakata and H. Morishima, Phys. Stat. Sol. A 33, 633 (1976)
324. H. Nakata and H. Morishima, Phys. Stat. Sol. A 43, 417 (1977)
325. S. Namba, K. Masuda, K. Gamo, S. Ohba and K. Murakami, Scientific Papers Institute of Physical & Chemical Research 65, 22 (1971)
326. S. Namba and K. Masuda, in "Advances in Electronics and Electron Physics", ed. C. Marton (Academic Press, New York, 1975) pp. 263-330
327. H. Neubrand, Phys. Stat. Sol. A 17, 459 (1973)
328. H. Neubrand, Phys. Stat. Sol. B 86, 269 (1978)
329. M. Nisenoff and H.Y. Fan, Phys. Rev. 128, 1605 (1962)
330. Y. Nishi, Japanese J. Appl. Phys. 10, 52 (1971)
331. Y. Nishi, K. Tanaka and A. Ohwada, Jap. J. Appl. Phys. 11, 85 (1972)
332. C.B. Norris, K.L. Brower and F.L. Vook, Rad. Effects 18, 1 (1973)
333. Y. Ochiai and E. Matsuura, Phys. Stat. Sol. A 38, 243 (1976)
334. Y. Ochiai and E. Matsuura, Phys. Stat. Sol. A 45, K101 (1978)
335. I. Ohdomari, M. Ikeda, K. Ohno and T. Itoh, Journ. of the Vacuum Society of Japan 18, 300 (1975)
336. I. Ohdomari, M. Ikeda, K. Ohna and T. Itoh, J. Appl. Phys. 46, 5065 (1975)
337. D.W. Olson and T.G. Castner, Phys. Rev. B 17, 3318 (1978)
338. J.R. Onffroy, Phys. Rev. B 17, 2062 (1978)
339. Y. Ootuka, F. Komori, Y. Monden, S. Kobayashi and W. Sasaki, Solid State Comm. 33, 793 (1980)
340. R.R. Parsons, J.A. Rostworowski, R.E. Halliwell and R. Barrie, Can. J. Phys. 55, 1349 (1977)
341. J.C. Phillips, Comments on Solid State Physics 3, 67 (1970)
342. J.H. Pifer, Phys. Rev. B 12, 4391 (1975)
343. G.E. Pikus and G.L. Bir, Phys. Rev. Lett. 6, 103 (1961)
344. J.D. Quirt and J.R. Marko, Phys. Rev. Lett. 26, 318 (1971)
345. J.D. Quirt and J.R. Marko, Phys. Rev. B 5, 1716 (1972)
346. J.D. Quirt and J.R. Marko, Phys. Rev. B 7, 3842 (1973)
347. A.G. Revesz and B. Goldstein, Surf. Sci. 14, 361 (1969)
348. C. Robert, C.R. Acad. Sci. Paris 260, 6337 (1965)
349. M. Rosso, J. Phys. Soc. Japan 38, 780 (1975)
350. M. Rosso, in "Physics of Semiconductors, 1978", ed. B.L.H. Wilson (Institute of Physics, London, 1979) pp. 939-942
351. L.M. Roth in "Semiconductor Physics" (Academic Press, New York, 1961) pp. 592-596
352. D. Schechter, Phys. Rev. B 9, 1751 (1974)
353. U. Schmidt, E. Weber, H. Alexander and W. Sander, Solid State Comm. 14, 735 (1974)
354. E. Schultz-Dubois, M. Nisenoff, H.Y. Fan and K. Lark-Horovitz, Phys. Rev. 98, 1561 (1955)
355. F. Schwiderfsky, Thin Solid Films 18, 45 (1973)
356. I.T. Semenov and M.S. Fogel'son, Sov. Phys.-Semicond. 9, 1264 (1975)
357. I.T. Semenov, M.S. Fogel'son and L.A. Shul'man, Sov. Phys.-Solid State 16, 2127 (1975)
358. I.T. Semenov and M.S. Fogel'son, Sov. Phys.-Semicond. 9, 1264 (1976)
359. K. Seshan, Radiation Effects 40, 29 (1979)
360. N.J. Shevchuk and W. Paul, J. Non-Cryst. Solids 16, 55 (1974)
361. T. Shimizu and M. Nakayama, J. Phys. Soc. Japan 19, 1829 (1964)
362. T. Shimizu, J. Phys. Soc. Japan 20, 1096 (1965)
363. T. Shimizu, J. Phys. Soc. Japan 25, 1021 (1968)
364. T. Shimizu, S. Hasegawa and H. Karimoto, in "Ion Implantation in Semiconductors", ed. S. Namba (Plenum Press, New York, 1975) pp. 525-531
365. T. Shimizu, M. Kimeda, I. Watanabe and K. Kamono, Journ. of Non-Cryst. Solids 35 & 36, 303 (1980)

366. T. Shimizu, M. Kimeda and Y. Kiriyama, Solid State Comm. 37, 699 (1981)
367. I. Shiota, N. Miyamoto and J. Nishizawa, Sur. Sci. 36, 414 (1973)
368. I. Shiota, N. Miyamoto and J. Nishizawa, Jap. J. Appl. Phys. 2, 417 (1974)
369. I. Shiota, N. Miyamoto and J. Nishizawa, J. Appl. Phys. 48, 2556 (1977)
370. E.G. Sieverts and C.A.J. Ammerlaan, in "Radiation Effects in Semiconductors, 1976" (Institute of Physics, London, 1977) pp. 213-220
371. E.G. Sieverts, S.H. Muller and C.A.J. Ammerlaan, Solid State Comm. 28, 221-225
372. E.G. Sieverts, S.H. Muller and C.A.J. Ammerlaan, Phys. Rev. B 18, 6834 (1978)
373. E.G. Sieverts, J. Phys. C: Solid State Physics 14, 2217 (1981)
374. E.G. Sieverts, S.H. Muller, C.A.J. Ammerlaan, R.L. Kleinhenz and J.W. Corbett, Phys. Stat. Sol. (b) 109, 83 (1982)
375. C.P. Slichter, Phys. Rev. 99, 479 (1955)
376. N.A. Sobolev, G. Götz, W. Karthe and B. Schnabel, Rad. Effects 42, 23 (1979)
377. I. Solomon, "11th International Conf. on the Physics of Semiconductors Proceedings" (Elsevier Publishing Co., New York, 1972) pp. 27-37
378. H.J. Stein, Appl. Phys. Lett. 15, 61 (1969)
379. R.A. Street and D.K. Biegelsen, Solid State Comm. 33, 1159 (1980)
380. F.V. Strnisa and J.W. Corbett, Cryst. Latt. Defects 5, 261 (1974)
381. K. Sugihara, J. Phys. Chem. Solids 29, 1099 (1968)
382. Y. Sugiura, J. Phys. Soc. Japan 20, 1272 (1965)
383. Y. Sugiura, J. Phys. Soc. Japan 20, 2098 (1965)
384. K. Suzuki and N. Mikoshiba, J. Phys. Soc. Japan 27, 1207 (1969)
385. P. Swarup and P.L. Trivedi, Sov. Phys.-Solid State 18, 190 (1976)
386. A. Taloni and W.J. Rogers, Surface Science 19, 371 (1970)
387. S. Tanaka, M. Kobayashi, K. Uchinokura and A. Koma, J. Phys. Soc. Japan 21, 566 (1966)
388. M. Taniguchi, M. Hirose, Y. Osaka, S. Hasegawa and T. Shimizu, Japanese Journ. of Appl. Phys. 19, 665 (1980)
389. D.R. Taylor, J.R. Marko and I.G. Bartlet, Sol. Stat. Comm. 14, 295 (1974)
390. P.A. Thomas, D. Lepine and D. Kaplan, in "Tetrahedrally Bonded Amorphous Semiconductors", eds. M.H. Brodsky, S. Kirkpatrick and D. Weaire (American Institute of Physics, New York, 1974) pp. 47-52
391. P.A. Thomas and D. Kaplan, in "Structure and Excitations of Amorphous Solids" (American Institute of Physics, New York, 1976) pp. 85-90
392. K.B. Tolpygo and V.A. Telezhkin, Phys. Stat. Sol. B 81, K139 (1977)
393. Y. Toyoda and Y. Hayashi, J. Phys. Soc. Japan 29, 247 (1970)
394. Y. Toyoda and Y. Hayashi, J. Phys. Soc. Japan 30, 1511 (1971)
395. Y. Toyoda and Y. Hayashi, J. Phys. Soc. Japan 33, 718 (1972)
396. Y. Toyoda, N. Kishimoto, K. Murakami and K. Morigaki, J. Phys. Soc. Japan 43, 114 (1977)
397. Y. Toyoda and K. Morigaki, J. Phys. Soc. Japan 43, 118 (1977)
398. S. Toyotomi, J. Phys. Soc. Japan 37, 130 (1974)
399. H. Ue and S. Maekawa, Phys. Rev. B 3, 4232 (1971)
400. I. Ursu, V. Lupei and S.V. Nistor, Atom. Energ. Rev. 5, 97 (1967)
401. R.H. Van der Linde and C.A.J. Ammerlaan, Semiconductors and Insulators 4, 139 (1978)
402. V.S. Vavilov, Yu.P. Koval' and O.G. Koshelev, Sov. Phys.-Solid State 8, 1908 (1967)
403. V.S. Vavilov, Ya.G. Klyava and O.G. Koshelev, Sov. Phys.-Semicond. 3, 169 (1969)

404. V.S. Vikhnin, I.M. Zaritskii, A.A. Konchits and B.K. Krulikovskii, Sov. Phys.-Solid State 17, 2348 (1975)
405. V.S. Vikhnin, M.F. Deigen, Yu.G. Semenov and B.D. Shanina, Sov. Phys.-Solid State 18, 1293 (1976)
406. B.I. Vikhrev, N.N. Gerasimenko, A.V. Dvurechenskii and L.S. Smirnov, Sov. Phys.-Semicond. 8, 871 (1975)
407. U. Voget-Grote, J. Stuke and H. Wagner, in "Structure and Excitations of Amorphous Solids" (American Institute of Physics, New York, 1976) pp. 91-96
408. U. Vogel-Grote, W. Kummerle, R. Firscher and J. Stuke, Phil. Mag. B 41, 127 (1980)
409. A.F. Volkov, Sov. Phys.-Solid State 7, 1304 (1965)
410. H. Vollmer and D. Geist, Phys. Stat. Sol. B 62, 367 (1974)
411. F.L. Vook, "US-Japan Seminar on Ion Implantation in Semiconductors", ed. S. Namba (Japan Soc. for the Promotion of Science, 1972) pp. 9-17
412. T. Wada, T. Mizutani, M. Hirose and T. Arizumi, J. Phys. Soc. Japan 22, 1060 (1967)
413. G.K. Walters and T.L. Estle, J. Appl. Phys. 32, 1854 (1961)
414. S.S. Wang, J. Shirafuji and Y. Inuishi, Osaka Univ. Techn. Rep. 23, 469 (1973)
415. G.D. Watkins, J.W. Corbett and R.M. Walker, J. Appl. Phys. 30, 1198 (1959)
416. G.D. Watkins and J.W. Corbett, Disc. Faraday Soc. 31, 86 (1961)
417. G.D. Watkins, J. Phys. Soc. Japan 18, 22 (1963)
418. G.D. Watkins and J.W. Corbett, Phys. Rev. 134, A1359 (1964)
419. G.D. Watkins and J.W. Corbett, Phys. Rev. 138, A543 (1965)
420. G.D. Watkins, in "Symposium on Radiation Effects on Semiconductor Components", ed. F. Cambou (Journees d'Electronique, Toulouse, 1967) pp. A1-0 to A1-13
421. G.D. Watkins, Phys. Rev. 155, 802 (1967)
422. G.D. Watkins, in "Radiation Damage in Semiconductors", ed. P. Baruch (Dunod, Paris, 1967) pp. 97-113
423. G.D. Watkins, in "Radiation Effects in Semiconductors", ed. F.L. Vook (Plenum Press, New York, 1968) pp. 67-81
424. G.D. Watkins, IEEE Trans. Nucl. Sci. NS-16, 13 (1969)
425. G.D. Watkins and F.S. Ham, Phys. Rev. B 1, 4071 (1970)
426. G.D. Watkins, in "Radiation Damage and Defects in Semiconductors" ed. J.E. Whitehouse (Institute of Physics, London, 1973) pp. 228-237
427. G.D. Watkins, in "Lattice Defects in Semiconductors, 1974", ed. F.A. Huntley (Institute of Physics, London, 1975) pp. 1-22
428. G.D. Watkins, Solid State Comm. 17, 1205 (1975)
429. G.D. Watkins, Phys. Rev. B 12, 4383 (1975)
430. G.D. Watkins, Phys. Rev. B 12, 5824 (1975)
431. G.D. Watkins and K.L. Brower, Phys. Rev. Lett. 36, 1329 (1976)
432. G.D. Watkins, Phys. Rev. B 13, 2511 (1976)
433. G.D. Watkins, Phys. Rev. B 13, 4644 (1976)
434. G.D. Watkins, Chinese Journal of Physics 15, 92 (1977)
435. E. Weber and H. Alexander, in "Radiation Effects in Semiconductors, 1976" (Institute of Physics, London, 1977) pp. 266-271
436. E. Weber and H.G. Riotte, Appl. Phys. Lett. 33, 433 (1978)
437. E. Weber and H.G. Riotte, Journ. of Applied Physics 51, 1484 (1980)
438. R.M. White and J.F. Gouyet, Phys. Rev. B 16, 3596 (1977)
439. J.K. Wigmore, Phys. Lett. 26A, 15 (1967)
440. E.G. Wikner and H. Horiye, IEEE Trans. Nucl. Sci. NS-13, 18 (1956)
441. J.M. Winter, J. Phys. Soc. Japan 21, 213 (1966)
442. F.D. Wohler, H. Alexander and W. Sander, J. Phys. Chem. Solids 31, 1381 (1970)
443. H.H. Woodbury and G.W. Ludwig, Phys. Rev. 117, 102 (1960)
444. H.H. Woodbury and G.W. Ludwig, Phys. Rev. 117, 1287 (1960)

445. H.H. Woodbury and G.W. Ludwig, Phys. Rev. Lett. 5, 96 (1960)
446. H.H. Woodbury and G.W. Ludwig, Phys. Rev. 126, 466 (1962)
447. T. Wosinski and T. Figielski, Phys. Stat. Sol. B 71, K73 (1975)
448. T. Wosinski, T. Figielski and A. Makosa, Phys. Stat. Sol. 37, K57 (1976)
449. T. Wosinski and T. Figielski, Phys. Stat. Sol. B 71, K73 (1976)
450. Y. Yafet, Solid State Physics 14, 1 (1963)
451. Y. Yafet, J. Phys. Chem. Solids 26, 647 (1965)
452. M. Yamaguchi and T. Kirayama, Japanese J. Appl. Phys. 14, 365 (1976)
453. G. Yang and A. Honig, Phys. Rev. 168, 271 (1968)
454. A. Yoshida, Oyo Buturi 45, 1018 (1976)
455. K. Zellama, P. Germain, S. Squelard, J. Monge and E. Ligeon, Journ. of Non-Cryst. Solids 35 & 36, 225 (1980)
456. B.G. Zhurkin and N.A. Penin, Sov. Phys.-Solid State 6, 879 (1964)
457. B.G. Zhurkin and N.A. Penin, JETP Lett. 2, 13 (1965)
458. B.G. Zhurkin, N.A. Penin and B.A. Volkov, Sov. Phys.-Solid State 6, 2040 (1965)
459. B.G. Zhurkin and N.A. Penin, Sov. Phys.-Solid State 7, 2592 (1966)
460. B.G. Zhurkin, N.A. Penin and P. Swarup, Sov. Phys.-Solid State 8, 2839 (1967)
461. B.G. Zhurkin and N.A. Penich, JETP Lett. 6, 321 (1967)
462. B.G. Zhurkin, N.A. Penin and N.N. Sibel'din, Sov. Phys-Semicond. 2, 688 (1968)
463. J.W. Corbett, Surface Sci. 90, 205 (1979)
464. G. Feher, R.C. Fletcher and E.A. Gere, Phys. Rev. 100, 1784 (1955)
465. T.G. Castner, Jr., E.B. Hale and R. Craven, in "Proc. Tenth Int'l Conf. on Physics of Semiconductors, 1970" (USAEC, Washington, 1970) pp. 613-619
466. T.G. Castner, Jr.,Phys. Rev. B 2, 4911 (1970)
467. E.B. Hale and R.L. Mieher, Phys. Letters 29A, 350 (1969)
468. J.L. Ivey and R.L. Mieher, Phys. Rev. B 11, 849 (1975)
469. Zh.S. Takibaev, Yu.V. Gorelkinskii, V.F. Grishchenko and N.N. Gerasimenko, Sov. Phys. Semicond. 9, 1303 (1976)
470. N.F. Mott, Advances in Phys. 21, 785 (1972)
471. N.F. Mott, Metal-Insulator Transitions (Taylor & Francis, London, 1974)
472. C. Kittel and A.H. Mitchell, Phys. Rev. 96, 1488 (1954)
473. J.M. Luttinger and W. Kohn, Phys. Rev. 97, 869 (1955)
474. W. Kohn and J.M. Luttinger, Phys. Rev. 98, 915 (1955)
475. W. Kohn, in "Solid State Physics", eds. F. Seitz and D. Turnbull (Academic Press, New York, 1957) Vol. 5, p.257
476. M. Breitenecker, R. Sexl and W. Thirring, Z. Phys. 182, 123 (1964)
477. A.M.K. Müller, Solid State Comm. 2, 205 (1964)
478. A.M.K. Müller, Z. Naturforsch. A20, 1476 (1965)
479. R.A. Falukner, Phys. Rev. 184, 713 (1969)
480. A. Baldereschi, Phys. Rev. B 1, 4673 (1970)
481. T.H. Ning and C.T. Sah, Phys. Rev. B 4, 3468 (1971)
482. T.H. Ning and C.T. Sah, Phys. Rev. B 4, 3482 (1971)
483. S.K. Pantelides and C.T. Sah, Phys. Rev. B 10, 621 (1974)
484. S.K. Pantelides and C.T. Sah, Phys. Rev. B 10, 638 (1974)
485. S.K. Pantelides, in "Festkörperprobleme", ed. H.J. Queisser (Pergamon/Vieweg, Braunschweig, 1975) Vol. XV, pp. 149-181
486. D. Schechter, Phys. Rev. B 8, 652 (1973)
487. D. Schechter, Phys. Rev. B 9, 1751 (1974)
488. D. Schechter, Phys. Rev. B 11, 5043 (1975)
489. See also Hale and Mieher, ref. 165
490. R.L. Aggarwal, P. Fisher, V. Mourzine and A.K. Ramdas, Phys. Rev. 138, A882 (1965)
491. E.E. Haller, B. Jóos and L.M. Falicov, Phys. Rev. B 21, 4729 (1980)

492. J.W. Corbett, S.N. Sahu, T.S. Shi and L.C. Snyder, to be published
493. Vijay A. Singh, C. Weigel, J.W. Corbett and L.M. Roth, Phys. Stat. Sol. (b) 81, 637 (1977)
494. L.C. Snyder and J.W. Corbett, to be published
495. W. Kaiser, H.L. Frisch and H. Reiss, Phys. Rev. 112, 1546 (1958)
496. W. Kohn and D. Schechter, Phys. Rev. 99, 1903 (1955)
497. G.E. Pikus and G.L. Bir, Sov. Phys. Solid State 1, 136 (1959)
498. G.E. Pikus and G.L. Bir, Sov. Phys. Solid State 1, 1502 (1959)
499. D. Schechter, J. Phys. Chem. Solids 23, 237 (1962)
500. K.S. Mendelson and H.M. James, J. Phys. Chem. Solids 25, 729 (1964)
501. N.O. Lipari and A. Baldereschi, Phys. Rev. Lett. 25, 1660 (1970)
502. N.O. Lipari and A. Baldereschi, Phys. Rev. B 9, 1525 (1974)
503. T. Shimizu and K. Minami, Phys. Stat. Solidi (b) 62, 585 (1974)
504. T.G. Castner, Jr., ref. 466, p. 4911
505. T.G. Castner, Jr., ref. 466, p. 4934
506. J.-M. Spaeth, private communication
507. C.A. Coulson and M.J. Kearsley, Proc. Roy. Soc. (London) 241A, 433 (1957)
508. R.A. Swalin, J. Phys. Chem. Solids 18, 290 (1961)
509. A. Scholz and A. Seeger, Phys. Stat. Sol. 3, 1480 (1963)
510. R.P. Messmer and G.D. Watkins, in "Radiation Effects in Semiconductors", eds. J.W. Corbett and G.D. Watkins (Gordon and Breach, N.Y. 1970) p. 23
511. G.T. Surratt and W.A. Goddard, Solid State Commun. 22, 314 (1977)
512. A. Pêcheur, E. Kauffer and M. Gerl, in "Defects and Radiation Effects in Semiconductors, 1978", ed. J.H. Albany (Institute of Physics, Bristol-London, 1979) pp. 174-179
513. C.O. Rodrigues, S. Brand and M. Jaros, in "Defects and Radiation Effects in Semiconductors, 1978", ed. J.H. Albany (Institute of Physics, Bristol-London, 1979) pp. 193-199
514. J. Bernholc and S.T. Pantelides, Phys. Rev. B 18, 1780 (1978)
515. G.A. Baraff and M. Schlüter, Phys. Rev. B 19, 4965 (1979)
516. N.O. Lipari, J. Bernholc and S.T. Pantelides, Phys. Rev. Lett. 43, 1354 (1979)
517. J. Bernholc, N.O. Lipari and S.T. Pantelides, Phys. Rev. B 21, 3545 (1980)
518. G.A. Baraff, E.O. Kane and M. Schlüter, Phys. Rev. Lett. 43, 956 (1979)
519. G.A. Baraff, E.O. Kane and M. Schlüter, Phys. Rev. B 21, 3563 (1980)
520. R.L. Kleinhenz, S.N. Sahu, Vijay A. Singh and J.W. Corbett, Phys. Lett. 83A, 291 (1981)
521. G.D. Watkins and J.R. Troxell, Phys. Rev. Lett. 44, 593 (1980)
522. P.W. Anderson, Phys. Rev. Lett. 34, 953 (1975)
523. M. Kastner, D. Adler and H. Fritsche, Phys. Rev. Lett. 37, 1504 (1975)
524. J.W. Corbett and G.D. Watkins, Phys. Rev. Lett. 7, 314 (1961)
525. R.H. van der Linde and C.A.J. Ammerlaan, in "Defects and Radiation Effects in Semiconductors, 1978", ed. J.H. Albany (Institute of Physics, Bristol-London, 1979) pp. 242-247
526. E.G. Sieverts and J.W. Corbett, to be published
527. T.F. Lee and T.C. McGill, J. Phys. C: Solid State Physics 6, 3438 (1973)
528. J.G. DeWit, Ph.D. Thesis, University of Amsterdam, 1975
529. C.A.J. Ammerlaan and C. Weigel, in "Lattice Defects in Semiconductors, 1976", eds. N.B. Urli and J.W. Corbett (Institute of Physics, Bristol-London, 1977) pp. 448-457
530. C.A.J. Ammerlaan and J.C. Wolfrat, Phys. Stat. Solidi (b) 89, 85 (1978)
531. E.G. Sieverts, Ph.D. Thesis, University of Amserdam, 1978

532. C. Wigel and C.A.J. Ammerlaan, Phys. Stat. Solidi (b) 94, 505 (1979)
533. J.W. Corbett, J.P. Karins and T.Y. Tan, Nucl. Inst. and Methods 182/183, 457 (1981)
534. Teh-Yu Tan, Phil. Mag. A 44, 101 (1981)
535. J.W. Corbett, R.L. Kleinhenz and N.D. Wilsey, in "Defects in Semiconductors", eds. J. Narayan and T.-Y. Tan (North-Holland Press, Amsterdam, 1981) pp. 1-19
536. B.J. Masters, Solid State Comm. 9, 283 (1971)
537. R.L. Kleinhenz, Y.H. Lee, Vijay A. Singh, P.M. Mooney, A. Jaworowski, L.M. Roth, J.C. Corelli and J.W. Corbett, in "Defects and Radiation Effects in Semiconductors, 1978", ed. J.H. Albany (Institute of Physics, Bristol, 1979) p. 200
538. H. Stein, J. Electr. Materials 4, 159 (1975)
539. M.H. Brodsky, M. Cardona and J.J. Cuomo, Phys. Rev. B 16, 3556 (1977)
540. N.N. Gerasimenko, M. Rollê, L.-J. Cheng, Y.H. Lee, J.C. Corelli and J.W. Corbett, Phys. Stat. Sol. (b) 90, 689 (1978)
541. J.C. Knights, G. Lukovsky and R.J. Nemanich, Phil. Mag. B37, 467 (1978)
542. S.T. Picraux, F.L. Vook and H.J. Stein, in "Defects and Radiation Effects in Semiconductors, 1978", ed. J.H. Albany (Institute of Physics, Bristol-London, 1979) p. 31-44
543. G. Lucovsky, R.J. Nemanich and J.C. Knights, Phys. Rev. B 19, 2064 (1979)
544. H.J. Stein, Phys. Rev. Letters 43, 1030 (1979)
545. Cui Shu-fan, Ge Pei-wen, Zhao Ya-qin, Wu Lan-sheng, Acta Physica Sinica 28, 791 (1979)
546. B.N. Mukashev, K.H. Nussupov and M.F. Tamendarov, Phys. Lett. 72A, 381 (1979)
547. H. Wieder, M. Cardona and C.R. Guarnieri, Phys. Stat. Sol. (b) 92, 99 (1979)
548. C.J. Fang, K.J. Gruntz, L. Ley, M. Cardona, F.J. Demond, G. Müller and S. Kalbitzer, J. Non-Cryst. Solids 35/36, 255 (1980)
549. H. Shanks, C.J. Fang, L. Ley, M. Cardona, F.J. Demond and S. Kalbitzer, Phys. Stat. Sol. (b) 100, 43 (1980)
550. J. Tatarkiewicz and K. Wieteska, Phys. Stat. Sol. (a) 66, K101 (1981)
551. B.N. Mukashev, K.H. Nussupov, M.F. Tamendarov and V.V. Frolov, Phys. Lett. 87A, 376 (1982)
552. A. Hiraki, T. Imura, K. Mogi, M. Tashiro, in "The Ninth Int'l Conf. on Amorphous and Liquid Semiconductors, 1981", in press
553. Mai Zen-hong, Cui Shu-fan, Ge Pei-wen, Tsien Ling-chao, in "The XIIth Conf. of the Int'l Union of Crystallography, 1981" in press
554. S.N. Sahu, T.-S. Shi, P.-W. Ge, A. Hiraki, T. Imura, M. Tashiro, Vijay A. Singh and J.W. Corbett, J. Chem. Phys., in press
555. T.-S. Shi, S.N. Sahu, G.S. Oehrlein, A. Hiraki and J.W. Corbett, Phys. Stat. Sol., in press
556. K.L. Brower, private communication
557. J.W. Corbett, S.N. Sahu, T.-S. Shi and L.C. Snyder, to be published
558. S.T. Picraux and F.L. Vook, Phys. Rev. B 18, 2066 (1978)
559. G.D. Watkins, R.P. Messmer, C. Weigel, D. Peak and J.W. Corbett, Phys. Rev. Lett. 27, 1573 (1971)
560. C. Weigel, D. Peak, J.W. Corbett, G.D. Watkins and R.P. Messmer, Phys. Rev. B 8, 2906 (1973)
561. C. Weigel and J.W. Corbett, Z. Physik B 23, 233 (1976)
562. J.W. Corbett and J.C. Bourgoin, in "Point Defects in Solids", eds. J.H. Crawford, Jr. and L.M. Slifkin (Plenum Press, New York, 1975) Vol. 2, p. 1

REVIEW OF THE POSSIBILITIES OF ELECTRON MICROSCOPY IN THE IDENTIFICATION OF DEFECT STRUCTURES

J. Heydenreich

Akademie der Wissenschaften der DDR,
Institut für Festkörperphysik und Elektronenmikroskopie
4020 Halle, Wittenberg, DDR

The strong dependence of the properties of semiconducting materials and semiconductor devices on the presence of defects in the used crystalline materials (and in devices) - silicon in microelectronics, compound semiconductors in optoelectronics - implies the necessity of using diagnostic methods which are extremely sensitive to the defect detection and which facilitate a reliable defect characterization. The electron microscopy with its high lateral resolution is especially suitable for this purpose. It covers the transmission electron microscopy (TEM) as well as the surface scanning electron microscopy (SEM) and also the modern technique of scanning transmission electron microscopy (STEM).

With respect to the investigation of defects in the crystal volume there are two main tasks:

1) Identification of defects by TEM (or STEM).
2) Detection of the electrical activity of defects by the combined application of TEM and special SEM techniques.

The wide field of defects in semiconductors surfaces will not be considered within this review.

1.) The identification of defects by TEM (or STEM) is related to defects in the as-grown material (swirls, microdefects, for review see [1]) as well as to defects induced during the processes of device technology (heat treatment, diffusion processes, implantation processes). Investigations of this kind should necessarily be directed to a complex detection of the defect behaviour taking into account the interaction of defects of the same kind, or of different defects with one another. Here are involved processes of the clustering of point defects, of the formation of precipitations, planar defects and dislocations as well as their possible transformations. The decoration of dislocations by impurities, e.g. is of topical interest. A complex arrangement of defects is imaged e.g. in Fig. 1, which is related to Czochralski grown silicon. The micrograph shows clearly the presence of defects in the neighbourhood of precipitates, which could be identified as SiO_x (α-cristobalite). In the majority of cases the de-

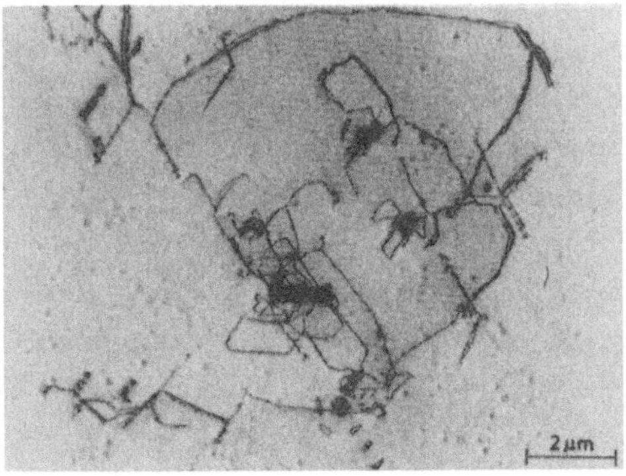

Fig. 1. Complex of crystal defects in Czochralski-grown silicon

termination of the defect type by using diffraction contrast electron microscopy [2,3] as e.g. in the case of Fig. 1, and by interpreting the micrographs using standard rules [4,5] and/or computer simulations [6-8] does not suffice for the investigations mentioned above. If detailed statements on the nature of defects, e.g. the core structure of dislocations, are required, higher resolving imaging techniques have to be applied, i.e. weak beam electron microscopy [9,10] or high-resolution phase contrast methods (see e.g. [11]) including lattice fringe imaging [12,13] and many-beam imaging (for silicon, see [14,15]).

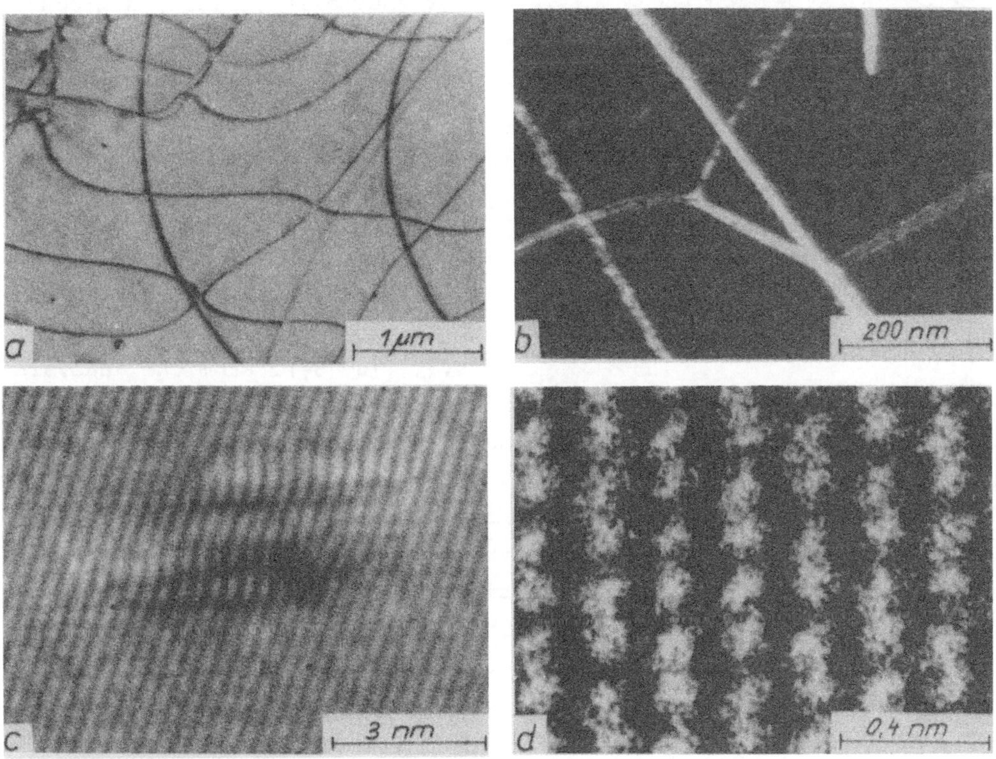

Fig. 2. Crystal defects in silicon. (a) dislocations (diffraction contrast electron micrograph); (b) dissociated dislocations (weak beam electron micrograph); (c) dislocation dipole (lattice fringe image); (d) <110>-oriented silicon (many-beam image)

Figure 2 shows the possibilities of the mentioned techniques to image defects in silicon, i.e. in a) the conventional diffraction contrast micrograph of dislocations, b) the weak beam micrograph showing the state of dissociation of dislocations into partials, c) the lattice fringe image of a dislocation dipole (distance of the lattice fringes: 0.316 nm), and d) the many-beam image of a <110>-oriented Si foil of about 10 nm thickness. The latter micrograph shows contrasts which are related to pairs of closely spaced columns of atoms lying in the beam direction and having a distance of only 0,14 nm.

Fig. 3. Rod-like defects in electron-irradiated silicon (1000 keV, 450°C, 10'), lying on different {113} planes (weak beam micrograph: (10 $\bar{1}\bar{0}$ 0)/(2 $\bar{2}$ 0))

For both weak beam and lattice imaging one further example shall be given . Figure 3 is a weak beam micrograph of rod-like defects typical of radiation treated (in some cases also only for thermally treated) semiconductors. These defects are dipole-like closed dislocation loops extended in <110> directions surrounding stacking faults. In the meantime these defects could be identified as intermediate defects in the process of dislocation nucleation from the condensation of point defects (see e.g. [16-18]). From the analysis of the micrograph taken from silicon which was irradiated with 1000 keV electrons at 450°C for 10', the planes of the stacking faults belonging to the rod-like defects are {113} planes, which are in agreement with the investigations of other authors. Different directions of the fringes of the stacking fault contrast within the same defect point to the fact, that two different {113} planes participate, the intersection line of which is identical with the <110> direction, in which the rod-like defect is elongated. In places marked by arrows a transformation into complete dislocation loops on {113} planes took place according to a well-known dislocation reaction.

The high-resolution (9-beam) electron micrograph of Fig. 4 shows a coherent precipitate in a copper-doped CdTe crystal. The analysis of the image taken in (110) orientation, considering the special fact that from the (111) planes only every second plane is imaged, leads to the

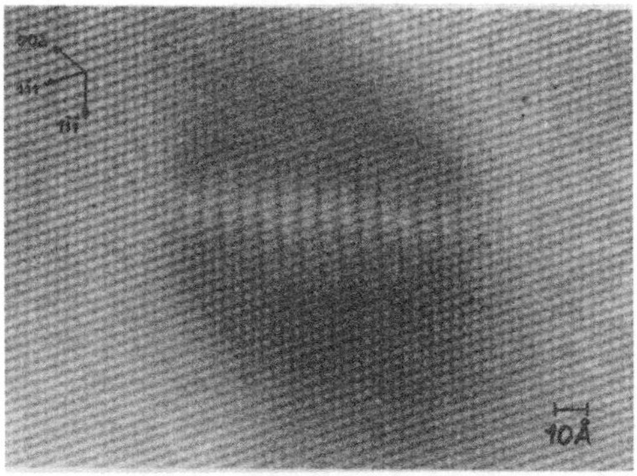

Fig. 4. High-resolution image of a coherent precipitate in copper-doped CdTe (nearly (110)-orientation)

assumption that the precipitate consists of cadmium copper telluride [19].

High-resolution electron microscopy is especially suitable also for the detection of the structure interfaces between different crystalline regions. In the investigations of interfaces between different materials (MOS,SOS) and besides that of thin specimen foils with the interesting interface parallel to the foil plane (perpendicular to the electron beam), the use of cross-sectional techniques which allows the observation of an interface "edge-on" (see e.g. [20,21]) gains increasing importance.

Apart from the defect identification with a sufficiently high resolution the observation of the behaviour of defects under external influences (e.g. thermal or radiation treatments) is very important and requires the realization of electron microscopic in-situ observations [22]. With respect to the fact that specimen foils used for TEM have to be sufficiently thick to draw conclusions on the bulk material behaviour,the use of the high-voltage electron microscope (HVEM) is necessary for such a purpose. By the help of a 1000 kV HVEM it is possible to investigate specimen foils up to thicknesses of some µm (compared to some 0.1 µm in the usual 100 kV EM),and it was shown [23] that in specimens with thicknesses of some µm the influence of the surfaces can be neglected so that they might be regarded as representatives for the bulk material.

One example for the in-situ observation of a (111) oriented Ge crystal (n-Ge, 10 Ωcm, pretreatment: deformation in <110> direction at 800°C) carried out in a 100 keV HVEM is given in Fig. 5. The initial state at room temperature is shown in a). After irradiation with an electron beam (current density: 1.5 A/cm^2) at 200°C for 2 and 5 min resp., micrographs b) and c) were obtained. After annealing the specimen at 700°C for 45' micrograph d) was taken. The series of micrographs shows convincingly that the originally present dislocations are very strongly influenced by the action of point defects which causes the existing dislocations to climb (helical dislocations) furthermore, new

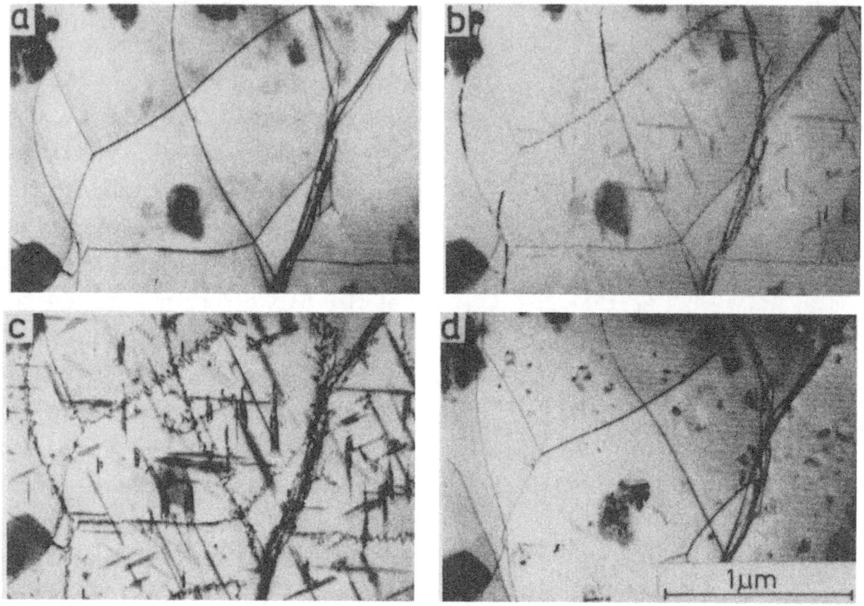

Fig. 5. HVEM in-situ observation of defects in <111> oriented Ge: a) initial state, b,c) after electron irradiation (1.5 A/cm²) at 200°C for 2' and 5', resp., d) after annealing at 700°C for 45'

defects (lying in <110> directions - rod-like defects) are created. Disregarding some change in the position of dislocations the annealing has brought out the initial state.

2.) The detection of the electrical activity of crystal defects may be carried out indirectly as well as directly. On one hand, in the indirect method the defect structure of different microregions is characterized and statistically evaluated, on the other hand the detection of properties is carried out by integral electrical measurements on extended solids. The direct method increasingly used facilitates the detection of the structure and its properties in a submicroscopic region, thus yielding a 1:1 correlation between structure and properties of a single crystal defect, which requires the combined application of TEM and special SEM techniques. For detecting the electrical or optical properties of semiconducting materials the following interaction processes between the electron probe and the specimen may be used in a SEM: a) the electron beam induced current (EBIC), which reveals the recombination efficiency (or the multiplication efficiency) of a defect for beam-induced charge carriers, b) the emission of cathodoluminescence radiation, which is related to radiative recombination processes in the material, and c) the thermal ionization of charge carriers from deep level defect centers by the action of a pulsed electron beam (scanning deep level transient spectroscopy - SDLTS), which facilitates the local detection of electron states in the band gap. In the cases mentioned the achievable information is laterally restricted to the micron region. For characterizing an acting defect reliably the higher resolving TEM has to be used additionally for this kind of direct correlation between defect structure and property, and the following combined methods are widely used: SEM(EBIC)/TEM, SEM(CL)/TEM and SDLTS/TEM [24-28]. Considering the special fact that thin crystal foils can be used throughout the experiments, STEM should be applied instead of TEM, so that the combined investigations may be carried out in the same instrument in a relatively simple way.

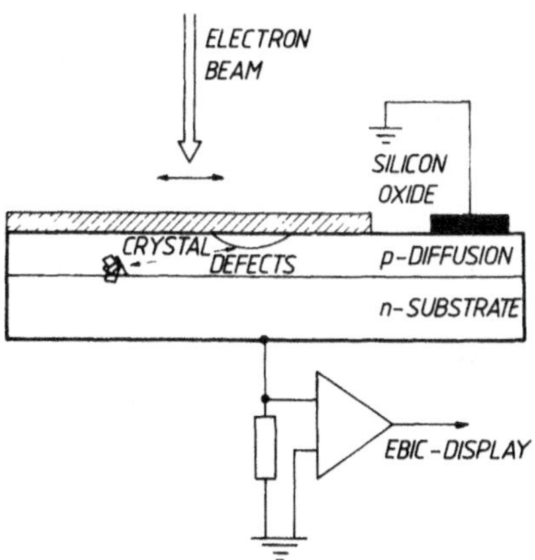

Fig. 6. Scheme of the EBIC technique

The method of detecting crystal defects by using the SEM(EBIC) technique (for review see [29]) is based on the fact that the action of the electron probe (see Fig. 6) of the SEM produces electron-hole pairs in a semiconducting specimen, which usually recombine quickly. In regions with a sufficiently high electrical field, e.g. at a p-n junction or at a Schottky barrier, a separation of the electron-hole pairs is reached, thus rendering it possible to use the produced current and conductivity, resp. to form the SEM image. Defects present in the specimens influence the recombination behaviour, as sometimes also the generation behaviour of the electron-hole pairs. Thus, the SEM technique is most suitable for the local detection of this defect-induced recombination or generation behaviour of the defects. In evaluating the electrical activity of defects one has to take into account that the resulting EBIC contrast is dependent on the position of the defect with respect to the junction plane and the specimen surface (see e.g. [30,31]). Taking this into account, the results of many SEM(EBIC)/TEM investigations e.g. show that primarily it is not the type of a dislocation (edge, mixed, screw) which determines its electrical activity, but rather its core structure having a strong influence. By combining TEM (weak beam) investigations with

SEM(EBIC) studies Booker et al. [32] showed for the first time that dissociated segments of edge dislocations in silicon have a higher electrical activity than those not dissociated. The same applies as well to 60° dislocations in silicon [33], as it is shown in Fig. 7. The example is related to investigations of diffusion-induced dislocations in a boron-doped silicon layer. The micrographs (a) TEM, b) SEM/EBIC show a screw dislocation (A), 60° dislocations (B,C) and a 60° dislocation (D) inside the field region. Besides the almost zero contrasts of the screw dislocation (A) and the dislocation in the field region (D), the micrographs display the very different EBIC contrasts of segments of 60° dislocations, which could be stereoscopically identified to lie in the same depth position. The high-contrast segments (B), however, can be attributed to dissociated parts of the 60° dislocations, thus verifying their high electrical activity. Contrary to this, the segments C, which are attributed to not-dissociated parts of the 60° dislocations, exhibit only a low electrical activity.

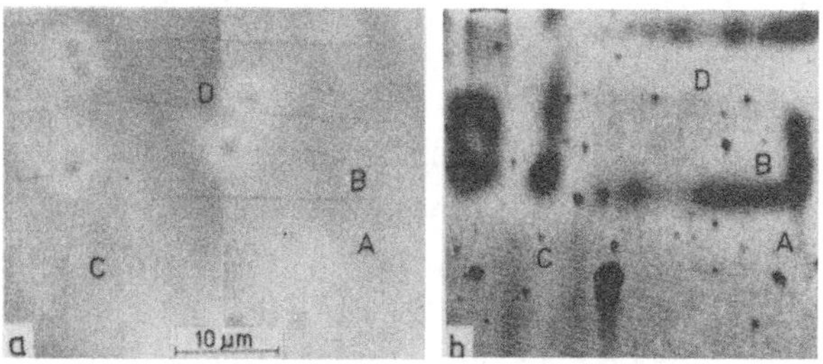

Fig. 7. Screw dislocation (A) and 60° dislocations (B,C,D) in <001> oriented silicon. a) HVEM micrograph (1000 kV, $\vec{g}$ = 220); b) SEM/EBIC micrograph (10 kV)

Up to now the question is not yet completely solved whether the changed core structure of the dissociated dislocation itself has a higher recombination efficiency, or whether the core of a dissociated dislocation strongly attracts impurity atoms from the surrounding, thus resulting in higher recombination efficiency because of its special state of decoration. According to the observed strong gettering action of defects (see e.g. [34-36]) the latter assumption might be valid. This is supported by the example of Fig. 8 demonstrating the decisive influence of the decoration on the recombination efficiency [37]. The micrographs (a) TEM, (b) SEM/EBIC show a stacking fault with a strong gettering efficiency, which can be concluded from the bright area in the surroundings of the stacking fault visible in the SEM/EBIC micrograph. In the region of this gettering zone the recombination efficiency of two neighbouring 60° dislocations is drastically reduced.

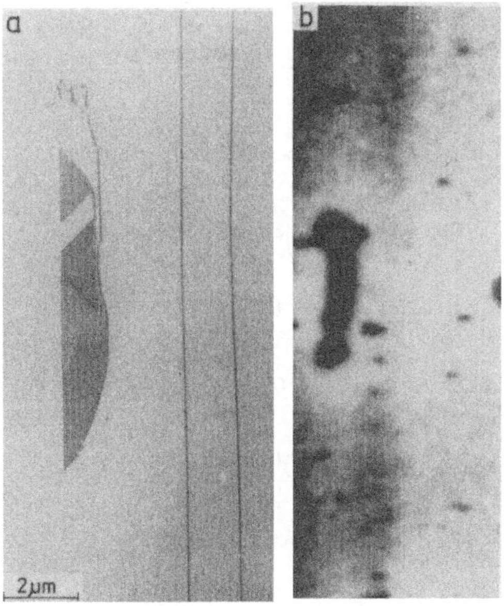

Fig. 8. Influence of a strongly gettering stacking fault (with a bright gettering zone) on the EBIC contrast of neighbouring 60° dislocations. (a) HVEM micrograph; (b) SEM/EBIC micrograph

Finally, a topical investigation carried out by using combined SEM/EBIC, TEM and EPR measurements [38] should be mentioned, in which the influence of point defects on dislocations of different types could be shown by investigating samples with and without dangling bond EPR centres.

The great variety of interaction processes between the electron beam and the specimen in the SEM offers advantageous possibilities of defect investigations, not only with respect to the imaging of crystal defects, but also concerning rough defects in the operation of devices (interruption of conducting paths etc.), which are of eminent practical importance in semiconductor technology (microelectronics) making use especially of the secondary electron voltage contrast. The latter one, however, is also used in the stroboscopic technique, which allows the dynamics of switching states to be investigated.

Acknowledgement

The shown electron micrographs are related to the research work of the IFE Halle. The author is indebted to the following colleagues engaged in this work: Dr. R. Gleichmann, Dr. H. Johansen, Dr. M. Pasemann, Mr. D. Baither, Mr. H. Bartsch, Mr. H. Blumtritt, Mr. D. Hoehl.

References

1. J. Chikawa, S. Shirai, Jap. J. Appl. Phys. 18, Suppl. 18-1, 153 (1979)
2. P.B. Hirsch, R.W. Horne, M.J. Whelan, Phil. Mag. 1, 677 (1956)
3. W. Bollmann, Phys. Rev. 103, 1588 (1956)

4. P.B. Hirsch, A. Howie, R.B. Nicholson, M.J. Whelan, D.W. Pashley, Electron Microscopy of Thin Crystals, London, Butterworths (1965)
5. G. Thomas, M.J. Goringe, Transmission Electron Microscopy of Materials, New York, Wiley & Sons (1979)
6. A.K. Head, Aust. J. Phys. 20, 557 (1967)
7. P. Humble, Aust. J. Phys. 21, 325 (1968)
8. K. Scheerschmidt, J. Heydenreich, phys. status solidi(a) 42, 47 (1977)
9. D.J.H. Cockayne, I.L.F. Ray, M.J. Whelan, Phil. Mag. 20, 1265 (1969)
10. D.J.H. Cockayne, A. Hons, J. Phys. (Paris) 40, C6-11 (1979)
11. J.C.H. Spence, Experimental High-Resolution Electron Microscopy, Oxford, Clarendon Press (1981)
12. J.W. Menter, Proc. Roy. Soc. (London) A236, 119 (1956)
13. J.C.H. Spence, J.M. Cowley, R. Gronsky, Ultramicroscopy 4, 429 (1979)
14. K. Izui, S. Foruno, H. Otsu, J. Electron Micr. 26, 129 (1977)
15. J.L. Hutchison, C.J. Humphreys, A. Ourmazd, P.B. Hirsch, Proc. 7th Europ. Reg. Conf. Electr. Micr., The Hague 1980, Vol.I, P.304
16. T.Y. Tan, H. Föll, S.M. Hu, Phil. Mag. A44, 127 (1981)
17. I.G. Salisbury, M.H. Loretto, Phil. Mag. A39, 317 (1979)
18. J.A. Lambert, P.S. Dobson, Phil. Mag. A44, 1043 (1981)
19. M. Pasemann, Z. Lilienthal, J. Microsc. Spectrosc. Electron. 6, 123 (1981)
20. I.L. Krivanek, J.H. Mazur, Appl. Phys. Lett. 37, 392 (1980)
21. G.W. Cullen, M.S. Abrahams, J.F. Corboy, M.T. Duffy, W.E. Ham, L. Jastrzebski, R.T. Smith, M. Blumenfeld, G. Harbeke, J. Lagowski, J. Cryst. Growth 56, 281 (1982)
22. Proc. Int. Symp. In-Situ High Voltage Electron Microscopy, Halle 1979; in: Cryst. Res. Technol. 14, 1169 (1979)
23. H. Fujita, Y. Kawasaki, E. Furubayashi, S. Kajiwara, H. Taoka, Jap. J. Appl. Phys. 6, 214 (1967)
24. C.J. Varker, K.V. Ravi, J. Appl. Phys. 45, 272 (1974)
25. A. Ourmazd, D.B. Darby, G.R. Booker, Inst. Phys. Conf. Ser. No. 36, 251 (1977)
26. P.M. Petroff, D.V. Lang, J.L. Strudel, A. Savage, Proc. 9th Int. Congr. El- Micr., Toronto 1978, Vol.I, p.130
27. J. Heydenreich, H. Blumtritt, R. Gelichmann, H. Johansen, J. Phys. (Paris) 40, C6-23 (1979)
28. D. Fathy, T.G. Sparrow, U. Valdrè, J. Micr. (London) 118, 263 (1980)
29. H. Leamy, J. Appl. Phys. 53, R51 (1982)
30. C. Donalato, Optic 52, 19 (1978)
31. L. Pasemann, Ultramicroscopy 6, 237 (1981)
32. G.R. Booker, A. Ourmazd, D.B. Darby, J. Phys. (Paris) 40, C6-19 (1979)
33. J. Heydenreich, H. Blumtritt, R. Gelichmann, H. Johansen, Scanning Electron Microscopy 1981/I (Eds. P. Becker, O. Johari) Chicago, SEM Inc. 1981, p.315
34. H. Blumtritt, R. Gleichmann, J. Heydenreich, H. Johansen, physica status solidi (a) 55, 611 (1979)
35. H. Menniger, H. Raidt, R. Gleichmann, physica status solidi (a) 58, 173 (1980)
36. M. Kittler, W. Seifert, Cryst. Res. Technol. 16, 157 (1981)
37. J. Heydenreich, H. Blumtritt, R. Gleichmann, H. Johansen, Cryst. Res. Technol. 16, 133 (1981)
38. A. Ourmazd, E. Weber, H. Gottschalk, G.R. Booker, H. Alexander, Inst. Phys. Conf. Ser. No.60, Sect.I, 63 (1981)

ELECTRICAL AND OPTICAL MEASURING TECHNIQUES FOR FLAW STATES*

J. S. Blakemore

Oregon Graduate Center
Beaverton, Oregon 97006, USA

1. Introduction

In reviewing here some electrical and optical techniques, useful for measurements relating to the localized states of deep-level centers (impurities, native defects, and/or complexes), it must be noted that such techniques have made major strides in the last decade. We should also acknowledge comprehensive reviews published in recent years. These include contributions by Grimmeiss [1], and by Mircea et al. [2], in a volume from a previous School in this series.

The declared topic of the present school is the study of "defect complexes". However, measuring techniques have much in common, whether the localized deep-lying states arise from an impurity (substitutional or interstitial), from a single site native defect, or from a local complex. Shockley and Last [3] introduced the convenient short word "flaw" to denote any of the above which provides localized states that communicate with the valence and/or conduction bands. The present writer uses this convenient name elsewhere [4] in the same context: a perturbation of the perfect lattice, which is spatially constrained in all directions.

Blood and Orton [5] have given a lengthy review on electrical characterization of semiconductors. This gave ample consideration to methods sensitive to flaws. These included techniques suitable for use with semiconductor material in "bulk" form, as well as with the depletion regions of P-N junctions, or of barrier ("Schottky") contacts.

A second major contribution to the recent literature on methods for

---

*Work supported by the U.S. National Science Foundation, through Grant DMR79-16454.

identification and diagnosis of states from deep-level flaws has been a review by Miller et al. [6] of various ways to conduct capacitance transient spectroscopy. Most of the techniques discussed therein were intended to deduce the emission coefficient characteristics of deep-lying flaws. These were outgrowths of the "emission rate window" procedure which Lang [7] popularized as "deep-level transient spectroscopy", or DLTS for short. That popular approach has engendered an extensive literature, which includes some more recent comments by Lang [8] on the experimental aspects.

A review by Grimmeiss [9] represented a different emphasis on the possible experimental methods for the investigation of flaw states and their capture/emission characteristics. This review recapitulated some ideas that Grimmeiss had expressed previously [10], concerning deep-level flaw analysis in Schottky barrier depletion layers by optical excitation. The techniques noted in the account which now follows includes both bulk material and barrier layer approaches. However, the usefulness of processes which are optically induced (and/or with a luminescent consequence) will be noted in each of the following sections.

2. The Relative Merits for Bulk and Barrier Layer Methods

Until the late 1960s, observations of deep-lying flaws in semiconductors, and in insulating photoconductors, were almost invariably made with the solid in a "bulk" form. While some measurements such as carrier lifetime *can* be made upon a complete ingot [11,12], this is not practical for obtaining *quantitative* information about deep level flaws in the crystal. Some nondestructive contactless (electrical as opposed to purely optical) measurement method were described by Miller et al. [13]. Also, lifetime can be sampled over regions of a complete semiconductor wafer by using temporary mercury Schottky contacts [14], or in a contactless arrangment that monitors the time evolution of Q for an adjacent tuned R.F. circuit following illumination [15]. However, the range of conditions over which a phenomenon must be explored to analyze a deep-level flaw quantitatively usually requires that a bulk sample be shaped specifically for that purpose.

Simple measurements of conductivity and magnetotransport, usually employed for enumeration of the *shallow* impurity dopant concentrations controlling n_o and p_o, tend to be useless even for the detection - let alone the characterization - of important deep-level flaws present in small concentrations. For we know that the performance of an active device may be profoundly affected by some deep flaw species with a concentration 10^5 or more times smaller than that of the shallow dopants. Such flaws can be detected (and hopefully both identified and characterized) only by non-equilibrium methods.

The 1960s-era books by Bube [16] and by Ryvkin [17] described various ways that photoconductive (and photoHall) methods can be used with a semiconductor sample in bulk form. The word "bulk" here may mean a separately cut slab, or a relatively thin conducting epitaxial layer on an inert (such as semi-insulating) substrate, or a deposited polycrystalline or amorphous thin film on an entirely alien mechanical support base. A "bulk" sample does not have to be large in all three dimensions; but the term is used here in the connotation that the flaws being monitored are not just those located within the depletion depth of a P-N junction or Schottky barrier contact. A "bulk" sample may be measured by a purely optical method which requires no contacts, or by an electrical method for which some current flow does occur through contacts. In the latter case, ohmic contacts are often desirable - but not always possible. And so "bulk" electrical measurements must often be arranged so that contacts with some blocking/rectifying characteristics can be tolerated, and with most of the flaw sites at which capture/emission processes occur lying in the relatively low-field region beyond contact depletion regions.

This question of contacts and their status and influence has long been known as a vexing one. The problem is not severe for work with a semiconductor such as Ge or Si, where well established (and reasonably foolproof) methods exist for making ohmic or barrier layer contacts - as desired - to samples of N-type, near-intrinsic, and P-type material. That happy status has not been achieved with most compound semiconductors. Complications which can arise when a "bulk" sample has non-ohmic contacts were discussed 20 years ago by Rose [18]. Such complications helped lay the groundwork for the 1968 proposal by Weisberg and Schade [19] that the thermally-stimulated-current (TSC) method be applied to flaws lying exclusively within the depletion layer depth of a diode structure.

Shortly thereafter, it was recognized that barrier layer capacitance, by being indicative of the bound charge in that layer, must also be sensitive to changes in the charge status for deep-level flaws in that specific volume of space. These potentialities for a diode structure were explored in great detail in a lengthy 1970 paper by Sah et al. [20]. They analyzed the uses of forward bias injecting pulses, reverse bias sweepout transients, and changes induced by intrinsic and/or extrinsic light; on electron or hole capture processes in the depletion layer, and also on the rate constants for electron or hole emission from flaws located in that layer. The consequences of this exposition by Sah et al. were a dramatic change in the balance between popularity of device structure measurements as contrasted with bulk *electrical* methods.

That balance was further changed when capacitative transient spectroscopy, DLTS, was proposed [6,7]. For this conveniently offered temperature as the independent variable, in determining the emission coefficient characteristics of deep-lying flaws. More recently, the emission rate window approach which DLTS popularized has been used in the analysis of quasi-bulk conductance transients [21-25]. This photoconductance transient method, valuable for high-resistivity materials, was named "optical DLTS" or "ODLTS" by Martin and Bois [22]; but the rather more distinctive name "PITS" (for photo-induced transient spectroscopy) was suggested by Fairman et al. [23] at about the same time.

Thus both bulk, and depletion layer electrical methods have their useful roles, and they can often usefully complement each other. Before returning to this, it is appropriate to comment on some purely optical methods, for which contacts (ohmic or not) are not required.

3. Optical Methods

Upwards optical transitions, between a flaw ground state and an excited state (photoexcitation) or a band state (photoionization, or photoneutralization), can be detected - in principle - from the decrease in optical transmission of a bulk sample, in the form of a polished parallel-sided slab. Among others, the present writer has used this method for deducing the magnitude and spectral shape of the photoionization cross-section $\sigma_I(h\nu)$ for flaw states in silicon [26] and GaAs [27]. However, the fractional transmittance for a sample of thickness t:

$$T \equiv (I_t/I_o) = (1 - R)^2 \exp(-\alpha t)\left[1 - R^2 \exp(-2\alpha t)\right]^{-1}$$

falls from a maximum of $(1 - R)/(1 + R)$ when $\alpha t \ll 1$, towards zero when $\alpha t \gg 1$, in a manner that provides satisfactory discrimination for the absorption coefficient (and thus for the photoionization cross-section σ_I) only for the (approximate) range $1 < \alpha t < 5$. Thus a sample thick enough for a reliable determination of σ_I(max) is *too thin* for any adequate exploration of the rapid fall-off of $\sigma_I(h\nu)$ near threshold.

Fortunately, there are other methods for which the techniques of data retrieval and analysis permit tracing $\sigma_I(h\nu)$ far down the threshold slope. One of these is the electrical procedure of photoconductivity; for which the "constant photocurrent" technique [28] provides one variant of the more usual procedure [29]. One alternative is given by photo-luminescence excitation (usually called PLE; an example is shown in Fig. 3), while photocapacitance provides another. Examples of the latter are shown in both Figs. 3 and 4, with the curve of Fig. 4 displaying the results of what Chantre et al. [30] call "DLOS". First, however, we need to get back to the topic of optical *absorption*.

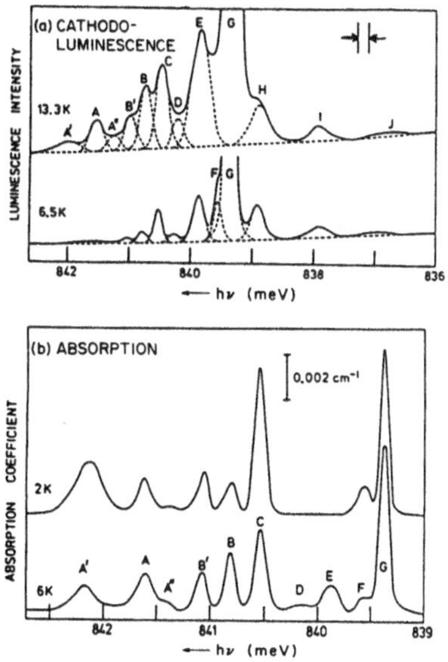

Fig. 1. Results of Lightowlers et al. [31] for (a) cathodoluminescence, and (b) optical absorption, associated with the fine structure of the 0.84 eV zero phonon line of the (111)-oriented Cr-related complex in GaAs. It should be noted that the absorption shown here was superimposed on a much stronger broad-spectrum absorption

For optical absorption, as deduced from transmission, can be valuable in the analysis of bound-to-bound radiative transitions. This requires low temperatures, and high resolution. Part (b) of Fig. 1 exemplifies such data, for Cr-doped GaAs near 0.84 eV, the location of the "zero-phonon" line of the (111)-oriented Cr-related complex. The work of Lightowlers et al. [31] is shown in Fig. 1, with cathodoluminescence induced by 50 keV electrons in part (a), and the upwards transitions of optical absorption in part (b). The temperature dependences for the various fine structure components permitted a deduction of ground state splittings (in the range 1 to 3 meV) for this complex.

Optical *absorption* can be detected in ways other than simply by a reduction of transmission. In one recently described and slightly unconventional approach that Bubenzer and Bimberg [32] call "CAS" (signifying calorimetric absorption spectroscopy), GaP was excited with pulsed monochromatic light. Then the absorption strength (at a temperature $T \simeq 1.5$ K) was detected by means of sensitive carbon thermistors, which were thermally coupled to the GaP sample. In this way, Bubenzer and Bimberg were able to register a weak excitonic transition, which remained undetected by optical transmission methods (though evident in luminescence).

Luminescence, already demonstrated by part (a) of Fig. 1, is a powerful technique for low temperature investigation of bulk material. Figure 2 illustrates some features of the photoluminescence near 1.03 eV,

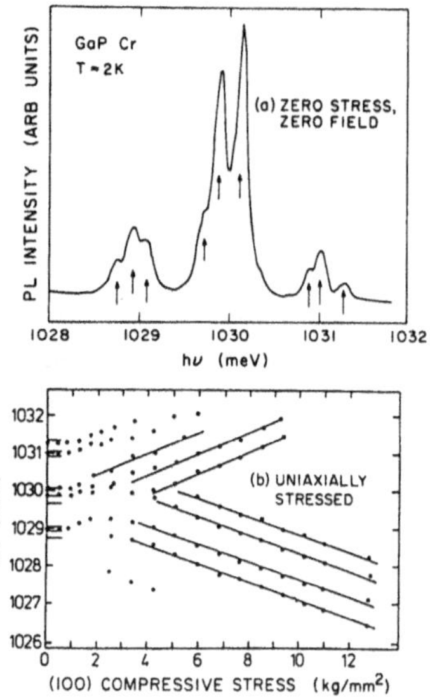

Fig. 2. Results of Eaves et al. [33] for photoluminescence near 1.03 eV of the (100)-oriented Cr-related complex in GaP. (a) The PL trace with no (externally applied) stress, and zero magnetic field. (b) Splittings created by a (100)-oriented compressive stress

for Cr-related complexes in GaP investigated by Eaves et al. [33]. Even with no magnetic field and no externally applied stress, the zero-phonon "line" has several components, as shown in part (a). Eaves et al. studied the Zeeman splitting for this line system, and also the effects of stress, with the consequences of (100) compressive stress shown in part (b). One may note that compressive uniaxial stress such as this is easy to apply (with a weighted rod bearing on the sample from outside the cryogenic part of the system).

Whereas cathodoluminescence (Fig. 1) and conventional photoluminescence (Fig. 2) use an excitation source that creates electron-hole pairs, and examine the consequences with a high resolution monochromator, the technique of photoluminescence excitation (PLE) varies hν for sub-bandgap exciting light, and collects a broad range of lower-energy photons which are emitted as a consequence of the excitation. The PLE method, advocated by Dean [34], has been used to good advantage by Monemar and Samuelson for the deep oxygen donor in GaP, in studying donor-acceptor transitions [35], and bound-to-free transitions [36]. Figure 3 illustrates one result of the latter, the deduction of Monemar and Samuelson from PLE of the photoneutralization cross-section spectrum for $\sigma^{o}_{pl}(h\nu)$ of GaP:O_p. It may be noted that these data extend in the threshold region to some 10^{-4} of the maximum response.

Figure 3 also includes, as a dashed line, the slope of the edge deduced by Henry [37] from a barrier layer (photocapacitance) method. Note that this has a distincly less rapid drop-off, which Monemar and

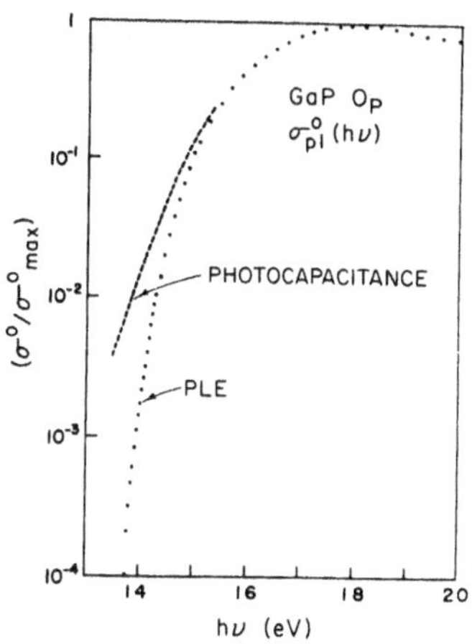

Fig. 3. Spectral dependence of the photoneutralization cross-section for GaP:O_P, after Monemar and Samuelson [36]. Their PLE data are shown as the dotted curve, compared with the results of Henry [37] obtained by a barrier layer method (photocapacitance)

Samuelson surmised could be a perturbation caused by the large electric field in a depletion layer. That large electric field can have other perturbing consequences also, in terms of field-induced phonon-assisted tunneling [38,39], when emission coefficients are being measured. This matter will be raised again later, when Fig. 7 is presented.

4. Thermally Stimulated Methods

It has long been known that a crystal containing deep flaws, with their occupancies perturbed by illumination at a low temperature, can then be heated at a constant (dT/dt) to permit restoration of equilibrium. A trace of conductance (TSC) or luminescence (TSL) versus time (and hence of temperature) can then show a series of peaks as successively deeper sets of flaw states become thermally accessible on the time scale of the experiment, and the theory of these effects was described 40 years ago [40]. TSC was discussed in detail in the books of Bube [16] and Milnes [41], and a more sophisticated differential analysis version has recently been described by Castagne et al. [42]. However, the form of a TSC trace can be affected by many of the experimental conditions; and several writers have given cautions [43,44] in recent years concerning pitfalls in TSC measurements that could lead to incorrect deductions concerning flaw depth, etc.

Whereas Weisberg and Schade [19] had described measuring TSC for the depletion layer of a P-N junction, Buehler [45] and Sah et al. [46] subsequently pointed out the value of monitoring the diode capacitance during the thermally stimulated readjustment of flaw occupancies, the procedure that Sah suggested be called "TSCAP". There has been a substantial ensuing literature concerning TSCAP measurements; but here again, there are many opportunities for pitfalls in correctly interpreting what one sees. (The writer still remembers TSCAP measurements in his own laboratory that did not work out happily.) Thus we should consider now capacitance measured under isothermal conditions; or with temperature varied in a manner that does not affect the results.

5. Isothermal Capacitance and Photocapacitance

The simplest measurement (in principle, at least!) for a depletion layer structure is of C-V at constant temperature, for a profile in integral form of the bound charge. Grimmeiss et al. [47] have warned that this is reliable only if the time scale of the reading exceeds the flaw refilling time, *not* a trivial consideration for deep-lying flaws at a low temperature.

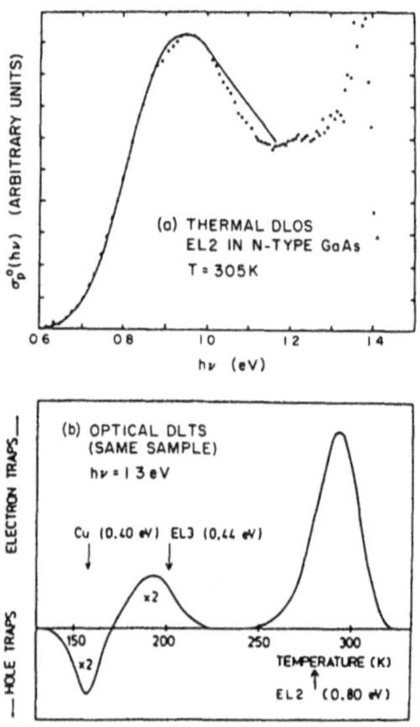

Fig. 4. Data of Chantre et al. [30] for N-type VPE GaAs ($n_o = 7 \cdot 10^{14}$ cm^{-3}) with a Schottky barrier contact. (a) The spectrum of "Thermal DLOS", transient photocapacitance, conducted isothermally at 305 K. (b) An optically excited (constant reverse bias) capacitance transient analysis versus temperature: "Optical DLTS", using short pulses of 1.3 eV photons

Photocapacitance has been used to study deep-level flaws in GaAs in a number of investigations over the years [48-50], and a differential signal form can be highly sensitive [50]. The deep-level-optical-spectroscopy, or "DLOS", group of procedures described by Chantre et al. [30] is isothermal transient photocapacitance, from which the spectral dependence of a flaw's optical cross-sections can be deduced. Part (a) of Fig. 4 shows DLOS for hole emission (photoneutralization) of EL2 in GaAs, as carried out near room temperature. The two parts of Fig. 4 are shown together to emphasize how useful it can be to make several kinds of measurement on the same sample. For part (b) of Fig. 4 shows the data of Chantre et al. of that GaAs sample (measured by the optically-stimulated DLTS method to be noted further in Section 6). Such optical DLTS traces, for a succession of

$h\nu$ values, can demonstrate when a DLOS spectral feature is controlled by just one optical transition process, and when several competing processes are involved.

6. Emission Rate Window Methods

Suppose flaw occupancies are disturbed by some stimulus: a flash of light, a forward biased pulse for a diode, etc., and then allowed to return towards equilibrium at constant temperature. How long this takes depends on that temperature, as well as on flaw depths, capture cross-sections, etc. If thermal emission from some flaw species provides the rate-determining step, then the decay rate speeds up with increasing temperature, as sympolized in Fig. 5. Thus a series of measurements at

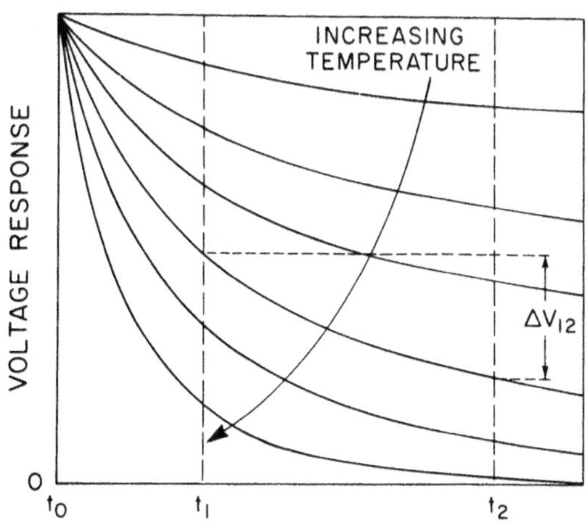

Fig. 5. Increase of recovery rate with temperature when this is controlled by a flaw emission coefficient. Thus for set times t_1 and t_2, the voltage signal differential ΔV_{12} rises, maximizes, and falls, with increasing T

progressively higher temperature will show ΔV_{12} to rise, maximize, and fall, if only one emission process is important. For a real sample with several flaw species, a curve of ΔV_{12} versus T can be expected to demonstrate these complexities. And upon this premise, DLTS was proposed by Lang [7], with the signal in his case coming from the relaxation of barrier layer capacitance. The emission coefficient selected by the choices for t_1 and t_2 controls the temperatures for the various peaks; and the variation of T_{max} with t_1, t_2, permits a determination of emission coefficient versus temperature - the "signature" of the flaw. Figure 6, taken from the classic paper of Lang and Logan [51], exemplifies temperature-scanned DLTS traces for various GaAs samples.

Note that temperature was *scanned* to obtain the curves of Fig. 6, yet the scan rate is slow enough so that each relaxation is an iso-

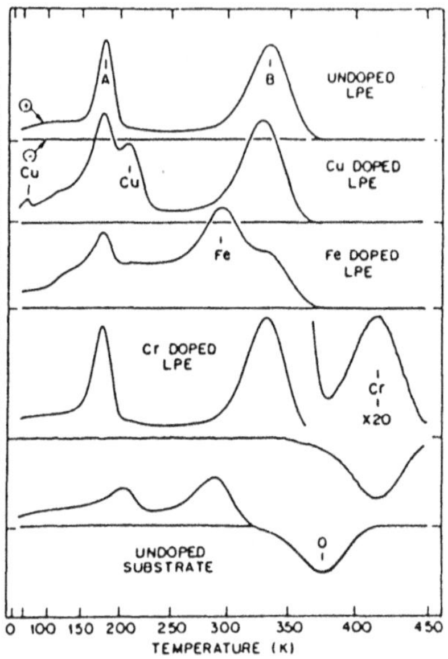

<u>Fig. 6.</u> Typical electrical DLTS traces, as reported by Lang and Logan [51] for variously prepared and doped GaAs samples, each made into the form of a P^+N junction for capacitance spectroscopy

thermal one. DLTS results do not (or should not!) depend on the temperature scan rate, and the temperature of each maximum can be established more accurately by means of isothermal point-by-point measurements.

The sensitivity and versatility of DLTS have resulted in numerous proposals for variations in the experimental or analytical procedure [6, 52,53, etc.], including constant capacitance versions [54], isothermal scans of the rate window [55], etc. The transient *current* response of a diode can provide an alternative to capacitance monitoring for DLTS-type experiments [56], and of course the pulse which perturbs flaw occupancies can be of photons [57,58] rather than of electrical bias. The extensive account of Chantre et al. [30] draws attention to the wide range of related and complementary techniques that can be applied to a diode structure, of which two were illustrated in Fig. 4.

Some words of caution are in order concerning DLTS and its ever-growing variant spectroscopic techniques. One is that non-exponential transients are often required by the laws of physics. This does not make analysis impossible, but recognition of this possible complication [59,60] is necessary to avoid a major source of error and misinterpretation.

A second caution is that the depletion layer of a P-N junction, or of a Schottky barrier, inevitably has a large electric field strength; even for zero bias, and more so for the reverse bias of most DLTS experiments. As we all know from elementary device physics, that field strength increases with doping. Figure 7 shows an example of the increase of electron emission coefficient with field strength, from the work of Makram-Ebeid [39] in studying EL2 in GaAs for various temperatures. He showed that this increase arises from phonon-assisted tunneling, and that it is prone to happen with many deep-level flaw species. Thus,

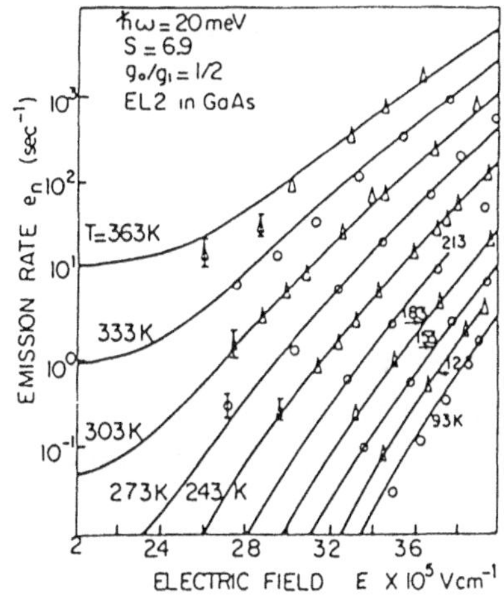

Fig. 7. Results of Makram-Ebeid [39] for the increase in electron emission coefficient with electric field strength, for EL2 in GaAs at various temperatures. This increase arises from phonon-assisted tunneling

Makram-Ebeid has shown [38] a similar trend of e_n with field for the E3 level in GaAs.

Such trends of emission coefficient with field can have severe consequences for DLTS results, unless recognized and averted (such as by the use of lightly doped material, to maximize W_o and minimize field). The present writer has discussed elsewhere [61] the contrast between *early* DLTS measurements of e_n for GaAs:Cr [51,62], and the values some 200 times smaller obtained by *subsequent* DLTS measurements [63] with *lightly* doped N-type GaAs containing chromium.

For all its other inconveniences, a bulk sample, with a reasonable volume of material between its contacts (of possibly questionable status), does enable the experimenter to study phenomena in a low-field regime. This has certainly contributed to the growth of the PITS technique [21-25, etc.], especially for high resistivity materials such as semi-insulating GaAs. Here one can vary temperature as an independent variable, in examining the response of a bulk or quasi-bulk conductive sample; relative to an emission coefficient window set by sampling times t_1 and t_2 following an exciting light pulse.

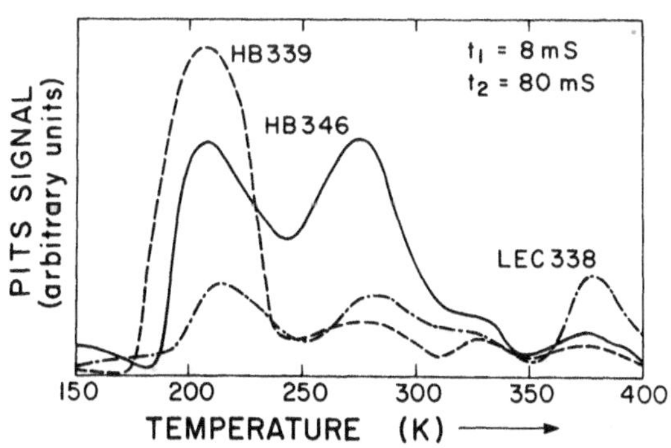

Fig. 8. Typical examples of PITS spectra. These are for semi-insulating Cr-doped GaAs samples [64]. The response of the deep chromium acceptor is actually the rather modest peak near 380 K. The other, larger, maxima result from shallower flaw species, including a prominent peak (centered on 210 K) arising from the E3 flaw species

Figure 8 shows typical examples of PITS traces, for GaAs:Cr samples measured in this laboratory [64]. Such curves clearly show that the material investigated contained deep-level flaws *other* than chromium, such as the EL6 center which contributes (for these values of t_1 and t_2) a peak near 210 K. As with DLTS, one can vary the values of t_1 and t_2 to scan for the response to a different emission rate window; and can consequently construct a flaw "signature" in terms of $e_n(T)$ or $e_p(T)$.

Thus, in emission rate window analysis, and in various other forms of experimental analysis, bulk material techniques and barrier structure techniques both have their uses. And these often can complement one another.

Acknowledgement

Acknowledgement is made to the U.S. National Science Foundation for their support of this work, through Grant DMR-79-16454.

References

1. H.G. Grimmeiss, in "New Developments in Semiconductors Physics; Lecture Notes in Semiconductor Physics, Vol. 122", ed. by F. Beleznay, G. Ferenczi and J. Giber (Springer-Verlag, Berlin, 1980) p. 50
2. A. Mircea, D. Pons and S. Makram-Ebeid, in Ref. 1, p. 69
3. W. Shockley and J.T. Last, Phys. Rev. 107, 392 (1957)
4. J.S. Blakemore, "Semiconductors Statistics", 2nd Edition (Dover Publications, New York, 1983)
5. P. Blood and J.W. Orton, Repts. Prog. Phys. 41, 157 (1978)
6. G.V. Miller, D.V. Lang and L.C. Kimmerling, Ann. Rev. Mater. Sci. 7, 377 (1977)
7. D.V. Lang, J. Appl. Phys. 45, 3023 (1974)
8. D.V. Lang, in "Thermally Stimulated Relaxation in Solids, Topics in Applied Physics, Vol. 37", ed. by P. Bräunlich (Springer-Verlag, Berlin, 1979) p. 93
9. H.G. Grimmeiss, Ann. Rev. Mater Sci. 7, 341 (1977)
10. H.G. Grimmeiss, in "Metal Semiconductor Contacts", Inst. of Phys. Conf. Ser. No. 22. (Inst. of Physics, London, 1974) p. 187
11. F.G. Vieweg-Gutberlet, in "Lifetime Factors in Silicon", A.S.T.M., Philadelphia (1980) p. 7
12. A.R. Gerhard and C.W. Pearce, in Ref. 11, p. 161
13. G.L. Miller, D.A.H. Robinson and S.D. Ferris, in "Semiconductor Characterization Techniques", ed. by P.A. Barnes and G.A. Rozgoni (Electrochemical Society, Princeton, 1978) p. 1
14. H. Reichl and H. Bernt, Solid State Electron. 18, 453 (1975)
15. H.W. Curtis and R.L. Verkuil, in Ref. 11, p. 210
16. R.H. Bube, "Photoconductivity of Solids" (Wiley, New York, 1960)
17. S.M. Ryvkin, "Photoelectric Effects in Semiconductors" (Consultants Bureau, New York, 1964) (English translation edition)

18. A. Rose, "Concepts in Photoconductivity and Allied Problems" (Wiley-Interscience, New York, 1963)
19. L.R. Weisberg and H. Schade, J. Appl. Phys. 39, 5149 (1968)
20. C.T. Sah, L. Forbes, L.L. Rosier and A.F. Tasch, Solid State Electron. 13, 759 (1970)
21. C. Hurtes, M. Boulou, A. Mitonneau and D. Bois, Appl. Phys. Lett. 32, 821 (1978)
22. G.M. Martin and D. Bois, in Ref. 13, p. 32
23. R.D. Fairman, F.J. Morin and J.R. Oliver, in "Gallium Arsenide and Related Compounds 1978", Inst. of Physics Conf. Ser. No. 45. (Inst. of Physics, London, 1979) p. 134
24. B. Devaud and B. Toulouse, in "Semi-Insulating III-V Materials", ed. by G.J. Rees (Shiva, Orpington, 1980) p. 241
25. T. Itoh and H. Yanai, in "Gallium Arsenide and Related Compounds 1980", Inst. of Physics Conf. Ser. No. 56. (Inst. of Physics, London, 1981) p. 537
26. R.A. Messenger and J.S. Blakemore, Solid State Commun. 9, 319 (1971)
27. W.J. Brown, D.A. Woodbury and J.S. Blakemore, Phys. Rev. B 8, 5664 (1973)
28. H.G. Grimmeiss and L-Å. Ledebo, J. Appl. Phys. 46, 2155 (1975)
29. M.A. Amato, M.C. Arikan and B.K. Ridley, in Ref. 24, p. 249
30. A. Chantre, G. Vincent and D. Bois, Phys. Rev. B 23, 5335 (1981)
31. E.C. Lightowlers, M.O. Henry and C.M. Penchina, in "Physics of Semiconductors, Edinburgh 1978", Inst. of Phys. Conf. Ser. No. 43, ed. B.L.H. Wilson (Inst. of Physics, London, 1979) p. 307
32. A. Bubenzer and D. Bimberg, J. Phys. Soc. Japan 49 (Suppl. A), 255 (1980)
33. L. Eaves, T. Englert, C. Uihlein and P.J. Williams, J. Phys. Soc. Japan 49 (Suppl. A) 279 (1980)
34. P.J. Dean, Phys. Rev. 168, 889 (1968)
35. B. Monemar and L. Samuelson, J. Luminescence 12/13, 507 (1976)
36. B. Monemar and L. Samuelson, Phys. Rev. B 18, 809 (1978)
37. C.H. Henry and D.V. Lang, Phys. Rev. B 15, 989 (1977)
38. S. Makram-Ebeid, Appl. Phys. Lett. 37, 464 (1980)
39. S. Makram-Ebeid, in "Defects in Semiconductors", ed. by J. Narayan and T.Y. Tan (North-Holland, New York, 1981) p. 495
40. J.T. Randall and M.H.F. Wilkins, Proc. Roy Soc. Lond. A 184, 366 (1945)
41. A.G. Milnes, "Deep Impurities in Semiconductors" (Wiley, New York, 1973)
42. M. Castagne, J. Bonnafe, J. Romestan and J.P. Fillard, J. Phys. C 13, 5555 (1980)
43. P.O. Fagerström and H.G. Grimmeiss, Solid State Commun. 26, 261 (1978)
44. C. De Blasi, S. Galassini, C. Manfredotti, C. Micocci and A. Tepore, Solid State Commun. 30, 453 (1979)
45. M.G. Buehler, Solid State Electron. 15, 69 (1972)
46. C.T. Sah, W.W. Chan, H.S. Fu and J.W. Walker, Appl. Phys. Lett. 20, 193 (1972)
47. H.G. Grimmeiss, L-Å. Ledebo and E. Meijer, Appl. Phys. Lett. 36, 307 (1980)
48. Y. Furukuwa, Japan J. Appl. Phys. 6, 675 (1967)
49. T.Okumura and T. Ikoma, Appl. Phys. Lett. 25, 572 (1974)
50. A.M. White, P.J. Dean and P. Porteous, J. Appl. Phys. 47, 3230 (1976)
51. D.V. Lang and R.A. Logan, J. Electron Mater. 4, 1053 (1975)
52. D.S. Day, M.Y. Tsai, B.G. Streetman and D.V. Lang, J. Appl. Phys. 50, 5093 (1980)
53. C.D. Wang and H.C. Lin, J. Appl. Phys. 52, 546 (1981)
54. N.M. Johnson, D.J. Bartelink, R.B. Gold and J.F. Gibbons, J. Appl. Phys. 50, 4828 (1979)

55. H. Okushi and Y. Tokumaru, Japan J. Appl. Phys. 19, L335 (1980)
56. J.A. Borsuk and R.M. Swanson, IEEE Trans. Electron Dev. ED-27, 2217 (1980); J. Appl. Phys. 52, 6704 (1981)
57. A. Mitonneau, G.M. Martin and A. Mircea, in "Gallium Arsenide and Related Compounds, Edinburgh 1976", Inst. of Phys. Conf. Ser. No. 33a. (Institute of Physics, London, 1977) p. 73
58. R. Brunwin, B. Hamilton, P. Jordan and A.R. Peaker, Electronics Lett. 15, 349 (1979)
59. A.M. White, B. Day and A.J. Grant, J. Phys. C 12, 4833 (1979)
60. G. Ferenczi, at Third "Lund" Conference on Deep-Level Impurities (1981)
61. J.S. Blakemore, J. Appl. Phys. 53, 520 (1982)
62. G.M. Martin, A. Mitonneau and A. Mircea, Electronics Lett. 13, 191 (1977)
63. G.M. Martin, A. Mitonneau, D. Pons, A. Mircea and D.W. Woodard, J. Phys. C 13, 3855 (1980)
64. J.S. Blakemore, S.G. Johnson and S. Rahimi, in "Semi-Insulating III-V Materials, Evian 1982", ed. by S. Makram-Ebeid and B. Tuck (Shiva, Nantwich, 1982)

PART II

THEORY

THEORY OF DEFECT COMPLEXES

Sokrates T. Pantelides

IBM Thomas J. Watson Research Center
Yorktown Heights, New York 10598, U.S.A.

Most methods that are used to describe point defects become cumbersome and impractical for defect complexes. Here, we review these methods and describe a simple alternative, which can provide both qualitative and semiempirical quantitative descriptions of the localized states associated with defect complexes. It is similar in spirit with the "defect-molecule" model which has sometimes been used for qualitative work, but is defined in a way that allows quantitative calculations. The method is particularly powerful for classes of complexes such as aggregates of identical point defects when the electronic structure of the primary point defect is known and for defects in wide-gap materials. We review results obtained for multivacancies in Si and for several defects in SiO_2.

1. Introduction

Defect complexes are aggregates of point defects which form during crystal growth, during processing (diffusion of dopants, ion implantation, annealing, etc.), or after a crystal has been irradiated with energetic particles or X-rays. Defect complexes often play a significant role in determining the characteristics and performance of devices by affecting carrier lifetimes, degradation, breakdown, etc. Microscopic theories, however, generally focus on point defects or, sometimes, pairs of point defects. For example, effective-mass theory has provided detailed understanding of the properties of the so called shallow donor and acceptor impurities which are used to dope semiconductors n-type or p-type, respectively. These impurities introduce very weak perturbation potentials which induce hydrogenic bound states in the band gap very near the band edges. Other impurities and most defect complexes generally introduce strong perturbing potentials. As a result, they induce highly localized states, some of which may lie deep in the band gap. Accurate theoretical descriptions of such deep defects have been lacking for many years. Recently, however, several techniques have been developed and implemented, and have provided detailed descriptions of several point defects. With the exception of some point-defect pairs, theoretical calculations for defect complexes are still lacking. In this paper, we shall review the merits and limitations of two methods which have been used widely for the study of point defects and point-defect pairs. We shall then describe a very simple method which can provide both qualitative understanding of the localized states and other related properties of defect complexes.

2. Methods for Point Defects

One method that has been used extensively for deep defects is based on so called cluster approximation [1]. Instead of treating an infinite crystal, one retains only a small cluster of atoms. Most calculational techniques can handle only a small number of atoms, typically 20-60. Thus, the selection of boundary conditions becomes very important. For example, a free cluster has states which are localized on the outer atoms ("surface" states). These states can interact with states that are localized about the defect in the "bulk" of the cluster and affect the results in unpredictable ways. For semiconductors, these cluster "surface" states are particularly bothersome, because they lie in the vicinity of the energy gap. One popular way to avoid this problem is to terminate the cluster with hydrogens which saturate the surface dangling bonds and move the surface states away from the energy-gap region. Sometimes, the strength of the hydrogen-host interactions is adjusted to optimize a particular bulk property such as the energy gap. An alternative way to avoid the surface dangling bonds has been to impose periodic boundary conditions. In effect, one is then treating a superlattice of defects in an infinite solid. Problems then arise from the fact the localized states in adjacent superlattice cells interact, giving rise to a band. Self-consistent calculations using a superlattice with a 54-atom cell [2] have shown that the vacancy in Si gives rise to a band with a dispersion of about 1 eV, spanning the entire energy gap. Semiempirical calculations using free clusters have also shown that the gap level changes significantly as a function of cluster size for clusters up to 64 atoms [1]. By using continued-fraction techniques to handle very large matrices, it has been found that one needs clusters of approximately 500 atoms to get acceptable convergence of the gap states [3].

During the last few years, Green's-function techniques have been developed and implemented which can describe an isolated defect in an otherwise infinite and periodic solid [4-7]. One first calculates an energy-band structure and the corresponding Green's functions of the perfect solid and relies on the localization of the defect potential to calculate the defect-induced <u>changes</u> in the electronic structure. The method has so far been successful in providing detailed descriptions of several point defects at the same level of accuracy as the corresponding descriptions of the perfect host crystals [4-10]. Even these calculations, however, become time-consuming and cumbersome when defect complexes are considered. If one forgoes self-consistency, and, in particular, uses semiempirical tight-binding Hamiltonians [11], the Green's-function method becomes practical for complexes involving a modest number of point defects [12]. The calculations again become time-consuming and involve manipulation of huge data bases when more extended defects need to be described.

In comparing the cluster and Green's-function approaches, we note that the key difference is the following: In the cluster approach, one attempts to calculate directly the eigenstates of the crystal with the defect, whereas, in the Green's-function approach, one first calculates the eigenstates of the perfect crystal (exploiting its periodicity), and then the changes induced by the defect. Since these changes are localized in space, no new approximation is introduced. In contrast, in the cluster approach, the termination of the infinite crystal is an additional approximation.

3. The Minimal-Basis Approach

As we saw in Sec. 2, small clusters (of order 20-60 atoms) do not generally yield convergent results and Green's-function methods become cumbersome for extended defects. It would, therefore, appear that large-cluster calculations involving more than 500 atoms in conjunction with a continued-fraction technique [3] are the most promising approach to describe extended defects. That approach, however, also involves substantial computational effort and has so far been used only for point defects and point-defects pairs. In this section, we will formulate a semiempirical theory which is isomorphic to a small-cluster calculation, but does not have any of the convergence problems faced by conventional cluster calculations. In fact, in the present approach, one must use a truly minimal basis set. In particular, one must use a number of basis orbitals which is precisely equal to the number of localized states one wishes to describe. The approach is thus isomorphic to the well-known "defect-molecule model", which has been widely used for <u>qualitative</u> determinations of the energy-level structure of many defects [13]. The defect-molecule model has, however, been discredited for quantiative calculations because both experiments and theory showed that the localized wavefunctions actually extend over several shells of neighbours. Indeed, it was this realization that led to semiempirical small-cluster calculations using atomiclike orbitals [14], in an effort to capture more of the localized wavefunction. We will show in this section, however, that a method can be defined in the spirit of the defect-molecule model which does not have its serious drawbacks and can be made quantitative. The central idea of this approach is contained in work by the author [15] on defects in SiO_2 and by Ivanov and the author [12] on multivacancies in Si. Here, we give a general discussion of the approach and draw general conclusions.

We start by redefining the concept of a cluster as follows. Instead of retaining a small number of atoms to simulate the infinite solid, we keep the infinite solid intact. The conventional cluster approximations can then be recovered by trunkating the basis set. The difference may, at first glance, appear insignificant. As we shall see, however, it has important consequences for semiempirical theories. We proceed by assuming the existence of an infinite basis set which is suitable for describing the infinite crystal. We divide this set into a small finite subset A and a subset B which contains the rest of the basis functions. The corresponding Hamiltonian matrix for the infinite crystal has the schematic form

$$\begin{pmatrix} AA & AB \\ BA & BB \end{pmatrix} \tag{1}$$

We can now recover all the conventional cluster approaches and identify their approximations:

(a) A free cluster corresponds to choosing the subset A to consist of the orbitals suitable for a finite cluster of atoms. The approximation amounts to dropping the AB matrix elements. This approximation can be justified if the AB matrix elements are small compared with the average separation between the energies of AA and of BB. In general, that is not true because AB are the matrix elements responsible for bandwidths and the eigenvalues of AA and BB are intermixed, spanning both valence and conduction bands. Thus, conventional free clusters are justified only if the clusters are large enough so that the neglect of AB affects the nature of the solutions only in the periphery of the cluster, i.e., gives rise to "surface" states without altering the "bulk" solutions, including the defect-induced states. All available studies so far point to the conclusion that several hundred atoms are needed for such convergence to be achieved.

(b) A periodic cluster corresponds to replacing BB by a periodic succession of AA's. The result is a defect-induced band of levels. As we mentioned already, clusters (i.e. unit cells) of 54 atoms in Si produce unacceptably large dispersions [2].

(c) A saturated cluster corresponds to replacing BB, AB, and BA with small finite matrices which simulate the role of the actual matrices. The only way to test the reliability of this procedure is to compare results with other more reliable calculations. Such comparisons, when possible [16], indicate that the method can produce reliable information. The approach has been used for point defectsand point-defect pairs, but would tend to become cumbersome for large complexes.

We are now able to identify the major intrinsic shortcoming of conventional cluster calculations: They are expected to produce both localized states and the "bulk bands". Indeed, the basis set, calculational technique, etc. are usually chosen to be suitable for the calculation of bulk properties. There is, therefore, little support for the often stated argument that cluster calculations ought to be good for deep defects since one is interested in localized states. This realization suggests that tremendous advantages can be gained if a cluster calculation can be designed to yield only localized states. Our formulation of the problem in terms of an infinite crystal and an infinite basis set consisting of two subsets A and B allows us to do just that: We can choose an optimal subset A which is sufficient to represent the defect-induced localized states and is such that the AB matrix elements can be neglected. One might surmise that such a choice would be virtually impossible, but that is not true. As we shall see, one can easily choose basis orbitals which, in principle, yield an exact expansion of the localized states and have AB matrix elements which are exactly zero! The only problem is that, though these basis orbitals exist in principle, they are too difficult to construct and hence not convenient for practical first-principles calculations. They can be used in semiempirical theories, however, where one requires only a suitable prescription for determining matrix elements. There is an important analogy here that is worth noting. In perfect crystals, Wannier functions exist in principle, but are difficult to construct and hence not very useful for practical first-principles calculations. Yet, they form the basis for semiempirical tight-binding theories in which only matrix elements are needed [17]. We will return to this analogy later on.

In order to choose the relevant basis orbitals for a given defect, we use the defect-molecule model as a guide. Take, for example, the isolated vacancy in Si. These states are known to arise primarily from the so called dangling bonds or dangling sp^3 hybrids of the four nearest neighbours. If we denote these four orbitals as h_1, h_2, h_3, and h_4, then symmetry allows us to write

$$\phi(A_1) = h_1 + h_2 + h_3 + h_4$$

$$\phi(T_2) = \begin{cases} h_1 - h_2 - h_3 + h_4 \\ h_1 + h_2 - h_3 - h_4 \\ h_1 - h_2 + h_3 - h_4 \end{cases} \tag{2}$$

In a conventional cluster calculation, these four basis orbitals would not be enough for quantitative work since the true vacancy localized wavefunctions are known to extend substantially beyond the nearest neighbours. We can, however, use eq.(2) to actually <u>define</u> the four h's in terms of the exact vacancy wavefunctions $\phi(A_1)$ and $\phi(T_2)$. When defined in this way, the h's have the symmetry of sp^3 hybrids. In fact, theylook like sp^3 hybrids near the vacant site, but also have whatever backbond structure is necessary to reproduce the exact vacancy wavefunctions when used in eq.(2). Recalling the analogy of these type of functions with the perfect-crystal Wannier functions, we refer to the h's as <u>Wannier hybrids</u> [12]. (In a more general case, when the principal basis orbitals do not have a dangling-hybrid character, we shall refer to them as <u>Wannier defect orbitals</u>.) It is now clear that we can describe an infinite crystal containing a single vacancy by choosing an infinite set which contains the four Wannier hybrids as subset A. The rest of the set (subset B) need not be specified further since, by construction, the AB matrix elements are zero. Thus the 4×4 matrix resulting from the expansion in terms of the four Wannier hybrids provides an exact description of the two main localized states of a single vacancy. By symmetry, only two matrix elements enter, namely

$$\varepsilon_h = \langle h_i|H|h_i\rangle \tag{3}$$

and

$$V_1 = \langle h_i|H|h_j\rangle, \quad i \neq j. \tag{4}$$

Thus, the method becomes very valuable if one has a reliable way of determining these two matrix elements. In general, the scheme is useful when applied to entire classes of defects (e.g. the single vacancy in a class of semiconductors) or defect complexes (e.g. multivacancies in a given semiconductor) when the energy-level structure of one or more numbers of the class is known from more parameter-free calculations or experiments. In the next Section we will briefly describe the multivacancies in Si, which have been studied by Ivanov and Pantelides [12], and a number of defects in SiO_2 which have been studied by Pantelides [15].

4. Applications

A. Multivacancies

We have already seen how the gap-region localized states of a single vacancy can be described by four Wannier hybrids corresponding to the four so-called dangling bonds or dangling hybrids of the four nearest neighbours. One can immediately extend the same discussion to any multivacancy. For example, the gap-region localized states of a divacancy can be described in terms of six Wannier hybrids corresponding to the six dangling bonds of the nearest neighbours. The resulting 6×6 matrix can yield, in principle, exact solutions, if all the matrix elements were known exactly. There are now four matrix elements, i.e. $\varepsilon_h = \langle h_i|H|h_i\rangle$; $V_1 = \langle h_i|H|h_j\rangle$, where $i \neq j$ and i, j are nearest neighbours of the same vacant site; and two different $V_2 = \langle h_i|H|h_j\rangle$, where $i \neq j$ and i, j are nearest neighbours of different vacant sites. We can now introduce an approximation and take ε_h and V_1 to be the same as the corresponding single-vacancy matrix elements, which in turn are

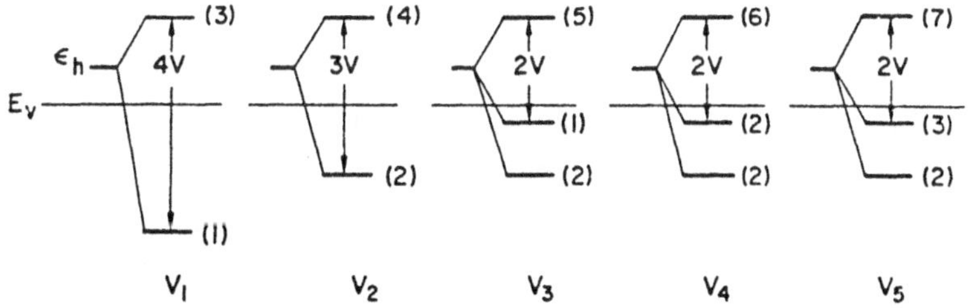

Fig. 1. Predictions for the gap-region localized energy levels of multivacancy chains in the (110) direction in Si using the minimal-basis approach. V_n denotes a chain of n vacancies and E_v is the top of the valence bands. The parameters ε_h and $V = V_1$ are fitter to reproduce the self-consistent results of the single vacancy (Refs. [5] and [6] from Ref. [12])

chosen by fitting the A_1 and T_2 levels obtained from an accurate Green's-function calculation [5,6]. The V_2 matrix elements may be taken as zero because they are between orbitals on more distant neighbours. In precisely the same manner one can describe any multivacancy by retaining only ε_h and V_1. Predictions for several multivacancies obtained in this manner by Ivanov and Pantelides [12] are shown in Fig. 1. The reliability of these predictions has been confirmed by thight-binding Green's-function calculations [12]. For even more accurate predictions one would need a reliable Green's-function calculation for both a single vacancy and a divacancy. One can then obtain values for ε_h, V_1 and the V_2's and predict the energy levels of higher-order multivacancies by neglecting only V_3's and higher-order matrix elements.

This very simple model of multivacancies has already been used to study lattice distortions and other properties of multivacancies [12]. It can be generalized to other defect complexes in a straightforward manner. The basic idea is to describe point defect in terms of minimal basis sets and determine the necessary matrix elements by fitting known accurate solutions. One can then build up defect complexes from these point defects and predict their electronic structure without needing any new parameter.

B. Defects in SiO_2

The scheme described above for multivacancies in semiconductors can in principle be used to describe defect complexes in SiO_2 or any other material. For SiO_2, however, no reliable self-consistent Green's-function calculations are available for any point defect. The minimal-basis approach is still useful, however, and has already been used by the author [15] to describe several point defects and small complexes in that material.

Let us start by describing the perfect crystal with a set of four sp^3 hybrids on every Si atom, pointing in the direction of the four nearest-neighbour oxygens, and three p orbitals on every oxygen atom. A bonding unit is then defined to consist of the three p orbitals on an oxygen and the two sp^3 hybrids pointing toward that oxygen atom (Fig.2). We can form linear combinations of these orbitals and construct bonding and antibonding orbitals giving us the energy-level structure shown in Fig. 3. By introducing matrix elements between orbitals belonging to neighbouring bonding units, we can then construct the energy bands [18]. If we now replace an oxygen with, say, a phosphorus atom, the only change in the basis set will consist of replacing the oxygen p

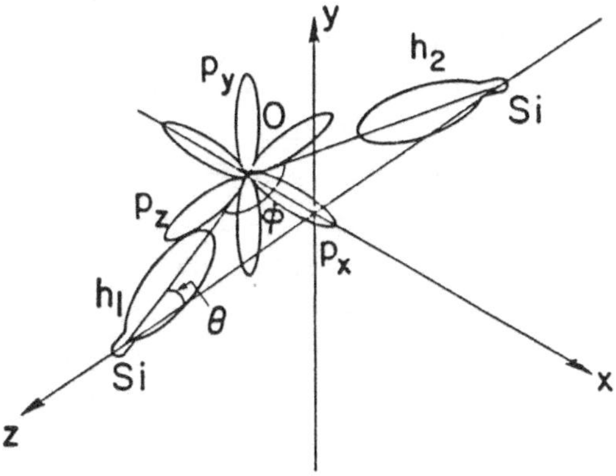

Fig. 2. Schematic representation of the basis orbitals in a bonding unit of SiO_2

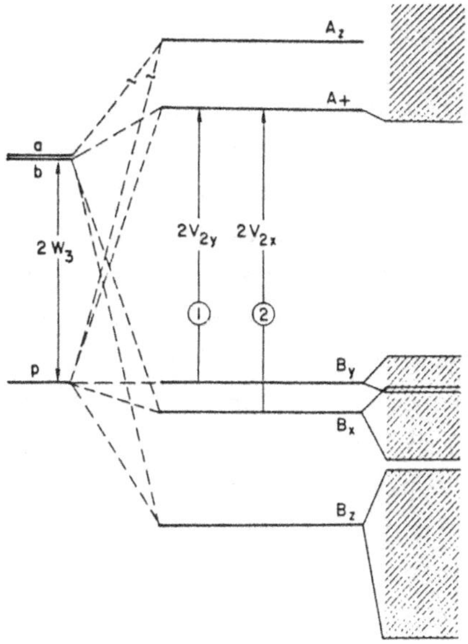

Fig. 3. Schematic energy levels in SiO_2. Left: the levels of the Si hybrids and the O p orbitals in a bonding unit; middle: the levels of the bonding and antibonding orbitals in a bonding unit; right: the energy bands

orbitals with phosphorus p orbitals. Thus, just one bonding unit is different from all the others. The "Wannier defect orbitals" are therefore chosen to correspond to the orbitals in the phosphorus bonding unit. In evaluating the necessary matrix elements we neglect contributions from orbitals in other bonding units. The resulting energy level structure is similar to that of the host bonding unit, but shifted upward in energy. Calculations for ten substitutional impurities have been reported and discussed in Ref. [16]. Similarly, predictions have been made for "dangling oxygens", SiH and SiOH defects. In all cases where experimental data are available, the theoretical predictions are in good agreement [15].

5. Summary

We have seen that the localized states of defects can be described by using only a minimal basis set of orbitals which are just adequate to describe the localized states. With such a choice, the coupling with the rest of the orbitals, which are primarily responsible for the propagating states, can be formally set to zero. The scheme is particularly valuable for describing defect complexes when the electronic structure of the component point defects is known from accurate calculations. One is then able to obtain reliable estimates of the matrix elements needed to describe the states associated with defect complexes. The scheme can also be used for point defects as well as complexes in wide-gap materials such as SiO_2, even if accurate calculations of point defects are not available.

References

1. For a review and critique of cluster methods, see S.T. Pantelides, Rev. Mod. Phys. 50, 797 (1978)
2. S.G. Louie, M. Schlüter, J.R. Chelikowsky and M.L. Cohen, Phys. Rev. B 13, 1654 (1976)
3. See, e.g., T.L. Reinecke and P.J. Lin-Chung, Solid State Comm. 42, 623 (1981)
4. M. Jaros, Adv. in Phys. 59 (1980), and references therein
5. J. Bernholc, N.O. Lipari and S.T. Pantelides, Phys. Rev. Lett. 41, 895 (1978); Phys. Rev. B 21, 3545 (1980)
6. G.A. Baraff and M. Schlüter, Phys. Rev. Lett. 41, 892 (1978); Phys. Rev. B 19, 4965 (1979)
7. U. Lindefelt and A. Zunger, Phys. Rev. B 24, 5913 (1981)
8. M. Scheffler, S.T. Pantelides, N.O. Lipari and J. Bernholc, Phys. Rev. Lett. 47, 413 (1981); Phys. Rev. B, to be published
9. J. Bernholc, N.O. Lipari, S.T. Pantelides and M. Scheffler, Phys. Rev. 26, Nov. 15 (1982);
10. G.O. Baraff, E.O. Kane and M. Schlüter, Phys. Rev. Lett. 47, 601 (1981); G.B. Bachelet, G.A. Baraff and M. Schlüter, Phys. Rev. B 24, 915 (1981)
11. J. Bernholc and S.T. Pantelides, Phys. Rev. 18, 1780 (1978)
12. I. Ivanov and S.T. Pantelides, to be published; S.T. Pantelides, I. Ivanov, M. Scheffler and J.P. Vigneron, Proceedings of the 12th International Conference on Defects in Semiconductors, Amsterdam, 1982 (in press)
13. See, e.g., the review article by J.W. Corbett and J.C. Bourgoin, in Point Defects in Solids, edited by J.H. Crawford, Jr. and L.M. Slifkin (Plenum, New York, 1975)
14. E.G., R.P. Messmer and G.D. Watkins, Phys. Rev. B 7, 1568 (1973)
15. S.T. Pantelides, Thin Sol. Films 89, 103 (1982)
16. For example, the gap-region localized states of the single vacancy in Si calculated by L.A. Hemstreet (Phys.Rev.B 15, 834 (1977)) are in good agreement with the results of Green's-function calculations.
17. See, e.g., S.T. Pantelides and J. Pollmann, J. Vac. Sci. Technol. 16, 1354 (1979)
18. S.T. Pantelides and W.A. Harrison, Phys. Rev. B 13, 2667 (1976)

CRITICAL COMPARISON OF THE THEORETICAL MODELS FOR ANOMALOUS LARGE LATTICE RELAXATION IN III-V COMPOUNDS

F. Beleznay

Research Institute for Technical Physics of the
Hungarian Academy of Sciences
Budapest, Újpest 1, P.O.Box 76, Hungary

Highly anomalous properties of some deep levels have recently been observed in many semiconductor materials, especially in III-V compounds and mixed crystals of them. Perhaps the most spectacular is the observation of the metastable effects such as persistent photoconductivity and photocapacitance quenching; accompanied with anomalous photoexcitation, absorption cross section and non-exponential time kinetics.

Most of the proposed theoretical models try to explain some of this properties neglecting some other important facts. Here a critical comparison between conflicting theoretical models is made and a new model is discussed which might explain the significant features of this complex problem.

1. Experimental Characteristics

Deep level experiments are using mostly one or closely related measuring technics. Hence here only those systems are reviewed in which anomalous properties measured by different technics are unabigously related to each other. These systems are the "EL2" level in GaAs (and more recently in $GaAs_{1-x}P_x$ and $Ga_xAl_{1-x}As$) and the "DX" level in $Ga_xAl_{1-x}As$: Te(Sn) mixed crystals.

Homogeneous samples of $Ga_xAl_{1-x}As$ doped with Te show persistent photoconductivity [1]; a typical concentration-temperature relation is shown in Fig. 1. In the same system the conductivity decay is highly temperature dependent and the time dependence is markedly non-exponential (Fig. 2). The activation energy estimated from this thermal dependence (∿120 meV) agrees well with the value measured by depletion layer methods [2] (DLTS and photocapacitance decay) with p-n junctions and Schottky barriers. Here both the emission and absorption cross sections can be determined which are both activated with the activation energies ∿280 meV and ∿180 meV, respectively. These anomalous features can only be found in a finite composition range ($0.4 \leq x \leq 0.8$), and from limited hydrostatic pressure experiments [3] the DLTS signal of this anomalous level behaves as if it were connected with the indirect conduction band minima (probably with those at points X). The optical

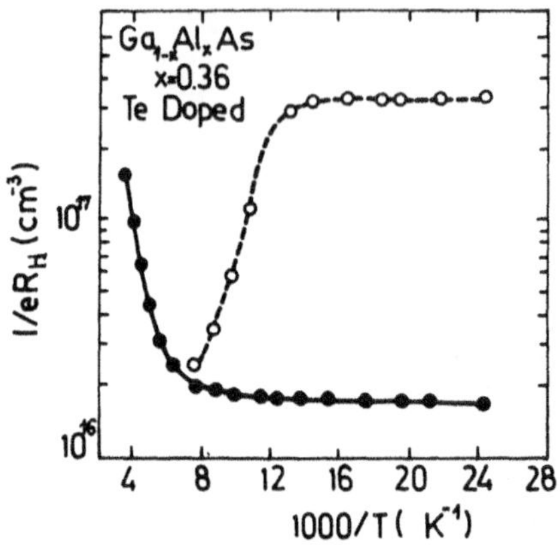

Fig. 1. Plot of the inverse Hall coefficient $1/eR_H$ versus temperature. The solid curve represents data taken with the sample in the dark, while the dashed curve is for data taken while the electron population is saturated under photoexcitations

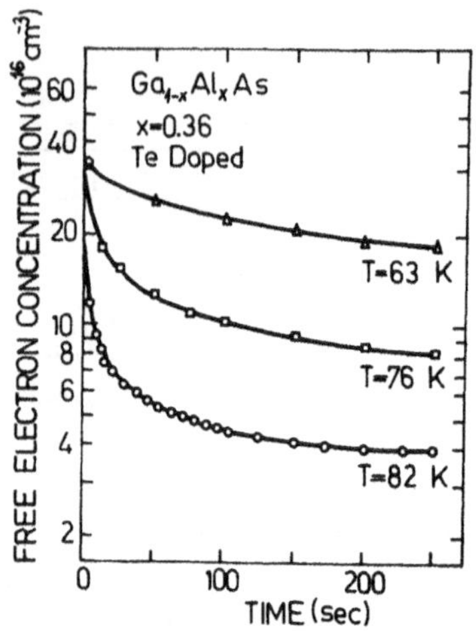

Fig. 2. Decay on the free-electron concentration $1/eR_H$ after photoexcitation for three different temperatures. In each case, $1/eR_H$ was saturated during a long excitation period

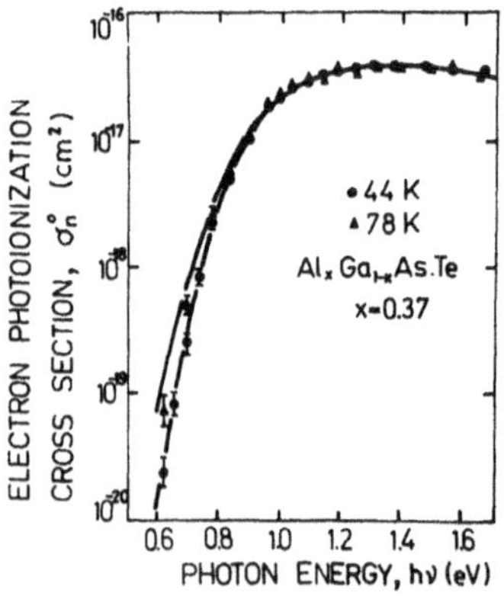

Fig. 3. Optical cross section for electron emission from DX center

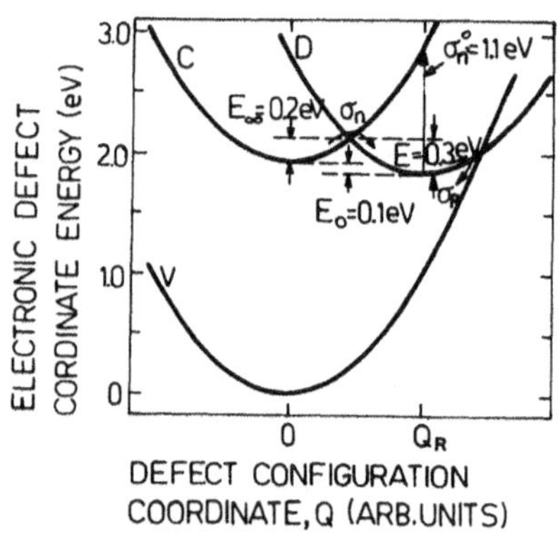

Fig. 4. C.c. diagram for DX center in $Ga_xAl_{1-x}As$ (phenomenological configuration coordinate diagram)

cross section is rather broad and temperature dependent (Fig. 3). If large lattice relaxation is the common cause of these features, a phenomenological configuration coordinate (c.c.) diagram can be derived which is shown in Fig. 4.

The "EL2" level, now identified as an antisite defect (or an antisite defect-vacancy complex) in as-grown undoped semiinsulating GaAs shows similar anomalies. The photo-capacitance quenching first observed by Vincent [4] can be explained if this system has two inequivalent state in GaAs, the higher one is metastable and the transition between them is an internal impurity transition [5]. Since this metastable state can not be excited optically with below band gap excitation (similar enormous Franck-Condon shift is observed in case of the DX level, too), the filling of the metastable state can be studied both with optical absorption measurements [6] in homogeneous material and with junction techniques. Once again independently on the actual geometries or the physical size of the investigated part of the sample, the experiments can be summarized in a configuration coordinate (c.c.) diagram [5].

More details of the experiments can be found in the original publications referred here and in two recent review papers [7,8].

2. Theoretical Models

Macroscopic space-charge barrier can inhibit excited charge carriers from getting recaptured as it was reviewed by Sheinkman [9] and investigated more recently by Queisser [10] in semiinsulating and n-type heterostructure of GaAs. In this latter case the space charge region is fixed by the value of the Debye screening length (about 1 μm). In the more microscopic cases having been previously discussed such a space charge region can hardly be imagined, especially in case of the junction techniques where the phenomena exist and independent of the searched part of the depletion layer which can be monitored by the reverse bias.

In addition, this model is intended to explain persistent conductivity and does not help to understand anomalous phonon scattering, temperature dependent optical properties among others.

A more microscopic model uses the potential fluctuation caused by microscopic disorder in mixed crystals. In mixed crystals compositional fluctuation is always present, and this local fluctuation can be visualized as some effective "local" medium [11] around the impurity or defect center with somewhat changed effective compositional ratio. For a highly localized electron level this fluctuation in the optical threshold energy can be calculated and it results in a broadened temperature dependent absorption curve. This is in full agreement with the experiments on transition metal impurities in mixed compound $GaAs_xP_{1-x}$. The magnitude of the fluctuation in the potential estimated by simple statistics is about 100 meV, and the fluctuation in the binding energy is roughly the same for a highly localized electron state such as the transition metal impurity state in III-V compounds.

However, this model, as far as I can see, is not relevant to the anomalous properties being discussed since a.) it can not occur in pure material ("EL2" in GaAs) and b.) it can hardly explain persistant behaviour. In fact the excited electron states are significantly more delocalized and the same sort of averaging as used for the localized deep level provides a microscopic barrier at least one order of magnitude less [12], much less than the experimental activation barriers (>100 meV). In case of excitons in $GaAs_{1-x}P_x$ which are expected somewhat more localized than a free electron state should be, the experimentally determined fluctuation broadening is about 5 meV [13] in good agreement with this estimation.

The most frequently used model for the anomalous properties listed in Section 1 is the large lattice relaxation (LLR) model combined with multiphoton processes proposed first by Henry and Lang in 1977 [14]. It is a semi-phenomenological model which can explain the activation barriers, the significant Franck Condon shift observed in the optical experiments and temperature dependences both in optical and thermal transitions. More recently the influence of the high electric field is also included with some success [15].

The difficulties, however, with this model are significant. First in lack of a microscopical basis the values of the c.c. data are purely empirical. It does not give any explanation why the Huang-Rhys factor is anomalously large. The dominance of the TA phonon-electron coupling is curious since LO a TO phonon couplings are expected in a partly ionic crystal and have usually observed in impurity level replicas. Are these phenomena connected with minima at X and if they are, why? In fact this is clearly demonstrated by Porowski et al. in a recent experiment [16] in InSb where similar persistent behaviour occurs under pressure.

Recent "a priory" deep level calculations [17] realize the significance of the inclusion of the lattice relaxation. In fact in a "realistic" calculation with relaxation for the vacancy state in Si about 100 atoms were moved from their equilibrium position. It is worth mentioning that the best results were obtained when the elastic model which is applied to calculate the relaxed lattice energy was the Keating model parametrized to fix the extremely low TA phonon energies at the points X.

3. Microscopic Model

In 1980 a new model has been proposed [18] which accounts the above features and with a relatively simple (perhaps oversimplified) calculation gives reasonable figures for the c.c. data.

The open structure zinc blende lattice is inherently unstable at high pressure. Every known zinc blende semiconductor crystal transforms into a more densely packed structure at high pressure, in the range of 30-150 kbars. Although the detailed atomic mechanism of this transformation has not been yet clarified [19], it is known that the shear elastic constant $c_s = (c_{11}-c_{12})/2$ and the TA phonon frequencies along the Σ and Δ lines and at point X are anomalously small and all soften at high pressure. In cubic materials the critical pressure for structural transition and the decrease of the elastic constant c_s are connected with each other. This structural stability criterion has been demonstrated for zinc blende semiconductors [20].

It is specific to semiconductors with a multi-valley indirect conduction band that the elastic constants depend on the electron concentration. Generally the electronic effect changes the shear elastic constant that splits the symmetry degeneracy of the valleys. The change of the shear elastic constant is negative because part of the free energy needed to strain the crystal is recovered by the transfer of electrons from valleys that are raised in energy by the strain to valleys that are lowered. An elementary symmetry argument shows that the electronic effect contributes to c_s in silicon-type materials which have energy minima along the Δ line, usually at point X. The corresponding symmetry related strain components are $\tilde{\varepsilon}_x = (2\varepsilon_{xx}-\varepsilon_{yy}-\varepsilon_{zz})/3$ and by the corresponding change of the indices one can obtain $\tilde{\varepsilon}_y$ and $\tilde{\varepsilon}_z$, respectively. In the high strain limit if the strain has only one $\tilde{\varepsilon}_x$ component, the energy at point X in the x direction is so much lowered that the repopulation is complete, hence the energy gain is linearly proportional to the electron density in the lowered minima ($\rho_x = \psi_x^* \cdot \psi_x$) and with the shift of these minima, usually given in eV's for unit strain (deformation potential C_u). Interestingly enough this picture is also locally true. A localized impurity wavefunction is a linear combination of the wavefunctions from degenerate minima. If the strain splits the energy degeneracy, also the measurable impurity transition energies will be splitted (and shifted correspondingly). At high strain the wavefunction has only components from the minima along the strain [21].

The proposed model describes impurity electrons which interact with the strain field. If the interaction between electrons and the strain field is strong enough,then by spontaneous symmetry breaking in the ground state the wavefunction has only components from minima which are lowered by the strain field. If local equilibrium for the strain is assumed, the following effective nonlinear Schrödinger equation can be derived

$$(-\nabla^2/2m^* - \frac{1}{\varepsilon_o r} - (C_u\rho_x^c/c_s)\tanh(\rho_x/\rho_x^c))\psi_x = E_e\psi_x \qquad (1)$$

Here the first two terms are the kinetic and potential energy terms for a single-charged impurity in the effective mass approximation and the third term describes the interaction with the (self-induced) strain field. In equilibrium $\tilde{\varepsilon}_x = C_u \cdot \rho_x/2c_s$. This cannot exceed the critical strain $\tilde{\varepsilon}_x^c$ at the zinc blende structural transition. Hence the interaction saturates at ρ_x^c corresponding to $\tilde{\varepsilon}_x^c$. In the highly strained (internal) region some contribution from the local "structural change" is also expected. In lack of a detailed microscopic treatment of the problem, however, this additive term is neglected.

Eq.(1) has to be solved with the condition that the total energy E_t should be minimal. The total energy contains the binding energy of the electrons E_e,and the energy of the strain field E_s $(= c_s\tilde{\varepsilon}_x^2/2)$. With this condition for a single-charged impurity one-electron and singlet two-electron bound states exist.

To illustrate the characteristics of this bound one-electron state some typical results are given [22]. If we used typical material parameters of the III-V compounds ($\varepsilon_o = 10$, $C_u = 6$ eV, $c_s = 5 \cdot 10^{10}$ dyn/cm^2, $m^* = 0.35$) the effective Bohr radius of the ground state is 25 Å, the corresponding electrical energy is 80 meV, rather similar to a hydrogenic level. The elastic distortion, however, is substantial. The range of the "transformed" critical domain is larger than 10 Å, and the stored elastic energy is 350 meV. It is seen that even in a reasonably delocalized state (where the use of the effective Hamiltonian, Eq.(1) a posteriory can be justified) an enormous lattice distortion can take place due to the nonlinearity of the electron-lattice interaction.

The difference between the present model and some more usual polaronic picture will be discussed elsewhere [22]. Here we only list some of the direct consequences of this model which are substantiated by the experimental facts:

- the anomalous state coupled to minima at X, it is strongly coupled to TA phonons, but it is not a simple electron-TA phonon model,
- always lowers the symmetry,
- pressure (stress) can induce,

- the lattice distortion (and the electron wavefunction) is delocalized,
- either soft phonons or fluctuations can introduce non-exponential time kinetics,
- in the former case in the long time limit $f(t) \sim e^{-(t/\tau)^{\alpha}}, (0 < \alpha < 1)$ and the log-log plot is linear,
- negativ U state can exists (second state of EL2?).

Naturally, as in every nonlinear calculation the results are rather sensitive for the value of c_s which depends on the initial stress. Moreover, an isotrop model can only approximate the highly anisotropic band minima at points X.

Keeping in mind these limitations it is remarkable that such a simple calculation can give reasonable figures with many characteristic features discussed earlier.

References

1. R.J. Nelson, Appl. Phys. Lett. 31, 351 (1977)
2. D.V. Lang and R.A. Logar, Phys. Rev. Lett. 39, 635 (1977)
3. D.V. Lang, R.A. Logan and M. Jaros, Phys. Rev. B19, 1015 (1979)
4. G. Vincent and D. Bois, Solid State Commun. 27, 431 (1978)
5. G. Vincent, D. Bois and A. Chartze, J. Appl. Phys. 53, 3643 (1982)
6. G.M. Martin, Appl. Phys. Lett. 39, 747 (1981)
7. J.M. Langer, Proc. 15th Conf. Phys. Semiconductors, Journal of the Phys. Soc. of Japan 49, 207 (1980)
8. D.V. Lang, ibid 49, 215 (1980)
9. M.K. Sheinkman and A.Ya. Shik, Fiz. Tekh. Poluprovodn. 10, 209 (1976) (Sov. Phys. Semicond. 10, 128 (1976))
10. H.J. Queisser and D.E. Theodorou, Phys. Rev. Lett. 43, 401 (1979)
11. L.A. Labedo, private communication, to be published
12. V. Sa-yakanit, Phys. Rev. B 19, 2266 (1979)
13. Shui Lai and M.V. Klein, Phys. Rev. Lett. 44, 1087 (1980)
14. C.H. Herry and D.V. Lang, Phys. Rev. B 15, 989 (1977)
15. S. Makram-Ebeid and M. Lannov, Phys. Rev. Lett. 48, 1281 (1982)
16. S. Pozowski, L. Konczewich, M. Konczykowski, R. Aulombard and J.L. Robert, Proc. 15th Int. Conf. Phys. Semiconductors, Journal of the Phys. Soc. of Japan 49, 271 (1980)
17. G.A. Baraff, E.O. Kane and M. Schlütez, ibid. 49, 231 (1980)
18. F. Beleznay, Recent Development is Condensed Matter Physics (Plenum Press, 1981) Vol. 1, p. 743
19. B. Okai and J. Yoshimoto, J. Phys. Soc. Japan 45, 1880 (1979)
20. D.S. Rimai and R.J. Sladek, Solid State Commun. 30, 591 (1979)
21. R.A. Cooke, R.J. Nicholas, R.A. Stzadling, J.C. Portal and S. Abkenazy, Solid State Commun. 26, 11 (1978)
22. The more detailed derivation of our model with different calculations will be published elsewhere. There a critical comparison with polaron-type calculations will also be given

VACANCY RELATED STRUCTURE DEFECTS IN SiO_2 - CYCLIC CLUSTER CALCULATIONS COMPARED WITH EXPERIMENTAL RESULTS

P. Deák and J. Giber
Physical Institute of the Technical University Budapest
H-1521 Budapest, P.O.Box 112, Hungary

1. Introduction

As it is well known, the defects and defect complexes of silica are of particular importance from the viewpoint of MOS device operation. A fundamental type of these centers are those connected with an oxygen vacancy. The vacancy in SiO_2 undergoes strong and different (symmetric or asymmetric) relaxations depending upon its charge state and moreover, it can easily form complexes with hydrogen atoms or protons. It is a general opinion in the literature [1] that one of these vacancy based structural defects should cause the most common center in silica: the so called E' center which appears both in crystalline and in amorphous forms with very similar experimental characteristics. Therefore, the deep levels of such centers may play a substantial role in an MOS device too. Our paper is devoted to the presentation of a theoretical method able to be applied for calculating the electronic structures of these defects on a level which is suitable for comparing results with experimental data.

The accurate calculation of deep levels of strongly relaxed point defects has ment a serious problem for long years [2]. Recently however, very efficient and elegant methods were elaborated and applied for semiconductor defects [3,4,5] based on the Koster-Slater approach [6]. In these methods the energies of the localized levels can be found by seeking the zeros of a determinant constructed from the perfect crystal Greenian and from the defect potential on a localized basis. Consequently, the order of the determinant is determined by the localization of the defect potential, while the environment of the defect is taken into account almost exactly via the perfect crystal Greenian. This means of course, that the latter has to be previously known, preferably from a band structure calculation carried out using the same localized basis. This condition can be easily fulfilled using pseudopotential methods for semiconductor materials [7]. As it is well known, however, the local pseudopotential approach fails for materials containing first row elements, as e.g. for SiO_2 [8]. Moreover, the nonlocal pseudopotential calculation carried out by Chelikowsky and Schlüter [6] have also left some problems unresolved in the band structure of α-quartz (e.g. a "gap" in the upper valence band was predicted at an energy where the photoelectron spectra show a minimum but not zero intensity; the position and width of the lower valence band do not agree with experimental data; the first peak of the Si $L_{2,3}$ spectrum could not be reproduced [9]).

The experimental photoelectron and emission spectra of vitreous silica show strong similarity with those of α-quartz [9,10]. This points to the fact that the electronic structure of both forms is determined primarily by short range interactions, leading to the consequence that a cluster type calculation using an advanced quantumchemical approach may yield valuable information regarding both the perfect and imperfect electronic structure. The quantumchemical cluster model posseses one more advantage. Namely, the various defects are usually identified by measuring the electronic transitions from or to their levels. Due to the defect-molecule nature of the environment of a vacancy, the energies of these transitions are strongly influenced by electronic relaxation and many-body effects. These can be straightforwardly accounted for in a quantumchemical approach.

Of course, the convergence properties of a cluster model are not as good as for the Koster-Slater type methods. They can be used on an equal footing only in those cases, when the defect wavefunction is not less localized than the defect potential [11].

2. Calculational Method

In recent years, we have developed a new method which does not suffer from the shortcomings of earlier cluster models [12,13]. This, so called *Cyclic Cluster Model* applies Born-Kármán boundary conditions to a "crystal" containing only a single unit cell. In this case the group of translations contains only the basis vectors of the unit cell and the only allowed $\underline{k}$ vektor is the $\underline{k}$ = 0. Then Bloch sums* belonging to

---

*"Bloch sums" can be formed by summing up equivalent atomic orbitals over all one-cell crystal pieces, ℓ filling in the space periodically.

$\underline{k} = 0$ can be formed:

$$\Phi_{\mu,\underline{k}=0} = \frac{1}{\sqrt{N}} \sum_{\ell=0}^{N-1} \mu_\ell ; \qquad N \to \infty \tag{1}$$

and the Crystalline Orbitals belonging to $\underline{k} = 0$ can be produced on this basis

$$\psi_{n,\underline{k}=0} = \sum_\mu C_{\mu n}(\underline{k}=0)\ \Phi_{\mu,\underline{k}=0} \tag{2}$$

The crystalline orbitals can be determined by solving the LCAO-CO equations [14] for $\underline{k} = 0$

$$\sum_v \sum_j \left[F^{oj}_{\mu v} - E_n(0) S^{oj}_{\mu v}\right] C_{vn}(\underline{k}=0) = 0 \tag{3}$$

where

$$F^{oj}_{\mu v} = H^{oj}_{\mu v} + \sum_{\lambda,\sigma} P_{\lambda\sigma}(0) \frac{1}{N} \sum_{k,\ell} \left[<\mu_o v_j|\lambda_k \sigma_\ell> - \frac{1}{2} <\mu_o \sigma_\ell|\lambda_k v_j>\right] \tag{4}$$

is the Foch matrix, $H^{oj}_{\mu v}$ is the core Hamiltonian matrix and

$$P_{\lambda\sigma}(0) = 2 \sum_m C^*_{\lambda m}(\underline{k}=0) C_{\sigma m}(\underline{k}=0) \tag{5}$$

is the density or bond order matrix. As eq.(4) shows, this procedure of obtaining $E_n(0)$ differs from the usual LCAO-CO band structure calculation in the fact, that all interactions of the state $\underline{k} = 0$ with other $\underline{k}' \neq 0$ states are neglected. The question is, how accurate the obtained $E_n(0)$ values can be.

At this point we apply the Large Unit Cell concept of Evarestov |15|. Namely, the unit cell (the cluster) is chosen to be a special manyfold of the primitive cell that way, that the $\underline{k}^{red} = 0$ state of the corresponding reduced Brillouin zone becomes equivalent with some special $\underline{k}$'s of the original B.Z. Choosing e.g. a $2\underline{a}_1 \times 2\underline{a}_2 \times 2\underline{c}$ unit cell in α-quartz (4 times the primitive cell), $\underline{k}^{red}$ is equivalent with $\underline{k}^{orig}_\Gamma$ and with all $\underline{k}^{orig}_\Gamma$ vectors (see Fig. 1). Consequently

$$\rho^{LUC}_\Gamma(\underline{r}) = \sum_{\lambda,\sigma}^{LUC} P_{\lambda\sigma}(\Gamma^{red}) |\lambda\sigma> = \sum_{\lambda',\sigma'}^{PUC} \left[P_{\lambda'\sigma'}(\Gamma^{orig}) + \right.$$

$$\left. + 3P_{\lambda'\sigma'}(M^{orig})\right] |\lambda'\sigma'> = \rho^{PUC}_\Gamma(\underline{r}) + 3\rho^{PUC}_M(\underline{r}) \tag{6}$$

However, since Γ and M in a hexagonal Brillouin zone satisfy the representative k point condition of Chadi and Cohen [16]:

$$\sum_{|\underline{R}|=C_m} (e^{i\underline{k}_\Gamma \underline{R}} + 3e^{i\underline{k}_M \underline{R}}) = 0 \; ; \qquad m = 1,2 \tag{7}$$

the crystalline electron density (normalized to the given unit cell) can be expressed by

$$\rho^{PUC}(\underline{r}) \cong \left[\frac{1}{4}\,\rho_\Gamma^{PUC}(\underline{r}) + \frac{3}{4}\,\rho_M^{PUC}(\underline{r})\right] = \frac{1}{4}\,\rho^{LUC}(\underline{r}) \tag{8}$$

So in eq.(4) we actually account for the interaction of a one electron state with the correct crystalline density approximated by the Chadi-Cohen averaging procedure.

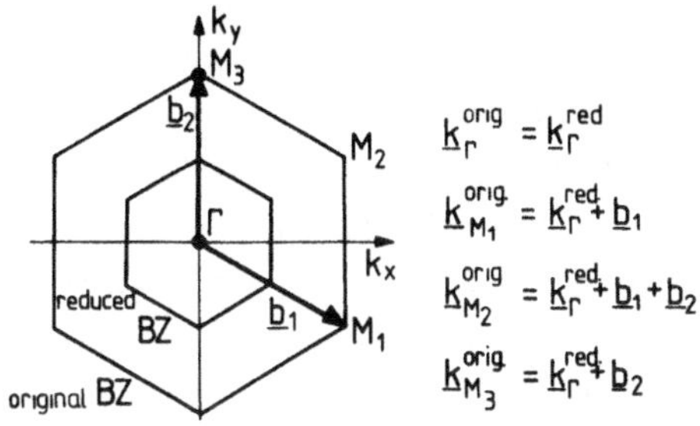

Fig. 1. Brillouin zones corresponding to the primitive cell (outer contour) and to the 4 time manyfold (inner contour). $\underline{b}_1$ and $\underline{b}_2$ are the primitive reciprocal lattice vectors belonging to the reduced Brillouin zone

So thus, the Cyclic Cluster Model is equivalent with a self-consistent LCAO-CO band structure calculation using representative k point theorem, with the restriction that only these representative states are calculated, but in one procedure.

3. The Electronic Structure of α-Quartz

The energy eigenvalues belonging to Γ and M in α-quartz were determined selfconsistently using a CNDO/S Fock matrix [17] reparametrized for saturated molecules on a (Si 3s,3p,3d; O 2s,2p) basis. Since the electronic structure is determined primarily by first neighbour Si-O and O-O interactions, N was set equal to ℓ in eq.(4). Gaps were determined by the lowest direct transitions calculated by a limited CI procedure. The band structure of α-quartz is highly dispersionless [8], what can be interpreted as a consequence of the importance of short range order. This may lead to the conclusion that $\rho(\underline{r})$ can be satisfactorily approximated by $\rho_\Gamma(\underline{r})$. Really, a calculation of the Γ and A^* states (an $\underline{a}_1 \times \underline{a}_2 \times 2\underline{c}$ unit cell) yielded $E_n(\Gamma)$ results very close to those obtained

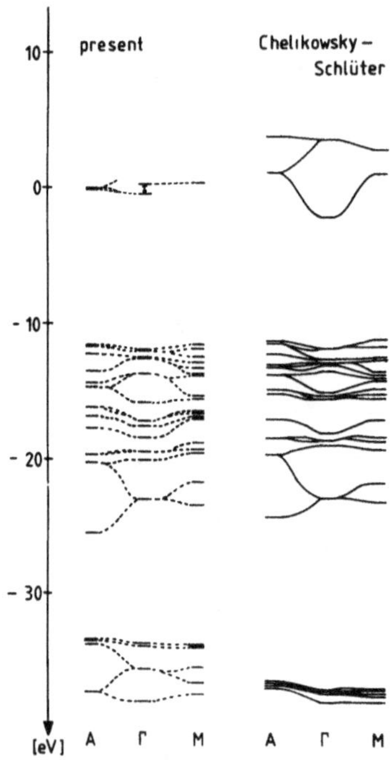

Fig. 2. Cluster electronic structure compared with the band structure [8]. Dotted lines are plotted corresponding to compatibility relations

by Γ and M^*, despite of the fact that Γ and A do not satisfy eq.(7). Since the differences were less than 0.2 eV, the two results are displayed together on Fig. 2, where cluster results are compared with the band structure reported in [8]. On the other hand, due to the different configurations, the electronic transitions obtained from the two different cluster differ by about 0.7 eV. This is indicated in the Figure by the arrow.

Corresponding to the above considerations, the results show no significant change relative to our preliminary calculation reported in ref. [18]. The overall agreement of the valence band with that of ref. [8] is striking. Differences occur just at the contraversial points mentioned in the introduction. The comparison of the total DOS histogram with experimental photoelectron spectra (Fig. 3) shows, that our results are closer to experimental findings than those of ref. [8]. As it can be seen from

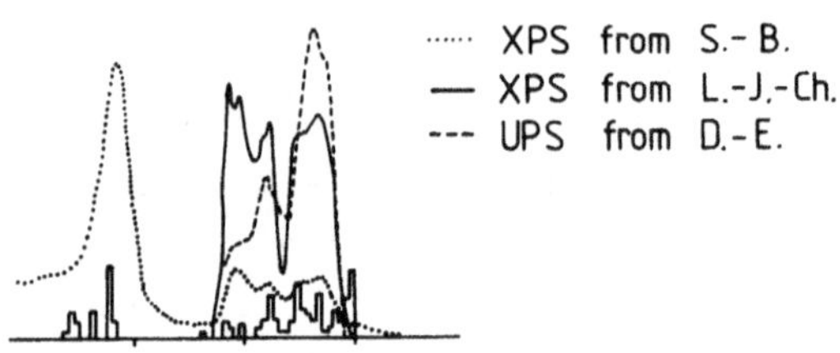

Fig. 3. The calculated DOS histogram compared with PES data. XPS results of Stephen and Binkovski (S-B) and UPS results of Distefano and Eastman (D-E) were taken from ref. [9], while XPS results of Laughlin, Joannopoulos and Schlüter (L-J-S) were taken from ref. [10]. The various spectra and the DOS histogram were joined at the place of the intensity minimum in the upper valence band

the comparison of partial DOS histograms with XES data (Fig. 4), the improvement is due to the inclusion of d orbitals. Namely, despite of their small participation in the upper crystalline orbitals (≤ 18 %), the weak d-π bonding causes a broadening in the top most part of the

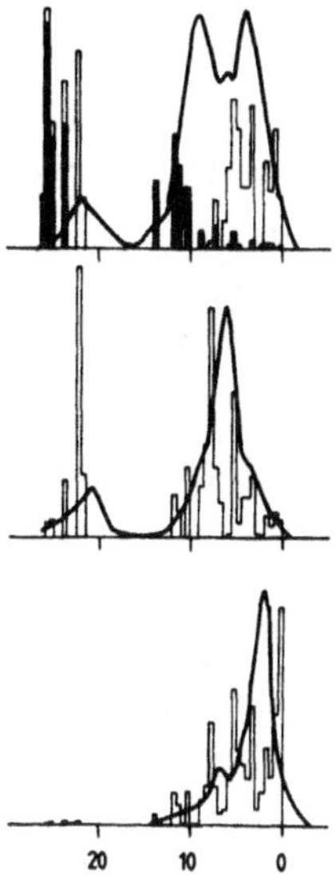

Fig. 4. Partial DOS histograms compared with XES data taken from ref. [9]. The zero energies of all experimental spectra are those given in ref. [9]

valence band. For more detailed discussion see ref. [18].

On the other hand, corresponding to the omission of O 3s orbitals from the basis, the lowest conduction band level is of Γ_3 symmetry, composed mainly of Si 3p orbitals. The energy of the first direct transitions are 11.42 eV at A and 11.84 eV at M, while 11.53 and 12.18 eV were obtained at Γ from the (Γ,A^*) and from the (Γ,M^*) calculation resp. These values fall well in the region of the first non-excitonic hump of the UV absorption spectrum peaked at 11.8 eV [9]. The width of the indirect gap is 11.21 eV from (Γ,A^*) while 11.79 eV from (Γ,M^*). These values are higher than those obtained in ref. [8], but agree well with the new interpretation of photoconductivity measurements [19] suggesting a band gap of about 11.5 eV for SiO_2.

The first ionization potentials (Koopmans' theorem) were calculated as 11.6003 eV at M, 11.6088 eV at A and 11.9932 eV or 12.0182 eV at Γ from (Γ,A^*) and from (Γ,M^*) resp. These agree very well with the 11.7 eV value deduced from experiments [9]. Therefore our energy values are correct on an absolute scale, and we might relate the defect levels also directly to the vacuum level.

4. The Electronic Structure of the V^+ Centre

The E' centre appearing both in amorphous and in crystalline silica shows characteristic UV transitions and ESR signals. These features are generally attributed to a positively charged, asymmetrically relaxed oxygen vacancy (V^+) in both forms [1]. The unpaired electron is placed on a dangling hybrid orbital belonging to the slightly relaxed, three coordinated silicon neighbour of the vacancy. If the latter is a ^{29}Si isotope, the orbital undergoes a "strong" hf interaction (a = 384 G), while "very weak" hf's (a = 0.3-1.2 G) occur with the other neighbouring ^{29}Si atom relaxed strongly toward the plane of the three oxygen surrounding it. Further "weak" hf's (a = 8 G) can be observed if proton impurities are present. The transitions connected to the above ESR features are 5.85 eV in vitreous silica, and 5.4 and 6.2 eV in α-quartz. The latter, so called E_1', E_2' splitting was not observed in the amorphous form, and no acceptable explanation could be given for it up to now [20].

The equilibrium geometry of the V^+ center was calculated by Yip and Fowler [21] resulting in also a correct orbital composition expected from g tensor data. Recently we reported [18] cyclic cluster calculations on a $Si_6O_{11}V^+$ cluster with the geometry given in ref. [21]. There was concluded, that for achieving acceptable results a $Si_{12}O_{23}V^+$ cluster is needed in order to prevent repeated defect interactions. Now, this calculation has been carried out.

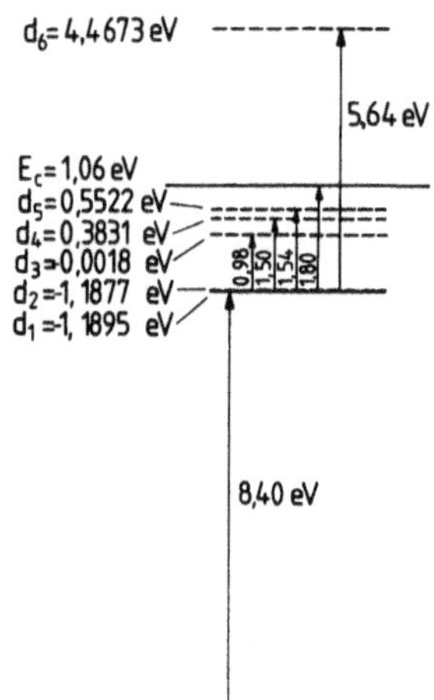

Fig. 5. Electronic structure of the $Si_{12}O_{23}V^+$ cluster. Levels are displayed according to Koopmans' theorem. Arrows indicate the transitions from or to the d_1 defect level

Our results on the V^+ centre (Yip-Fowler geometry) can be seen on Fig. 5. A singly occupied defect level, d_1, was found at -1.1895 eV. It belongs to an orbital localized in 85 % ($s^{0.31}p^{0.49}d^{0.05}$) on the slightly relaxed (Si_I) neighbour of the vacancy. Almost degenerated with this, an empty state, d_2, was found at -1.1877 eV localized in 66 % ($s^{0.26}p^{0.06}d^{0.34}$) on the other strongly relaxed (Si_{II}) neighbour. Further three d-like empty orbitals, d_3, d_4, d_5, localized up to 45 % on Si_{II} appear below the conduction band edge. Finally, a p like orbital ($s^{0.04}p^{0.68}d^{0.07}$) localized in 79 % on Si_{II} is placed at 4.4673 eV. Transitions in connection with these levels are also displaced in Fig. 5. The d_1-d_6 transition, 5.64 eV compares well with the experimental values attributed to E'. The near degeneracy of d_1 and d_2, on the other hand, shows the possibility of explaining the E_1', E_2' split by means of d_1-d_6 and d_2-d_6 transitions. Investigations in this direction are in progress. Transitions d_1-d_3, d_1-d_4, d_1-d_5 have oscillator strengths lower than that of d_1-d_6 by a factor of 20.

The transitions from the valence band to d_1 and from d_1 to the conduction band have energies of 8.40 and 1.80 eV resp. In the absorption spectrum of silica a shoulder can be detected splitting into two mean features at 7.6 and 8.2 eV [22]. The former can be related to a non-bridging oxygen with a trapped hole [29], but according to our results the latter might belong to the V^+ center. The 1.8 eV value of the d_1-E_c transition compares well with the characteristic red luminescence band of SiO_2 [22]. We have to note however, that the energies of these transitions depend strongly on the number of configurations included in the CI procedure.

The parameters of the hyperfine interactions calculated by the single centre approximation (and by the omission of d orbital contribution) used in ref. [24] are shown in Table I.

Table I. Parameters of the d_1-^{29}Si hf interactions in $10^{-4} cm^{-1}$ units. Exp. results are shown in parenthesis

	a		b
at Si_I	422.7 (384)		16.5 (20)
at Si_{II}	0.18		0.01
		(0.3-1.2)	
at Si neighbours	0.04		0.09
of Si_{II}	0.27		0.09
	0.74		0.14

The agreement with the experimental data is satisfactory. Regarding that the unpaired electron can easily be excited to the d_2 level the hf parameters were calculated also to this orbital. These are shown in Table II.

Table II. Parameters of the d_2-^{29}Si hf interactions in 10^{-4} cm^{-1} units. Exp. values are shown in parenthesis

	a		b
at Si_{II}	351.5 (384)		1.3 (20)
at Si_I	0.05		0.04
		(0.3-1.2)	
at Si neighbours	0.21		0.61
of Si_{II}	0.98		0.48
	8.89 (7.94)		0.74 (0.60)

The lowering of the "strong" hf parameters is due to the neglection of d contributions, which are here higher than at d_1. On the other hand, if the excitation would not change the d_2 state considerably, one of the "weak" hf's could be thus explained. Further work on this problem is also in progress.

The overall agreement of our calculated results with experimental data can be regarded as reliable only if the cluster size is large enough to prevent the interaction of the vacancy wavefunctions with the nearest repeated vacancy potential. The situation is displayed on a projection of the cluster onto the hexagonal plane in Fig. 6. The defect potential is characterized by the changes in the net atomic populations

$$\left(\sum_{\mu}^{atom} P_{\mu\mu} \right)$$

relative to the perfect case, caused by introducing the vacancy. Taking the highest value of this quantity (appearing at Si_{II}) to 100 %, the numbers in the figure show the changes at all atoms. As it can be seen the highest values on atoms with equal distances from two neighbouring vacancy are 12.8 %. On the other hand, the singly occupied state is localized up to 95 % in the region denoted by the solid contour, while the d_2 and d_6 empty orbitals are localized up to 95 % and 97 %, respectively in the region denoted by the dashed one. The d_3, d_4, d_5 states are localized in 77-89 % in the latter region. This means that the overlaps of the vacancy wavefunctions with neighbouring vacancy potentials are negligible.

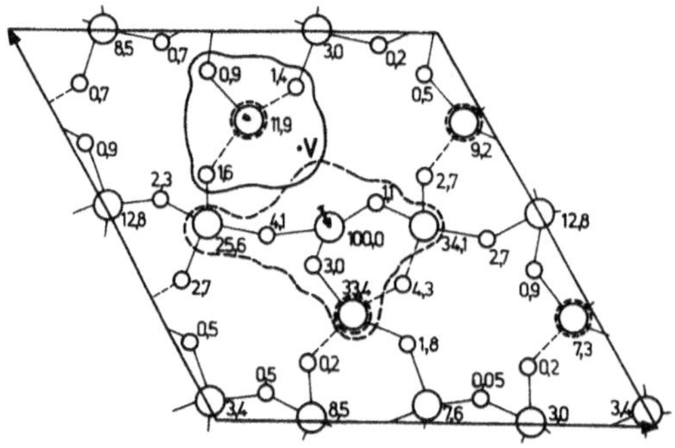

Fig. 6. Projection of the $Si_{12}O_{23}V^+$ cluster onto the hexagonal plane. Numbers show the changes in the net atomic populations due to the defect in the per cent of the highest value. Solid and dashed contours show the extents of the defect wavefunctions. For further explanation see the text

5. Conclusion

Our calculational results confirm the assumption that a neutral vacancy in SiO_2 can be regarded as a neutral hole trap. (The qualitative calculations of ref. [25] predict the doubly occupied level of a neutral vacancy at about 4 eV above the valence band. This corresponds to the neutral hole trap expected at this energy [26]. (Capturing a hole, the vacancy relaxes asymmetrically, and its occupied level is pushed upward to 2.25 eV below the conduction band (see Fig. 5). This level may play the role of a positively charged electron trap, characterized as a trapped hole site in ref. [26].

References

1. D.L. Griscom, Phys. Rev. B 22, 4192 (1980)
2. S.T. Pantelides, Rev. Mod. Phys. 50, 797 (1978)
3. N.O. Lipari, J. Bernholc and S.T. Pantelides, Phys. Rev. Lett. 43, 1354 (1979)
4. G.A. Baraff and M. Schlüter, Phys. Rev. Lett. 41, 892 (1978)
5. C.O. Rodriguez, S. Brand and M. Jaros, Inst. Phys. Conf. Ser. (1979)
6. G.F. Koster and J.C. Slater, Phys. Rev. 95, 1167 (1954)
7. D.J. Chadi, Phys. Rev. B 16, 3572 (1972)
8. J.R. Chelikowski and M. Schlüter, Phys. Rev. B 15, 4020 (1977)
9. D.L. Griscom, J. Non-Cryst. Solids 24, 155 (1977)
10. R.B. Laughlin, D. Joannopoulos and D.J. Chadi, Phys. Rev. B 20, 6228 (1979)
11. P. Deák, Acta Phys. Hung. 50, 247 (1981)
12. P. Deák, Phys. Lett. 83A, 39 (1981)
13. P. Deák, Thesis (1981)
14. J.M. André, L. Gouverneur and G. Leroy, Int. J. Quantum Chem. 1, 451 (1967)
15. R.A. Evarestov and V.A. Lovchikov, phys. stat. sol. (b) 79, 43 (1977)
16. D.J. Chadi, M.L. Cohen, Phys. Rev. B 7, 692 (1973)
17. P. Deák, Proceedings of the 4th ICQC, to be published in the Int. J. Quantum Chem.
18. P. Deák and J. Giber, Phys. Lett. 88A, 237 (1982)
19. R. Evrard and A.N. Trukhin, Phys. Rev. B 25, 4102 (1982)
20. D.L. Griscom, in "The Physics of SiO_2 and Its Interfaces", ed. S.T. Pantelides (Pergamon, New York 1978) p. 232
21. K.L. Yip and W.B. Fowler, Phys. Rev. B 11, 2327 (1975)
22. G.N. Greaves, in "The Physics of SiO_2 and Its Interfaces", ed. S.T. Pantelides (Pergamon, New York 1978) p. 268
23. D.L. Griscom, ibid p. 263
24. C. Gobsch, M. Haberlandt, J. Wechner and J. Reinhold, phys. stat. sol. (b) 90, 309 (1978)
25. A.J. Bennett and L.M. Roth, J. Phys. Chem. Solids 32, 1251 (1971)
26. D.J. DiMaria, in "The Physics of SiO_2 and Its Interfaces", ed. S.T. Pantelides (Pergamon, New York 1978) p. 160

A NEW MODEL FOR THE Si-A CENTER*

J.R. Leite, M.J. Caldas and A. Fazzio
Instituto de Fisica da Universidade de São Paulo
CP 20516, São Paulo, SP, Brasil

ABSTRACT

The multiple scattering-Xα molecular cluster model is used to study the A center in silicon. We show that this center can be created by Jahn-Teller effects over substitutional oxygen impurity in the lattice. Our defect model provides a consistent explanation for the EPR experimental results obtained for the Si-A center.

1. Introduction

The present work was undertaken in order to perform a rigorous theoretical investigation of the A centre in silicon. The Si-A center, also referred to as Si-B1 center, is a well known radiation-damage defect produced in "pulled" silicon by room temperature irradiation [1]. From the results of electron paramagnetic resonance (EPR) and infrared absorption (IR) experiments it was suggested that this center is an oxygen-silicon vacancy related defect [1,2]. According to the accepted microscopic model for the center, the oxygen atom bridges two of the four broken bonds associated to a vacancy making a molecular chemical bond with two silicon neighbours. The oxygen atom is then off the subsitutional site in a (100) direction, and the other two silicon atoms come close together to form a chemical bond. The microscopic model described above is currently adopted to interpret the experimental data on the Si-A center. In the neutral state of the defect two electrons are paired off in the Si-Si bonding orbital. The EPR signal and electrical activity of the center is explained in terms of an additional

*Work supported by CNPq and FAPESP, Brasil

trapped electron in the Si-Si antibonding state. This state should correspond to a deep level located at 0.17 eV below the bottom of the conduction band [3,4]. The occupancy of this level should give rise to a well known EPR spectrum that can be interpreted through a $S = \frac{1}{2}$ spin Hamiltonian. Hyperfine interaction with O nucleus is not observed and the symmetry for the stabilized center is C_{2v}. From the EPR spectrum it is observed that there is a strong interaction of the defect electron with two silicon nuclei, with hyperfine axis making an angle slightly smaller than the tetrahedral angle [1]. The defect is also associated to an IR absorption band at 12 u^2. Although the microscopic model presented, together with a linear combination of atomic orbital representation for the defect state describes correctly the main features of the Si-A center, a rigorous theoretical investigation of the electronic structure of this center has not yet been carried through.

The aim of this work is to show that the Si-A center can be created by Jahn-Teller (JT) distortions of the Si:O or $Si:O^-$ systems. In order to reach this goal we are using a molecular cluster model proposed by Fazzio, Leite and De Siqueira [5], within the framework of the self-consistent field multiple scattering-Xα method (MS-Xα) [6].

2. Electronic Structure of the Si-A Center

The cluster model used in this work comprises one central atom, four nearest neighbours and twelve next-nearest neighbours in a tetrahedral configuration. The "perfect" cluster representing the silicon crystal is denoted by (17Si). In order to simulate the oxygen substitutional impurity in the undistorted lattice a "defect" cluster was built through the replacement of the central silicon atom of the "perfect" cluster by an oxygen atom ($16Si+O^-$). The electronic structure of the clusters was obtained through the MS-Xα method within the framework of the muffin-tin approximation for the potential [7]. The radii of the atomic and outer spheres along with the assumed values for the spherical harmonic quantum numbers are listed in Reference[8].The exchange-correlation parameters used for the "perfect" cluster were the same for all regions, $\alpha = 0.72751$. When simulating the impurity we assume for the central sphere the value of α for oxygen [9].

In Figure 1 we show the self-consistent energy spectra of the clusters 17Si and $16Si+O^-$ used to represent the Si crystal and the $Si:O^-$ system, respectively. The energy spectrum of the cluster 17Si defines the crystal band edges. The uppermost occupied level, $3t_2$, and

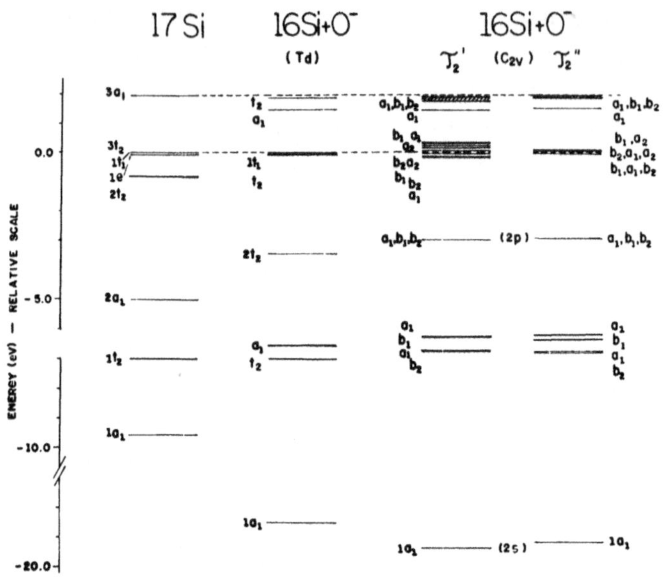

Fig. 1. Orbital energies of the clusters (17Si) and ($16Si+O^-$) representing the perfect silicon crystal and the oxygen substitutional impurity in the lattice

the lowermost occupied level, $1a_1$, define the top and the bottom of the valence band, respectively. The level $3a_1$ defines the bottom of the conduction band. The values for the gap, 1.96 eV, and for the valence bandwidth, 9.60 eV, obtained from the cluster calculation are in fairly good agreement with the experimental results, 1.15 eV and 12.4 eV, respectively [10,11].

The energy spectrum of the cluster $16Si+O^-$ in T_d symmetry represents the electronic structure of an ideal substitutional oxygen impurity, negatively charged, in the silicon lattice. The results for the ground state of the neutral impurity are similar to those obtained for the negatively charged state, therefore will not be shown here [12]. The first interesting feature of the energy spectrum of the cluster $16Si+O^-$ is that the introduction of the impurity gives rise to a hyperdeep level, $1a_1$, that lies at $\sim$9 eV below the bottom of the valence band. In Table I we make a comparison of charge distributions for the orbitals that receive the main contributions from the atomic orbitals of the central atom, for the 17Si and $16Si+O^-$ clusters. For the "perfect" cluster the charge distribution of the central atom and on the first shell shows the formation of the covalent chemical bonds between the central atom and the nearest neighbours. On the other hand, we see

Table I. Comparison of charge distributions for the orbitals that receive the main contributions from the atomic orbitals of the central atom, for the (17Si) and ($16Si+O^-$) clusters; units of one electron

Cluster	Orbital (occupancy)	Central atom	First shell	Second shell	Interatomic
17Si	$1a_1(2)$	0.48	0.30	0.01	0.21
	$2t_2(6)$	0.25	0.32	0.17	0.25
$16Si+O^-$	$1a_1(2)$	0.94	0.01	0.00	0.04
	$2t_2(6)$	0.72	0.14	0.04	0.10

from the entries displayed in Table I for the "defect" cluster that there is a strong concentration of charge on the central sphere corresponding to the 2s and 2p atomic orbitals of the oxygen atom. 94 % of the charge associated to the hyperdeep level is in the oxygen sphere. Therefore we conclude that there is no evidence for the formation of chemical covalent bonds between the oxygen atom and the two silicon neighbours.

The second feature of interest in the energy spectrum of the cluster $16Si+O^-$ is that the impurity introduces two levels in the band gap, one non-degenerate, a_1, and one three-fold degenerate, t_2. In the negative charge state of the center the a_1 level is fully occupied and the t_2 level accomodates one electron. Therefore we conclude that the $Si:O^-$ system in T_d symmetry is unstable with respect to normal JT distortions (permitted modes ε and τ_2). Indeed, even in the neutral charge state the proximity of the two levels in the band gap can force the system to undergo a second order JT (pseudo-JT) distortion in which case the only permitted mode is τ_2. Assuming that the center exists also in the neutral charge state we decided to investigate the distortions along the two τ_2 modes, without coupling. Since the MS-Xα method does not yield total energies reliable enough to determine equilibrium configurations, we decided to build clusters already at the "experimental geometry". Thus we assume for the angle between the hyperfine axis the experimental value [1]. Our calculations yield a C_{2v} symmetry for the stable configuration of the $Si:O^-$ system, that agrees with the symmetry detected for the Si-A center.

In Figure 2 we show a scheme of the z coordinate of the two independent normal modes of τ_2 symmetry for distortion of the central atom

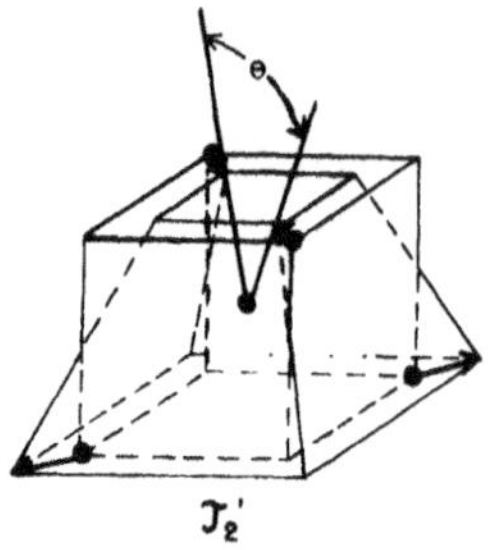

τ_2' τ_2''

Fig. 2. Scheme of the displacements of the central oxygen atom and four nearest silicon neighbours in the τ_2 mode of distortion. The symmetry of the distorted system is C_{2v}

and the nearest neighbours. The energy spectra of the distorted systems are shown in Fig. 1. The energies are now labelled according to the irreducible representation of the C_{2v} point group. We point out that the levels a_1, b_1, b_2, resulting from the splitting of the t_2 level in the gap remain degenerate. During the selfconsistency they interchange positions spanning an energy range indicated by the shaded regions in Fig. 1. This region is located at (0.10±0.04) eV and (0.05±0.02) eV below the bottom of the conduction band for the modes τ_2' and τ_2'' respectively, in fair agreement with the experimental value for the impurity level, 0.17 eV. According to our calculations when the b_1 orbital is occupied there is a charge concentration on the two silicon atoms which according to EPR experiments are supposed to come close together to form a chemical bond. Therefore we assume that this orbital accomodates the trapped electron.

We compare now our results with the magnetic resonance data, bearing in mind that the defect electron is occupying the b_1 orbital. The absence of hyperfine interaction with the oxygen nucleus is readily justified as this orbital presents a nodal plane between the two silicon atoms that come close together. Over these two nuclei, our calculations yield a hyperfine contact field of $267.1 \cdot 10^{-4}$ cm^{-1} and $258.8 \cdot 10^{-4}$ cm^{-1} for the τ_2' and τ_2'' modes, respectively. These results are in fair accord with the experimental value [1], $136.8 \cdot 10^{-4}$ cm^{-1}.

3. Conclusions

Using a 17-atom cluster model, with suitable boundary conditions and within the framework of the MS-X method, we demonstrate that the Si-A center can be created by JT or pseudo-JT effects over substitutional oxygen impurity. We found a hyperdeep level associated to this center. Our model for the A center gives results in good agreement with the experimental evidences for: the position of the impurity level in the band gap, the electronic spin, the hyperfine interaction with the two silicon neighbours and the absence of hyperfine interaction with the oxygen nucleus.

References

1. G.D. Watkins and J.W. Corbett, Phys. Rev. 121, 1001 (1961)
2. J.W. Corbett, G.D. Watkins, R.M. Cherenko and R.S. McDonald, Phys. Rev. 121, 1015 (1961)
3. G.K. Wertheim, Phys. Rev. 105, 1730 (1957); 110, 1272 (1958)
4. D.E. Hill, Phys. Rev. 114, 1414 (1959)
5. A. Fazzio, J.R. Leite and M.L. De Siqueira, J. Phys. C12, 513 (1979)
6. K.H. Johnson, Adv. Quantum Chem. 7, 143 (1973)
7. K.H. Johnson, Ann. Rev. Phys. Chem. 26, 39 (1975)
8. A. Fazzio, M.J. Caldas and J.R. Leite, Int. J. Quant. Chem. S13, 349 (1979)
9. K. Schwarz, Phys. Rev. B 5, 2466 (1972)
10. Data collected from CRC Handbook of Chemistry and Physics, 58th ed., Robert C. Weast, Ed., CRC Press, Cleveland, OH, 1977
11. W.D. Grobman and D.E. Eastman, Phys. Rev. Lett. 29, 1508 (1972)
12. M.J. Caldas, J.R. Leite and A. Fazzio, Phys. Stat. Solidi (b) 98, K109 (1980)

PART III

DEFECTS IN SILICON

DEFECT COMPLEXING IN IRON-DOPED SILICON

C.A.J. Ammerlaan

Natuurkundig Laboratorium, Universiteit van Amsterdam
Valckenierstraat 65, 1018 XE Amsterdam, The Netherlands

1. Introduction

Many impurities are solvable in silicon. In the crystal lattice they can occupy substitutional sites, interstitial sites, or new bond structures are locally formed involving the impurity and some silicon atoms. Impurities which are normally substitutional, such as Ge, B, P, and S are characterized by low diffusion coefficients. Interstitial impurities, examples are Li and Fe, have high mobilities. Oxygen, which occupies a bond-centered position, is an intermediately fast diffuser. With a mobile impurity species present, interactions between impurities become possible. Reaction rates are often diffusion-limited. Reaction products can vary in size from point-like defects via small complexes to macroscopic clusters. Impurity chemistry in silicon is rich in appearance. The possibilities are too numerous the be treated in any detail in limited space. Therefore, this review deals only with iron-doped silicon. First, the properties of isolated iron will be summarized. Subsequently, a discussion will be given on iron-impurity pairs, small iron complexes, and iron precipitation. Other 3d-transition metal impurities, for example chromium and manganese , show very similar behaviour in silicon. However, the most data are avilable for iron as it is the more common impurity.

2. Isolated Iron Impurity Atoms

Iron is a fast-diffusing element in silicon. The diffusion constant, measured over the range 1100-1300 °C is expressed by $D_{Fe} = 8.9 \cdot 10^{-3} \cdot \exp(-0.90\ eV/kt)\ cm^2/s$ [1]. The solubility at 1200 °C is about $1.5 \cdot 10^{16}$ atoms/cm³ [1-3]. Because of the high mobility, it is difficult to prevent iron contamination of crystals during treatments at high temperatures. Iron is easily introduced unintentionally [4-7] and has been identified as a thermal defect [8-14]. Iron diffuses interstitially [3]. After quenching to room temperature iron is found to occupy the tetrahedral interstitial lattice sites. A deep donor level in the silicon bandgap is associated with this impurity [15-17]. The level position at E_{vb} + (0.39 ± 0.02) eV was determined by Hall effect [2,18,19] and deep level transient spectroscopy (DLTS) [19-26]. Carrier capture cross-section, reported to be $\sigma_p^o \simeq 3 \cdot 10^{-16}$ cm² and $\sigma_n^+ \simeq 15 \cdot 10^{-16}$ cm² [27] are temperature dependent with an activation energy of 43 meV [19]. Both neutral Fe_i^o and positively charged Fe_i^+ are paramagnetic due to the partial filling of the 3d electron shell. The model of Ludwig and Woodbury accounts for the electronic structure [28]. Neutral iron has $(3d)^8$ configuration, spin S = 1, and isotropic Zeeman splitting factor g = 2.070 [29,30]. The coincidence of the two electron paramagnetic resonance (EPR) transitions is lifted when the symmetry of the centre is lifted by applying uniaxial stress [31-34]. Positively charged Fe_i^+ has a configuration $(3d)^7$. The ground state is an orbital triplet T_1 with effective momentum L' = 1, which couples to the spin S = 3/2 to give a total momentum J = 1/2. The isotropic g value is g = 3.524 [30,31]. The centre exhibits dynamical Jahn-Teller effects [38]. Spin-lattice relaxation rates were determined [35-37]. Photoluminescence from deep states associated with iron was observed [38]. Theoretically, the electronic structure has been described using the spin-restricted scattered-wave Xα method for a small cluster containing a central iron atom [39-41]. A deep bandgap donor level was found indeed. An electronic level at E_{cb}-0.55 eV, frequently reported [2,16,42-44] is probably not associated with a single iron impurity.

After 1.8 MeV electron irradiation of iron-doped silicon an EPR spectrum arising from a centre with one iron atom was observed [45,46]. The spectrum was ascribed to substitutional iron. Vacancies created by the irradiation may have enabled the interstitial to substitutional conversion. The centre has <111> axial symmetry.

3. Iron-Impurity Pairs

3.1. Iron-Acceptor Pairs

Keeping silicon crystals at room temperature interstitial iron is lost from solid solution. This process proceeds faster in p-type silicon; n-type material is more stable [47]. In p-type silicon iron atoms will release their donor electron to the acceptors, thereby becoming positively charged. The attraction between mobile positive iron ions Fe_i^+ and negative acceptor ions A^-, is sufficiently strong to favour associations at room temperature. The existence of the iron-acceptor pairs FeB, FeGa and FeIn, is confirmed by their EPR spectra [30,48]. The two chemical components in a pair are unambigously identified by a characteristic hyperfine structure. The FeB and FeGa pairs have <111> axial symmetry [30,49], consistently with the simplest pair model of a substitutional acceptor impurity with an iron ion on a nearest interstitial site. The angular dependence of the resonance of FeIn indicates rhombic symmetry [30,49]. Energy levels for the pairs have been measured by Hall effect and DLTS [24]. Donor levels are reported at E_{vb} + 0.10 eV for FeB [19,23,25,26,50], at E_{vb} + 0.19 eV for FeAl [19,50] and at E_{vb} + 0.24 eV for FeGa [19]. An additional level for FeB at E_{cb} - 0.55 eV is probably an acceptor level of the same pair [50]. Also for FeAl a second level, at E_{vb} + 0.13 eV, is found [50]. It is possibly related to a different pair structure. The formation of pairs proceeds faster at increased temperatures and at higher levels of acceptor doping [22]. This is consistent with a diffusion-limited reaction process and a time constant $\tau = (4\pi n_A D_{Fe} R)^{-1}$, with a Coulomb capture radius R [42]. The activation energy for the pairing process near room temperature, reported ≃ 0.83 eV, is slightly less than for interstitial diffusion at high temperatures [22,42]. Complicated kinetics have been observed [19]. To account for this a separation of the pair process into two phases has been proposed [50]. The pair interaction is an equilibrium process: pairs may associate and dissociate. The equilibrium constant $K = n_{Fe^+A^-}/n_{Fe^+} \cdot n_{A^-}$ will have a form $K(T) = K(\infty) \exp(-H/kT)$. Expressing concentrations as site fraction, the pre-exponential constant $K(\infty)$ must be close to the coordination number Z = 4 according to theory. The formation enthalpy ΔH may be approximated by $\Delta H = e^2/4\pi\varepsilon_0\varepsilon_r r_1$. Taking $\varepsilon_r = 11.8$ for the dielectric constant and $r_1 = 0.235$ nm, which is the distance between a lattice site and an interstitial T-site, one calculated ΔH = 0.52 eV. Accurate determinations of K(T) have not been performed. Published values for the binding energy of the FeB pair vary from 0.50 to 0.65 eV [22]. Illumination by intense white light causes dissociation of the FeB pair [50]. At temperatures above 150 °C pairs dissociate irreversibly as then iron is lost from the solution by precipitation. Information available for the iron-acceptor pairs is not yet complete.

3.2. Iron-Donor Pairs

Single interstitial iron is not a stable impurity in n-type silicon either. Also in this material the interstitial iron concentration decreases when kept at room temperature by a thermally activated process, with longer time constants than in p-type silicon. Interactions of neutral iron atoms with the group V donors, P, As, and Sb, either in the neutral or positive charge state have not been reported [11]. However, with sulphur, which can act as a double donor, iron forms an impurity pair. Four different such pairs were identified by means of EPR [51,52]. They exhibit the expected twofold splitting when the magnetic isotope ^{57}Fe (I=1/2) is used for diffusion. For one centre also the ^{33}S (I=3/2) hyperfine structure was observed. The g tensor of the spectra indicates orthorhombic symmetry for all centers. One of the pairs is nearly <111> axial with only a slight orthorhombic distortion. The geometric structure of these defects is therefore rather different from the iron-acceptor pairs. They may consist of an iron and a sulphur atom on a (110) plane. The electrical properties of these pairs were not studied. Information on iron-donor pairing is still incomplete.

3.3. Iron-Gold Pairs

Gold is an amphoteric impurity frequently used to control carrier lifetimes in silicon. After quenching of silicon doped with both gold and iron, the latter impurity sometimes unconsciously, an anisotropic EPR spectrum can be observed [11,53-57]. A fourfold splitting in the spectra indicates the presence of one gold atom (^{197}Au: abundance 100 %, I = 3/2). Additional twofold splitting is present due to hyperfine interactions with one ^{57}Fe atom, when this isotope is deliberately diffused [56]. The spectrum arises from a centre with trigonal symmetry. Its angular dependence is usually due to the large electric quadrupole moment of the ^{197}Au nuclei. The hyperfine structure, the anisotropy of the spectrum, and the known properties of gold and iron in silicon allow a reliable interpretation of the spectrum. The model for the centre is a neutral complex consisting of a substitutional gold atom with an iron atom on a nearest neighbour interstitial position.

The electronic structure of the Au_sFe_i complex was analysed using LCAO techniques [58]. The most satisfactory description was obtained by assuming the transfer of one electron from the iron to the gold impurity. The electronic configurations then are $(5d)^8 6sp^3$ and spin S_{Au} = 1 for the negative Au_s^- ion, and $(3d)^7$ with spin S_{Fe} = 3/2 for the positive Fe_i^+ ion. Antiferromagnetic coupling between the two localized spins is required to explain the experimentally observed spin S = 1/2. Due to covalency the quantities $\langle 1/r^3 \rangle$ for the d-orbitals on the gold and iron atoms are reduced considerably with respect to the free atoms. Also, the s-core polarizations of these impurities solved in silicon are smaller than in most other materials.

The Au_sFe_i pairs are formed to highest concentrations by heating the samples as quenched to temperatures of 200-250 °C for 10 minutes [56]. Their formation is accompanied by the disappearance of the interstitial iron component. By heating above 250 °C the pairs are irreversibly lost.

3.4. Iron-Iron Pairs

Also interactions between iron atoms themselves were observed in EPR [46]. To avoid competing processes dislocation-free silicon of high purity was used for these experiments. Pairs of iron atoms at small separation were produced by high-energy electron irradiation followed by modest annealing. Four EPR spectra related to iron pairs were reported. The participation of two iron atoms to form one centre is directly concluded from the characteristic hyperfine structure when using the magnetic isotope ^{57}Fe. The symmetry of the atomic arrangement is trigonal for one centre, monoclinic for one other, and orthorhombic, with slight distortion to lower symmetry, for the two remaining pairs. In all four centres the two iron atoms occupy sites which are equivalent by symmetry. Tentative atomic models for the iron-pair defects were proposed. These models also involve vacancies created by the irradiation. All iron pairs have spin S = 5/2. This could be explained in the Ludwig-Woodbury model by assuming the occupation of interstitial sites by the iron atoms and parallel alignment of the spins on the two atoms. Consisting of two iron atoms only, the complexes already act as ferromagnetic inclusions. Every pair is observed in a charged state. Deep bandgap levels are presumably associated with the pairs. Some of the iron-pairs are directly formed during the room temperature irradiation. This shows that the interstitial iron impurity is the subject to radiation-induced diffusion.

4. Iron Complexes

Upon anneal at temperatures above 120 °C the EPR signal of Fe_i^o decreases and is lost completely ultimately, even in pure silicon crystals. This is not due to a mere shift of the Fermi level as no resonance from Fe_i^+ becomes observable. Anneal in the temperature range 120-150 °C produces a new EPR spectrum [46]. Though this has much smaller intensity, its growth appears to be correlated with the decrease of the Fe_i^0 concentration. The EPR line width is too large for hyperfine interactions with ^{57}Fe, which are expected to be small, to be resolvable. Nevertheless, the formation kinetics and the g value of the spectrum strongly suggest an iron-related defect. Based on the observed spin S = 4 of the spectrum, a complex consisting of four interstitial iron atoms in the neutral state is proposed. Ferromagnetic coupling of the four S = 1 spins on the Fe_i^0 constituents explains the spin of the complex. The iron atoms are arranged in trigonal symmetry. The $(Fe_i^0)_4$ centre is reminiscent of $(Mn_i^0)_4$, though the latter centre has cubic symmetry and its structure is well established by resolved ^{55}Mn hyperfine interactions [59]. The spectrum of $(Fe_i^0)_4$ is lost by anneal at 250 °C. The identification of this four-iron-atom complex needs further confirmation.

5. Iron Precipitation

5.1. Formation of Clusters

After slow cooling of silicon following iron diffusion at high temperature, no interstitial component is present. Also, dissolved interstitial iron is lost by heating above 250 °C. After such treatments metallic precipitates are observable by transmission electron microscopy and by X-ray topography. The dimensions reported for these segregations vary from 10 nm to several µm's. The precipitates are not pure iron inclusions, but compounds of silicon and iron, possibly Fe_3Si [60]. They cause the so-called hazy appearance after preferential etching. A large density of minute etch pits, typically with a diameter of 1 µm, is visible on the surface. The density of these bulk clusters is larger in the middle of the crystals than close to the surfaces. The nature of the nucleation sites is not well defined yet. Oxygen can play a role as Czochralski silicon behaves differently from float-zone silicon.

5.2. Precipitation on Dislocations

In strongly dislocated silicon no interstitial iron impurity can be detected after doping and quenching. Though iron is introduced, as evidenced by nuclear activation analysis, the fraction present on interstitial sites, measured specifically by EPR, is much lower or entirely absent. In the range of 10^5 to 10^7 dislocations per cm², the interstitial iron component falls from normal to zero [61]. Around the dislocations there appear to exist depleted zones with the radius of about 2 µm [61]. Iron atoms within this radius are trapped by the dislocation strain field and are concentrated near the dislocation core already during the initial quenching. Iron atoms originally located outside the dislocation zones can diffuse to and be trapped in these zones in thermally activated anneal. The deposition of iron at the dislocations changes the shape of the etch pits [62].

5.3. Precipitation on Surfaces

When conditions of time and temperature are suitable for iron to migrate over macroscopic distances it can precipitate on surfaces. This has been demonstrated in cases where surfaces were carefully cleaned from iron before the heat treatment. Iron contamination of silicon crystals can thus be remedied by a proper gettering technique. Heating of the samples in a gas stream containing HCl will remove iron impurities, probably because metallic chlorides with high volatility are formed at the surface [12,63].

References

1. J.D. Struthers, J. Appl. Phys. 27, 1560 (1956)
2. C.B. Collins and R.O. Carlson, Phys. Rev. 108, 1409 (1957)
3. E. Weber and H.G. Riotte, J. Appl. Phys. 51, 1484 (1980)
4. G. Bemski and C.A. Dias, J. Appl. Phys. 35, 2983 (1964)
5. L. Elstner and W. Kamprath, Phys. Stat. Sol. 22, 541 (1967)
6. M.L. Swanson, Phys. Stat. Sol. 33, 721 (1969)
7. W. Leskochek, H. Feichtinger and G. Vidrich, Phys. Stat. Sol. (a) 20, 601 (1973)
8. Y.H. Lee, R.L. Kleinhenz and J.W. Corbett, Appl. Phys. Lett. 31, 142 (1977)
9. J.D. Gerson, L.J. Cheng and J.W. Corbett, J. Appl. Phys. 48, 4821 (1977)
10. E. Weber and H.G. Riotte, Appl. Phys. Lett. 33, 433 (1978)
11. Y.H. Lee, R.L. Kleinhenz and J.W. Corbett, Inst. Phys. Conf. Ser. 46, 521 (1979)
12. H.J. Rijks, J. Bloem and L.J. Giling, J. Appl. Phys. 50, 1370 (1979)
13. H. Feichtinger, A. Gschwandtner and J. Waltl, Phys. Stat. Sol. (a) 53, K71 (1979)
14. V.B. Voronkov, A.A. Lebedev, A.T. Mamadalimov, B.M. Urunbaev and T.A. Usmanov, Sov. Phys. Semicond. 14, 1217 (1980)
15. M.A. Abdugafurova, L.M. Kapitonova, L.S. Kostina, A.A. Lebedev and Sh. Makhkamov, Sov. Phys. Semicond. 9, 450 (1975)

16. A.I. Dyatlov, L.M. Kapitonova, A.A. Lebedev, M.A. Pogarskii and T.A. Shaposhnikova, Sov. Phys. Semicond. 11, 559 (1977)
17. H.R. Szawelska and H. Feichtinger, J. Phys. C: Solid State Phys. 14, 4131 (1981)
18. H. Feichtinger, J. Waltl and A. Gschwandtner, Solid State Comm. 27, 867 (1978)
19. K. Wünstel and P. Wagner, Appl. Phys. A 27, 207 (1982)
20. L.C. Kimerling, J.L. Benton and J.J. Rubin, Inst. Phys. Conf. Ser. 59, 217 (1981)
21. A.O. Evwaraye, Inst. Phys. Conf. Ser. 46, 533 (1979)
22. H. Lemke, Phys. Stat. Sol. (a) 64, 215 (1981)
23. K. Wünstel and P. Wagner, Solid State Comm. 40, 797 (1981)
24. K. Wünstel, O. Kumagai and P. Wagner, App. Phys. A 27, 251 (1982)
25. N.H. Sheng and J.L. Merz, Physica B, 1983 (to be published)
26. L.C. Kimerling and J.L. Benton, Physica B, 1983 (to be published)
27. E. Schibli and A.G. Milnes, Annu. Rev. Materials Science 10, 157 (1980)
28. G.W. Ludwig and H.H. Woodbury, Phys. Rev. Lett. 5, 98 (1960)
29. G.W. Ludwig, R.O. Carlson and H.H. Woodbury, Bull. Amer. Phys. Soc. Ser. II, 4, 22 (1959)
30. G.W. Ludwig and H.H. Woodbury, Solid State Phys. 13, 223 (1962)
31. H.H. Woodbury and G.W. Ludwig, Phys. Rev. 117, 102 (1960)
32. N.T. Bendik, L.S. Milevskii and E.G. Smirnov, Sov. Phys. Semicond. 5, 749 (1971)
33. M. Berke, E. Weber, H. Alexander, H. Luft and B. Elschner, Solid State Comm. 20, 881 (1976)
34. W. Gehlhoff and K.H. Segsa, Phys. Stat. Sol. (a) 41, K21 (1977)
35. A.A. Bugai, V.S. Vikhnin and V.E. Kustov, Sov. Phys. Solid State 22, 864 (1980)
36. M.F. Deigen, V.Ya. Bratus, B.E. Vugmeister, I.M. Zaritskii, A.A. Zolotukhin, A.A. Konchits and L.S. Milevskii, Sov. Phys. JETP 42, 1073 (1976)
37. A.A. Konchits, V.S. Yikhnin, I.M. Zaritskii and B.K. Knulikovskii, Sov. Phys. Solid State 20, 1348 (1978)
38. J. Weber and P. Wagner, J. Phys. Soc. Japan 49, Suppl. A, 263 (1980)
39. G.G. DeLeo, G.D. Watkins and W.B. Fowler, Phys. Rev. B 23, 1851 (1981)
40. G.G. DeLeo, G.D. Watkins and W.B. Fowler, Phys. Rev. B 25, 4962 (1982)
41. G.G. DeLeo, G.D. Watkins and W.B. Fowler, Phys. Rev. B 25, 4972 (1982)
42. W.H. Shepherd and J.A. Turnar, J. Phys. Chem. Solids 23, 1697 (1962)
43. N.T. Bendik, V.S. Garnyk and L.S. Milevskii, Sov. Phys. Solid State 12, 1340 (1970)
44. C. Schmidt, Appl. Phys. 17, 137 (1978)
45. S.H. Muller, H.G. Arnold, J.J. de Graaff, G.M. Tuynman and C.A.J. Ammerlaan, Radiaton Physics of Semiconductors and Related Materials (Tbilisi State University Press, Tbilisi, 1980) p. 288
46. S.H. Muller, G.M Tuynman, E.G. Sieverts and C.A.J. Ammerlaan, Phys. Rev. B 25, 25 (1982)
47. A.A. Lebedev and B.M. Urunbaev, Sov. Phys. Semicond. 15, 350 (1981)
48. W. Gehlhoff, K.H. Segsa and C. Meyer, Phys. Stat. Sol. (b) 105, K91 (1981)
49. G.W. Ludwig, R.O. Carlson and H.H. Woodbury, Bull. Amer. Phys. Soc. Ser. II, 4, 144 (1959)
50. K. Graff and H. Pieper, J. Electrochem. Soc. 128, 669 (1981)
51. G.W. Ludwig, Phys. Rev. 137, A1520 (1965)
52. O.F. Schirmer, Physica B, 1983 (to be published)
53. H. Feichtinger, Inst. Phys. Conf. Ser. 46, 528 (1979)

54. M. Höhne, Phys. Stat. Sol. (b) 99, 651 (1980)
55. M. Höhne and A.A. Lebedev, Phys. Stat. Sol. (b) 104, K79 (1981)
56. R.L. Kleinhenz, Y.H. Lee, J.W. Corbett, E.G. Sieverts, S.H. Muller and C.A.J. Ammerlaan, Phys. Stat. Sol. (b) 108, 363 (1981)
57. M. Höhne, Phys. Stat. Sol. (b) 109, 525 (1982)
58. E.G. Sieverts, S.H. Muller, C.A.J. Ammerlaan, R.L. Kleinhenz and J.W. Corbett, Phys. Stat. Sol. (b) 109, 83 (1982)
59. G.W. Ludwig, H.H. Woodbury and R.O. Carlson, J. Phys. Chem. Solids 8, 490 (1959)
60. E. Nes and J. Washburn, J. Appl. Phys. 42, 3562 (1971)
61. L.S. Milevskii, I.G. Varvarina and N.T. Bendik, Izv. Akad. Nauk SSSR Neorg. Mater. 12, 1382 (1976)
62. N.T. Bendik, V.S. Garnyk and L.S. Milevskii, Sov. Phys. Solid State 13, 1464 (1971)
63. M. Stojic, Phys. Stat. Sol. (a) 70, K39 (1982)

PHOTOLUMINESCENCE OF DEFECT COMPLEXES IN SILICON

R. Sauer and J. Weber

Physikalisches Institut (Teil 4), Universität Stuttgart
Pfaffenwaldring 57, D-7000 Stuttgart 80
Federal Republic of Germany

ABSTRACT

This article is concerned with three originally different species of defects in silicon which have only recently been recognized to be accessible to the photoluminescence (PL) technique: (i) Defects incorporating transition metal ions, (ii) defects related with oxygen and carbon doping the silicon and (iii) dislocation induced defects. We shall briefly discuss the state of the art from the PL point of view as well as present new data for all three kinds of defect complexes.

Introduction

Extrinsic photoluminescence (PL) related with shallow donor or acceptor doping has long been studied in silicon. The conception of bound excitons (BE) to the doping impurities usually the predominant origin of extrinsic PL has even been developed in this material [1]. Later works have brought detailed insight into the structure of BE's and in their recombination kinetics in silicon as well as in other semiconductors [2]. In particular, it is well established that the main recombination channel in silicon is non-radiative and is dominated by a phononless localized Auger process [3,4]. This process leads to non-radiative lifetimes which are roughly inversely proportional to the fourth power of the impurity ionization energy or, by virtue of Haynes' rule [1], of the exciton localization energy to the impurity, $\tau_{nonrad} \sim E_{loc}^{-4}$. As the comparatively much larger radiative lifetimes, $\tau_{rad} >> \tau_{nonrad}$, do not as much depend on the localization energy increases. Therefore, deep donors or acceptors are not expected to give rise to "normal" BE luminescence.

Actually, there has been no indication at all of PL from deep defects in silicon up to only a few years ago an exception to this statement being luminescence due to irradiation induced defects which is known since about 1965 ([5], and for further references, [6]). Since the second half of the seventies PL from deep defect complexes in silicon has been reported by several groups of workers in samples of quite different origin exhibiting a large variety of spectral features. There is also today an increasing number of PL line systems which are new detected and are due to defect complexes. The new optical centers comprise transition metal ions as well as oxygen/carbon related centers and dislocation induced defects. Important conditions needed for the PL to observe are, among others, suitable basic doping of the silicon, e.g. with oxygen, and appropriate sample treatments like heating, quenching, deforming or thermal annealing. As such processes are in a large scale necessary for device fabrication, the PL study of these defect complexes may also deserve great technological interest. With regard to the above mentioned kinetic properties of donor and acceptor BEs it is likely that all PL which is observed well below the bandgap of silicon, E_g (4.2 K) = 1.1695 eV, can be described in terms of centers of isoelectronic character involving in the luminescing state only one electron and one hole. This expectation is in a number of cases explicitly verified. Contrary to the BE's we encounter in the complex PL systems examples of high luminescence efficiencies despite the low spectral emission energies which correspond to large electron-hole binding energies. Whereas in the BE luminescence there was no doubt about the chemical nature of the optical centers since the early work, the chemistry and geometric configuration of the various complex, non-cubic centers need still to be identified in the current investigations, and in the majority of observed PL systems one has not yet arrived at detailed and generally accepted identifications.

1. Transition Metal Complexes

1.1. Copper related photoluminescence

In this section we shall discuss PL from centers incorporating the transition ions Cu, Fe and Cr. Copper gives rise to a characteristic PL spectrum (Fig. 1) around 1.014 eV exhibiting the main optical transition (Cu_o^o) along with local mode Stokes replicas (Cu_o^1,...) and, at higher temperatures, local mode Antistokes replicas (Cu_o^{-1},...) at a vibrational quantum energy of 7.0 meV. This spectrum was first observed by Minaev et al. [7] who observed it "accidentically" after some sample heating and could not identify the optical center. Weber and Wagner [8] demonstrated that Cu is necessary for the spectrum to observe, and Weber et al. [9] studied it comprehensively. The spectrum arises upon

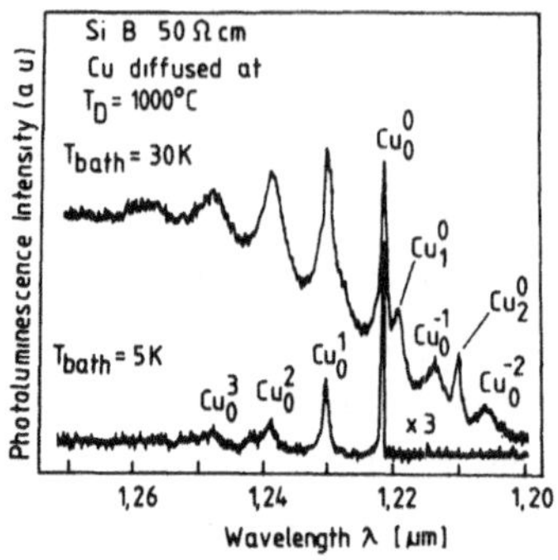

Fig. 1. Copper induced spectrum as recorded at 5 K and 30 K. Upper labels refer to the number of emitted (positive sign) or absorbed (negative sign) local vibrational modes, lower labels characterize electronic excited states

copper diffusion from Cu films evaporated on the surfaces of silicon slices. After the diffusion the samples are quenched in oil or distilled water as rapidly as possible to avoid precipitation of copper which is known to occur when the cooling rate is low. The PL intensities depend drastically on the cooling rate and the lines are only weakly seen in moderately slowly cooled samples at a given diffusion temperature. No dependence of the PL intensities exists on the concentration and species of shallow impurities in the starting material indicating that the optical center neither involves an intrinsic lattice defect nor a constituent other than copper. On the other hand, the incorporation of Cu in the defect is directly proven by an isotope shift of the principle optical transition upon ^{63}Cu or ^{65}Cu implantation and subsequent annealing and quenching of silicon samples (Fig. 2). More information towards the identification of the center is gained by studying the PL intensities

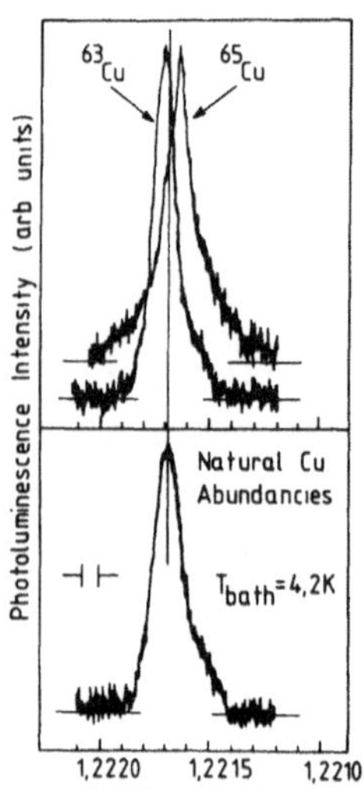

Fig. 2. Main optical transition (Cu_0^0) in naturally abundant copper doped samples and in ^{63}Cu or ^{65}Cu implanted silicon. The isotope splitting is 0.06 meV

at a fixed cooling rate as a function of the diffusion temperature. Since Cu is a very fast diffusor in silicon it is reasonable to assume that for each diffusion temperature the equilibrium value of the solid solubility is reached. The diffusion temperature can be eliminated from the measurements and replaced by the Cu concentration on this assumption using the dependence of the solid solubility on the diffusion temperature which is known from literature. The obtained relation of PL intensities versus Cu concentrations is quadratic [9], and an interpretation of this result in terms of second order reaction kinetics tells us that we are concerned with a Cu pair as the optical center. This is consistent with the analysis of uniaxial stress and Zeeman data [9] showing that the defect is non-cubic and posesses trigonal symmetry. A final answer to the local positions of the Cu atoms in the center cannot be given from the PL measurements at the present time. Following arguments based on the

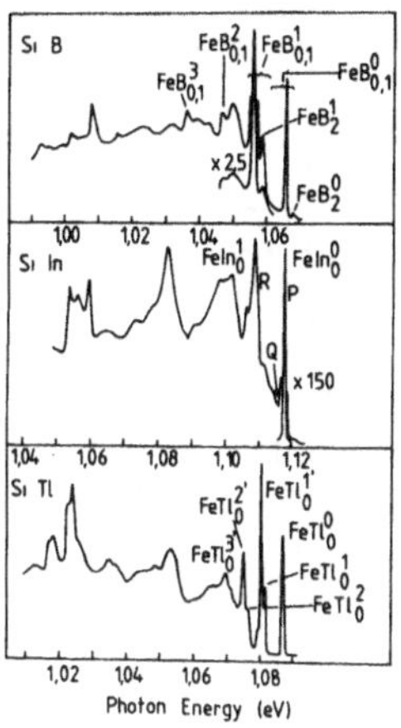

Fig. 3. Spectra of acceptor doped silicon arising upon Fe diffusion (Si:B, top) or Fe diffusion or heating and thermal quenching of silicon (Si:In,Tl). Labels of lines as in Fig. 1

thermalization features of the stress split Cu^0_0 line [9] it is not unlikely that one Cu atom occupies a substitutional lattice site whereas the other Cu atom resides in a nearest-neighbour interstitial site.

An excited state of the exciton localized at the Cu pair at 1.9 meV above the exciton ground state is evident from the PL spectra at higher temperatures (see Fig. 1, line Cu^0_1), and is consistent with the quantitative analysis of the temperature dependence of the Cu^0_0 line intensity. The exciton ground state splits fivefold in a magnetic field for H//[001] when orientational degeneracy effects due to the molecular Cu-Cu dumbbell along<111> are absent. This and the thermalization features of the Zeeman components show that the main optical line is a quintet-to-singlet transition. Further evaluation of the Zeeman data suggests that the luminescent state of the Cu center is an almost "classical" isoelectronic bound exciton (see, for fuller discussion, Ref. [6]): The exciton ground state corresponds to J = 2, and the 1.9 meV excited state split apart by the electron-hole electrostatic interaction corresponds to J = 1. The binding center is isoelectronic (J = 0), and though it represents an axial <111> configuration the related local strain field does not noticeably split the exciton states so that a classification of the states in terms of spherical angular momenta is possible. The isoelectronic nature of the Cu trap is currently supported by Watkins et al. [10] who measure long (radiatively determined) lifetimes of the Cu induced PL lines between 460 and 670 µsec. This is typical of an exciton at an isoelectronic trap having no possibility of recombining non-radiatively via an Auger process since a third Auger particle is not available.

It seems possible to establish a correlation between the optically deduced primary particle binding energy to the Cu center, 123 meV (see later discussion and Table 1), and a DLTS level, E_V + 102 meV, which is observed in similarly treated, copper diffused and quenched, silicon [9] or in the *same* samples as used for the PL studies.

Table 1 Summary of data, mostly from photoluminescence, on transition metal complexes in silicon [6]

Optical center	Cu-Cu	Fe-B	Fe-In	Fe-Tl	Cr-B
No-phonon transition (eV)	1.0145_5	1.0668	1.1179	1.0869	0.8436
Local mode energy (meV)	7.0	9.6	9.2	5.4/6.5	13.6
Defect symmetry Center configuration	trigonal <111>	trigonal <111>	monocl.II <100>	trigonal <111>	trigonal <111>
Spectroscopic localization energy (meV)	155.0	102.7	51.6	82.6	324
Thermal dissociation energy (meV)	32	≈15	36	≈20	≈20
Primary particle binding (meV)	123	≈88	-	-	≈304
DLTS level above E_V (meV)	102	100	160	-	290

1.2. Iron and chromium related photoluminescence

Novel sharp line spectra around 1.118 eV and 1.087 eV have been observed in silicon doped with In or Tl, respectively, by several groups. Part of the In related PL (Fig. 3) was first reported by Vouk and Lightowlers [11] and then studied by Mitchard et al. [12] who showed that the lines are long-lived (≈200 µsec) indicating an isoelectronic binding trap. The spectrum was in great detail studied [13,14] after the detection by Weber et al. [15] and Thewalt et al. [16] that the PL is dramatically enhanced (up to 10^4) by heating the In doped silicon to 800-1000 °C and then quenching it rapidly to room temperature. Thewalt and coworkers [17] showed that a similar heat treatment of Tl doped silicon results in a new spectrum around 1.087 eV which exhibits some features resembling the In related spectrum and is examined in detail in the forthcoming paper by Thewalt et al. [14]. In particular, the Tl related, low temperature PL lines (Fig. 3) have long lifetimes of 53 μsec as in the In case. A third spectrum of similar characteristic structure with sharp no-phonon lines around 1.067 eV was observed by Weber and Wagner [8] to arise in boron doped silicon upon Fe diffusion of the samples (Fig. 3) and is described by Sauer and Weber in more detail [6]. All three spectra posess several features in common:

- The principle no-phonon transition is replicated by local vibrational mode satellites whose vibration quantum energy is low and of similar size.
- The extended low energy structure reflects features of the phonon density-of-states indicating that the deep optical centers couple to phonons from the whole phonon spectrum.
- The lines are long-lived hinting to isoelectronic defects trapping the radiating excitons. For the In and Tl cases this was directly measured, for the boron case it is very likely considering the high luminescence efficiency in conjunction with the large exciton binding energy.
- The optical centers are non-cubic according to uniaxial stress and Zeeman measurements (In,Tl). For the boron case, the PL data reveal no symmetry lower than cubic, but when Fe - according to the sample treatment necessary to create the spectrum - is incorporated in the defect the statement is evident.

Another spectrum consistent in every respect with these criteria has very recently been reported by Conzelmann and Weber [18] to appear in boron doped, chromium diffused silicon near 0.844 eV. Important data of all four PL line systems are summarized in Table 1.

Can photoluminescence tell us what the nature of these optical centers is? The present authors believe that we are concerned with $Fe_i^+ - A_s^-$ pairs (where the acceptor ions A^- are B^-, In^- or Tl^-) or with analogous Cr-B pairs. This identification rests on four arguments:

- Sample heating for some minutes or even less to relatively low temperatures generates the In and Tl related spectra provided the starting material is appropriately acceptor doped. This hints to a fast diffusor which could be nearly any transition metal (except for Cu which never produces this kind of PL). The diffusor must then be present in the material as an unintentional contamination; however, considering the high PL efficiency near unity, the concentration does probably not exceed $10^{11}-10^{12}$ cm^{-3}.
- Fe diffusion generates the spectrum in the boron case. For In or Tl doped samples, Fe diffusion enhances the corresponding PL intensities substantially. This is also corroborated by work of Schlesinger and McGill [19].

- Symmetry of the optical centers. We may here preferentially refer to the In case of monoclinic II symmetry which corresponds in our identification to a <100> Fe-In pair configuration. Fe-In pairs are known from EPR studies [20] to have the same symmetry. This is striking as <100> centers have only seldom been found in PL as well as in EPR <111> configurations being abundant.
- A correlation which we make between the PL primary particle binding energies and DLTS levels found in the *same* samples used for the PL measurements. The primary binding energies result as the difference of spectroscopic and thermal dissociation energies if the corresponding numbers are sufficiently different (Table 1); the DLTS levels have with great certainty been associated with Fe-B, Fe-In and Cr-B pairs, respectively (see Ref. [6]). In the favourable cases, both methods yield energy levels which are very close to each other within the experimental errors of (estimated) <10-15 meV.

The local positions of the transition ions and their acceptor partners are in the present interpretation derived combining PL symmetry determinations with accepted EPR data: Fe-B is believed to be an interstitial-substitutional nearest-neighbour pair; Fe-In to be a next nearest-neighbour pair due to the bigger size of the In ion leading to the <100> interstitial-substitutional configuration; and Fe-Tl to be an overnext nearest-neighbour interstitial-substitutional <111> pair with a pair distance of half of the cube body diagonal due to the again increased size of the Tl ion.

The present identification suggests that the interstitial transition ions Cu, Fe or Cr are responsible for the size and the similarity of the local vibration mode quanta. It was shown [6] that consistency exists with the well-authenticated B_s-Li_i pairs in silicon where Li isotope shifts and the relative insensitivity of the infrared absorption spectra to the replacement of B by Al or Ga give evidence that the interstitial defect partner vibrates essentially independently of its substitutional counterparts.

An open question is at the present time why the shallow acceptors Al and Ga do not introduce similar novel spectra upon Fe diffusion or sample heating and quenching. The search for such spectra is being continued. Possibly, slight modifications in the sample treatment procedure are necessary for a success if such optical centers exist.

2. Oxygen and Carbon Related Photoluminescence

Novel luminescence features which correlate with oxygen and/or carbon doping the silicon have been found within the last three years. In an attempt to characterize the present state of the art one can roughly discern three principally different types of spectra which seem all to be associated with oxygen or carbon:

- In the comparatively near-band gap region ($h\nu \geqq 1.08$ eV) broad bands arise as no-phonon transitions as well as TA- and TO-momentum conserving phonon replicas when CZ wafers are subjected to 450 °C heating known to produce the oxygen thermal donor (TD). These bands are greatly modified upon heating the samples to 600 °C resulting in even broader overlapping bands which extend over a spectral range of about 80 meV. They are ascribed to "new donors". Finally, the PL features of as-received untreated material are essentially observed after a further anneal at 900 °C. The as-received PL and the PL after the thermal annealing stages have been labelled A-, B-, C- and E-type luminescence by Tajima and coworkers [21-23] who studied this class of new spectra. The 450 °C TD-bands (B-type luminescence) were also observed by Nakayama et al. [24,25] and by Yasutake et al. [26] in carbon lean CZ silicon and are there labelled O_{NP}, O_{TA} and O_{TO}. The present authors believe that it is difficult to obtain information on the nature of the optical centers in a simple and unambigous way from such broad bands and, therefore, shall no further discuss these spectra in the remainder of the article.
- Single sharp lines of considerable intensity sometimes superimposed on structureless broad bands are reported by Minaev and Mudryi [27] at energies well below the band gap (0.96-0.75 eV). They arise in CZ n-type silicon after heat treatment at 450 °C, and their intensities depend on the duration of the heating (1-300 hours). Most prominent are the lines P (0.7667 eV) and H (0.9259 eV). Higher carbon concentrations in CZ material bring new lines up (T: 0.9355 eV, I: 0.9652 eV) whose relative intensities depend again on the heating duration. Four sharp, weak lines (labelled 1 through 4) are most probably identical with particular lines of the many-lines spectrum to be mentioned in the following indent.
- Nakayama and coworkers [24,25] observed in silicon of high carbon *and* high oxygen concentrations, $[O_i] = 9.6 \cdot 10^{17}$ cm^{-3} and $[C_s] = 10.2 \cdot 10^{16}$ cm^{-3}, a characteristic spectrum consisting of many sharp lines of strongly different intensities which they labelled S_1 through S_{19}. This spectrum is independent of boron or phosphorus doping

and is strong for an annealing temperature of 500 °C and several hundred hours heating duration. Some of their experimental observations led these authors to argue that the optical centers are isoelectronic traps involving the intentional dopants oxygen and carbon.

In the PL experiments which we have performed we observe all three types of spectra in appropriately doped and treated silicon or in material lent to us by the original authors (in the last mentioned case). We report here new data on the P line and on the many-lines spectrum.

H. Nakayama, T. Nishino and coworkers were so generous as to provide some of their oxygen and carbon rich samples for experiments in the Stuttgart group. Using a Ge-detector instead of a S1-type cathode photomultiplier we detect many weak additional sharp lines in between the already known lines particularly in the low energy portion of the spectrum (Fig.4). The so completed spectral structure is reminiscent of discrete donor-acceptor pair line spectra as are long known to exist in direct and indirect semiconductors such as CdS, ZnSe or GaP, AlSb and cubic SiC, respectively [28]. Analogous line spectra have recently also been observed in silicon with a variety of donors and acceptors involved like Li-O and P or B, Al, Ga and In, respectively [29-31]. Our tentative interpretation of the present spectrum is in terms of donor-acceptor pair spectra and is consistent with the following observations: (1) The thermal dissociation energies of the stronger lines are, according to Nakayama et al., of similar size ranging roughly from 5 to 11 meV. This hints to transition initial states which are placed in the forbidden gap at not too different energies the wide spreading of the experimental photon energies being due to the final states which feel the Coulombic interaction strongly varying with the pair distance [2]. We find that all lines grow simultaneously with not too different activation energies when the temperature is raised from 4.2 K to about 10 K indicating some kind of supply channel from other traps. Especially, there are no two lines which can be associated with different initial states of *one* particular center being in thermal equilibrium with each

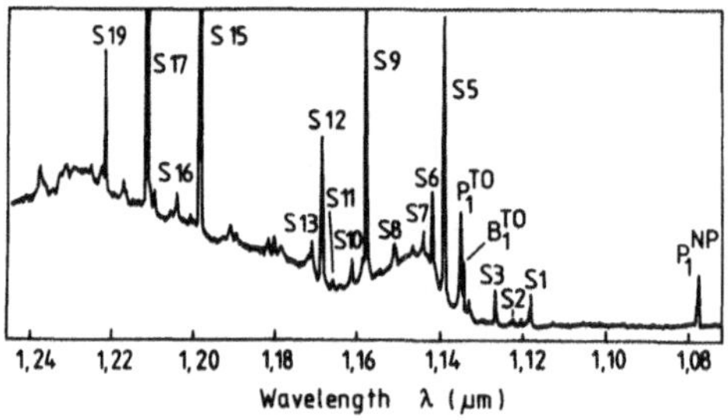

Fig. 4. Many-lines spectrum in oxygen *and* carbon rich silicon. S_1 through S_{19} have already been reported [24,25], unlabelled lines are new

other. (3) None of the lines splits in a magnetic field up to 5.3 Tesla, and no "forbidden" transitions show up by the perturbation of the field indicating that we are not concerned with close or nearest-neighbour donor-acceptor pairs which could be regarded as "molecular" isoelectronic traps in the approach usually made.

Quantitative fits to the spectrum on the basis of the well-developed theories on donor-acceptor pair spectra are in progress. Irrespectively of the possibly positive results we cannot content ourselves with a straight-forward interpretation in terms of oxygen-carbon associates as we failed to produce the many-lines spectrum in oxygen and carbon rich silicon annealed at 450-500 °C using starting material from other vendors than the Osaka group. It seems likely that a third chemical element is needed for the spectrum to observe.

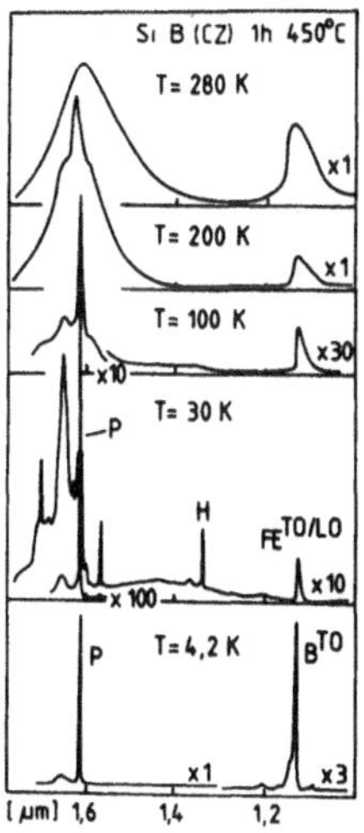

Fig. 5. Development of the P (0.7667 eV) line spectrum at increasing temperatures

First results on the P line at 0.7667 eV which we strongly observe in oxygen rich, carbon lean CZ silicon after heating at 450 °C for 1 hour as Minaev and Mudryi [27] are the following ones (Fig. 5): The P line (displacement from the band edge 402.8 meV) increases at fixed optical excitation from 4.2 K to 40 K by a factor of ≈300 and is then quenched with a thermal dissociation energy of ≈45 meV. It exhibits new structure in an intermediate temperature range both on the low and on the high energy side. From about 150 K on a broad band emerges near P which is stable up to the 600 K (!) regime and is even then strongly seen. At 280 K its intensity peak is displaced from E_g by ≈355 meV where the lower value than for the P line at 4.2 K at a comparable spectral position is due to the band shrinkage at higher temperatures. The broad hump (right hand side, Fig. 5) is the band-to-band recombination developing from the free exciton (FE^{TO}) peak. It is, however, much intenser than in intrinsic material under equivalent experimental conditions possibly due to a carrier confinement near the excitation spot by the oxygen related impurity associates. Similar effects are known to occur for the electron-hole drop (EHD) luminescence in doped silicon and germanium [32]. The P line quenching energy of 45 meV places one of the defect levels ≈45 meV below one band and accordingly, the other level ≈358 meV below the other band. The broad PL band may then be interpreted as a band-to-defect transition consistent with the spectral position and the temperature dependencies of the intensity and the halfwidth. The deep 358 meV defect level should be accessible to other analytical techniques; actually an acceptor level, E_V + 350 meV, has been observed by Hall effect measurements of the thermal oxygen donor in p-type, oxygen rich (> 10^{18} cm^{-3}) silicon after heat treatment at 430 °C for 140 min by Mordkovich [33].

3.Dislocation Induced Photoluminescence

Recombination radiation from dislocations was first observed by Drozdov et al. [34,35] who observed four lines D_1 through D_4 in deformed silicon crystals in the spectral region from 1.2 to 1.6 µm. Spatially resolved PL recordings of samples containing the highest dislocation density in the center of the bulk showed that the PL intensities are directly related with the dislocation concentration. The D lines were also observed after laser annealing of plastically deformed silicon [36].

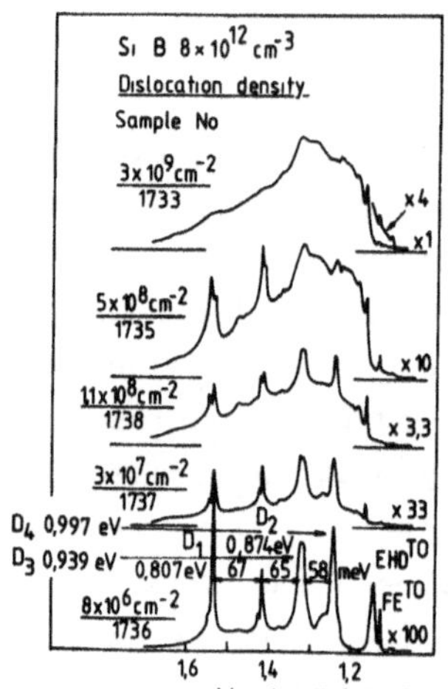

Fig. 6. Dependence of the D_1 through D_4 lines on the dislocation density (note intensity reduction factors)

We have performed PL measurements on deformed silicon containing well defined dislocations in a cooperation with E.R. Weber and H. Alexander (Universität Köln) who prepared the samples and characterized them by EPR and Transmission Electron Microscopy. There are four important observations which we make [37]:

- Independently of Drozdov's method we confirm that the D lines originate from dislocations. In a series of samples with increasing dislocation densities introduced at a deformation temperature of 650 °C the PL intensities are correspondingly enhanced (Fig. 6). The intrinsic near-band gap PL from free excitons (FE^{TO}) and electron-

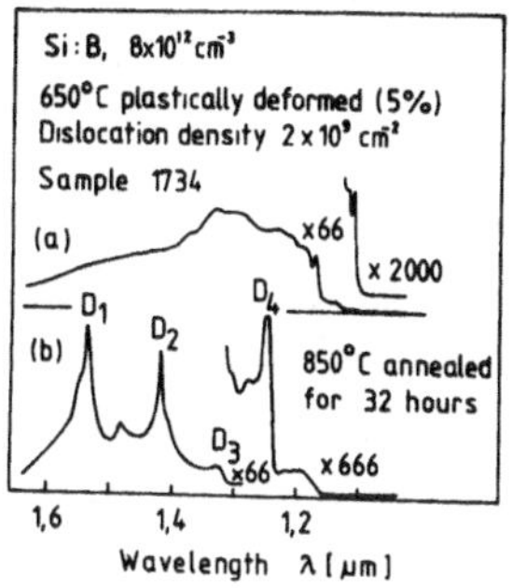

Fig. 7. Spectrum at 2.10^9 cm^{-2} dislocation density in as-deformed silicon (a) and after annealing at 850 °C (b)

hole drops (EHD^{TO}) is rapidly quenched by the dislocations.

- At higher dislocation densities a broad structured band emerges extending from 0.8 to 1.05 eV. This band disappears upon long-term annealing at 850 °C and is essentially replaced by the D lines (Fig. 7). The band exhibits production and annealing behaviour similar to EPR centers in deformed silicon; it is, therefore, likely due to point-defect like centers at dislocations or in the bulk of the crystal.

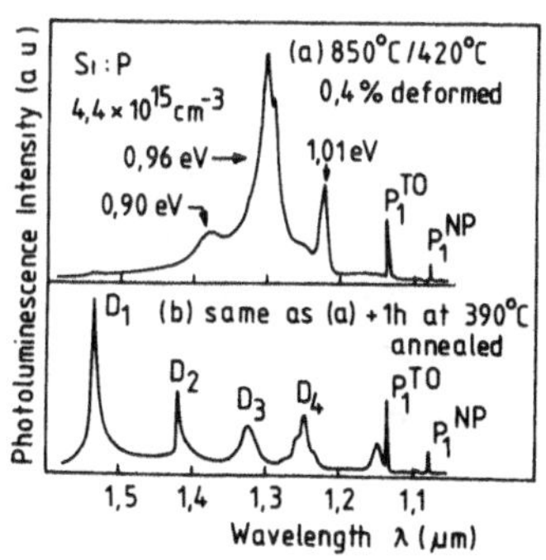

Fig. 8. New spectral features at 1.01 eV, 0.96 eV and 0.90 eV in silicon containing straight dislocations (a) and after annealing (b)

- New spectral features arise in samples deformed at low temperature (420 °C) under high compressive stress after a slight predeformation (850 °C) which produces straight 60° and screw dislocations throughout. The PL spectrum exhibits now a dominating, relatively sharp line at 0.96 eV together with a weak hump at 0.90 eV and a sharp line at 1.01 eV (Fig. 8). Stronger deformation enhances the dominating 0.96 eV band. Relaxation of the straight, widely dissociated dislocations (390 °C, 1 hour) leads to the reappearance of the D lines (Fig. 8).
- A similar high-stress deformation along a <2,1,11> axis which produces a large stacking fault density results in a spectrum (Fig. 9) which is dominated by the 1.01 eV line whereas only weak residual D_1 and D_2 luminescence is observed. Again, relaxation of the stacking faults in a 390 °C, 1 hour heat treatment restores the usual D line spectrum but now largely favoring the D_3 and D_4 lines over D_1 and D_2.

These findings allow to draw the following conclusions:

(1) The D lines originate from kinks or constrictions on curved dis-

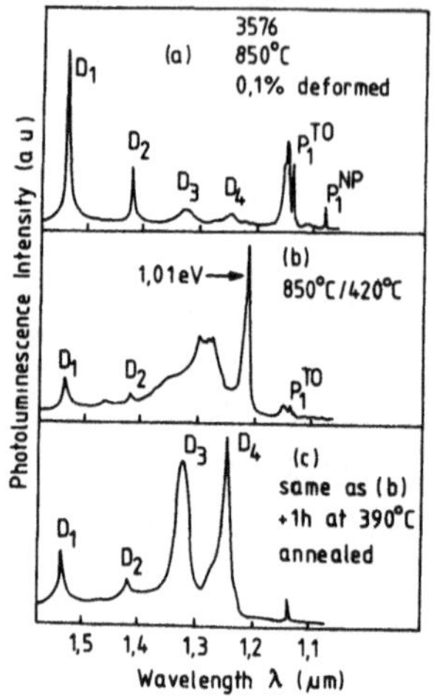

Fig. 9. Spectra after slight predeformation at 850 °C (a), after subsequent low temperature (420 °C) high-stress deformation (b) and after annealing (c)

locations rather than from perfect, straight dislocations.

(2) The broad band (0.8-1.05 eV) may be correlated with EPR point-defect like centers.

(3) The 0.96 eV and 0.90 eV bands are related with the presence of straight dislocations.

(4) The 1.01 eV line is associated with stacking faults.

Acknowledgements

It is a pleasure to thank E.R. Weber and H. Alexander for the fruitful and pleasant cooperation. Thanks are due to H. Nakayama, T. Nishino and their colleagues for the gift of some of their silicon samples. The financial support of the Bundesministerium für Forschung und Technologie (BMFT, contract NT 2555) is gratefully acknowledged.

References

1. J.R. Haynes, Phys. Rev. Lett. 4, 361 (1960)
2. P.J. Dean and D.C. Herbert, in "Excitons", ed. K. Cho (Springer, Berlin, 1979) p. 55
3. S.A. Lyon, G.C. Osbourn, D.L. Smith and T.C. McGill, Solid State Commun. 23, 425 (1977)
4. W. Schmid, phys. stat. sol. (b) 84, 529 (1977)
5. C.G. Kirkpatrick, J.R. Noonan and B.G. Streetman, Radiat. Eff. 30, 97 (1976)
6. R. Sauer and J. Weber, Physica B (Proc. Internat. Conf. Defects in Semiconductors, ICDS-12, Amsterdam 1982)
7. N.S. Minaev, A.V. Mudryi and V.D. Tkachev, Sov. Phys.-Semicond. 13, 233 (1979)
8. J. Weber and P. Wagner, J. Phys. Soc. Jpn 49, Suppl. A, 263 (1980)
9. J. Weber, H. Bauch and R. Sauer, Phys. Rev. B 25, 7688 (1982)
10. S.P. Watkins, U.O. Ziemelis, M.L.W. Thewalt and R.R. Parsons, to appear in Solid State Commun.
11. M.A. Vouk and E.C. Lightowlers, J. Lumin. 15, 357 (1977)
12. G.S. Mitchard, S.A. Lyon, K.R. Elliott and T.C. McGill, Solid State Commun. 29, 425 (1979)
13. J. Wagner, J. Weber and R. Sauer, Solid State Commun. 39, 1273 (1981)

14. M.L.W. Thewalt, U.O. Ziemelis, S.P. Watkins and R.R. Parsons, to appear in Can. J. Phys.
15. J. Weber, R. Sauer and P. Wagner, J. Lumin. 24/25, 155 (1981)
16. M.L.W. Thewalt, U.O. Ziemelis and R.R. Parsons, Solid State Commun. 39, 27 (1981)
17. M.L.W. Thewalt, U.O. Ziemelis and R.R. Parsons, Phys. Rev. B 24, 3655 (1981)
18. H. Conzelmann and J. Weber, Physica B (Proc. Internat. Conf. Defects in Semiconductors, ICDS-12, Amsterdam 1982)
19. T.E. Schlesinger and T.C. McGill, Phys. Rev. B 25, 7850 (1982)
20. G.W. Ludwig and H.H. Woodbury, Solid State Physics, ed. F. Seitz and D. Turnbull (Academic Press), Vol. 13, 223 (1962)
21. M. Tajima, A. Kanamori and T. Iizuka, Jpn. J. Appl. Phys. 18, 1401 (1979)
22. M. Tajima, S. Kishino and T. Iizuka, Jpn. J. Appl. Phys. 18, 1403 (1979)
23. M. Tajima, A. Kanamori, S. Kishino and T. Iizuka, Jpn. J. Appl. Phys. 19, L755 (1980)
24. H. Nakayama, J. Katsura, T. Nishino and Y. Hamakawa, J. Lumin. 24/25, 35 (1981)
25. H. Nakayama, T. Nishino and Y. Hamakawa, Appl. Phys. Lett. 38, 623 (1981)
26. K. Yasutake, M. Umeno, H. Kawabe, H. Nakayama, T. Nishino and Y. Hamakawa, Jpn. J. Appl. Phys. 21, 28 (1982)
27. N.S. Minaev and A.V. Mudryi, phys. stat. sol. (a) 68, 561 (1981)
28. P.J. Dean, in "Progress in Solid State Chemistry", ed. J.O. McCaldin and G. Somorjai (Pergamon Press, Oxford, 1973), Vol. 8, p. 1
29. U.O. Ziemelis, R.R. Parsons and M. Voos, Solid State Commun. 32, 445 (1979)
30. U.O. Ziemelis and R.R. Parsons, Can. J. Phys. 59, 784 (1981)
31. U.O. Ziemelis, M.L.W. Thewalt and R.R. Parsons, Can. J. Phys. 60, 1041 (1982)
32. J.C. Hensel, T.G. Philips and G.A. Thomas, Solid State Physics, ed. H. Ehrenreich et al. (Academic Press, New York, 1977), Vol. 32, p. 87
33. V.N. Mordkovich, Fiz. tverd. Tela 6, 847 (1964); cited after the listing of TD oxygen levels in P. Gaworzewski and K. Schmalz, phys. stat. sol. (a) 55, 699 (1979)
34. N.A. Drozdov, A.A. Patrin and V.D. Tkachev, JETP Lett. 23, 598 (1976)
35. N.A. Drozdov, A.A. Patrin and V.D. Tkachev, phys. stat. sol. (b) 83, K137 (1977)
36. R.H. Übbing, P. Wagner, H. Baumgart and H.J. Queisser, Appl. Phys. Lett. 37, 1078 (1980)
37. J. Weber, R. Sauer, E.R. Weber and H. Alexander, to be published

ELECTRON MICROSCOPICAL ANALYSIS OF THE STACKING FAULT BEHAVIOUR IN INERT-GAS ANNEALED CZOCHRALSKI SILICON

H. Bender, C. Claeys*, J. Van Landuyt, G. Declerck*, S. Amelinckx, R. Van Overstraeten*

Universiteit Antwerpen, R.U.C.A.
Groenenborgerlaan 171, B-2020 Antwerpen, Belgium
*K.U. Leuven, E.S.A.T.
Kardinaal Mercierlaan 94, B-3030 Heverlee, Belgium

1. Introduction

Lattice defect formation is observed during the inert-gas annealing of Czochralski silicon wafers [1-3]. The defects are characterized as partial dislocation loops associated with stacking faults, perfect dislocation loops and precipitate-dislocation loop complexes. It is demonstrated that a high interstitial oxygen supersaturation is necessary in order to obtain the defect nucleation and growth [4]. Therefore it is suggested to speak about oxygen induced stacting faults, rather than oxidation induced stacking faults. The symbol SF will be used. Depending on the experimental conditions an SF-denuded surface region can be obtained, beneath which the SF-length increases with depth under the surface to a maximal bulk value. It is pointed out that the denuded zone is not always free of other types of defects.

In this paper, the nucleation and growth of the stacking faults throughout the bulk of the silicon wafers will be discussed. Their time and temperature dependence is investigated. It is seen that the stacking fault density decreases with increasing anneal time, while the stacking fault length increases. A model to explain the transition from Frank-type stacking fault towards $\frac{1}{6}$<114>-type stacking fault and perfect dislocation loop will be discussed.

2. Experimental Procedure

This study is mainly performed on p-type, 7-21 Ωcm, 5 cm ∅, 250 μm thick, <111>-oriented Czochralski grown silicon wafers. Their initial interstitial oxygen concentration ($[O_i]$) is determined by means of IR-spectroscopy and lies in the range $1\text{-}2\cdot10^{18}$ atoms/cm³. The wafers received (from the manufacturer) a low temperature treatment in order to stabilize the thermal donors. The exact conditions of this pretreatment are unknown. Additionally also wafers from a different supplier are used. These are n-type <001>-oriented wafers with respectively 2-4 Ωcm and 3-5 Ωcm, and $[O_i] \sim 7 \cdot 10^{17}$ atoms/cm³.

The wafers are annealed in a nitrogen ambient at temperatures between 1050 °C and 1240 °C for times ranging between 5 and 1200 minutes. The SF-length and density distribution throughout the wafer bulk is analysed by means of optical microscopy combined with preferential etching at different depths under the surface or on cleavage planes. The defect structures are characterized by means of a high voltage electron microscope operating at 1000 kV.

3. Results and Discussion

The distribution of the SF-length versus the depth under the surface is determined throughout the whole described time and temperature region [3,4]. It is seen that a stacking fault denuded surface area is obtained after an N_2-annealing at temperatures in excess of 1150 °C. At 1240 °C, the width of this denuded zone increases very fast with anneal time, so that the wafers are completely SF-free after 8 h of treatment. At temperatures of 1050 °C and 1100 °C small SF's occur in the near surface region. As a function of the depth under the surface, larger SF's are observed. From a depth of about 50 µm onwards, the SF-sizes remain constant. At all depths, larger SF's occur with increasing anneal time. A maximum bulk value is reached after a certain anneal time, which depends on the temperature.

A significant variation of the SF-lengths and densities is observed between wafers from different ingots or crystal parts, although these have a comparable interstitial oxygen content. In order to examine this effect further, a number of wafers from different suppliers are partitioned in four quarters. These are annealed in N_2 at 1100 °C for respectively 1h, 3h, 6h and 10h. Figure 1 shows for the first type of silicon wafers (i.e. $[O_i] = 1\text{-}2 \cdot 10^{18}$ atoms/cm³), the results of the stacking fault densities and lengths as a function of the anneal time. It is seen that initially large density variations occur between the different wafers. The densities decrease with time to a value which is nearly equal for all wafers. The SF-lengths behave in an inverse way, so that larger SF's occur with increasing time. After 10h no further growth is observed [4]. In the two supplementary sets of wafers no defects are formed at all. Although they have a lower $[O_i]$, an interstitial oxygen supersaturation still exists at the anneal temperature. The absence of defects is therefore to be correlated with the conditions of the low temperature pretreatment given by the manufacturer

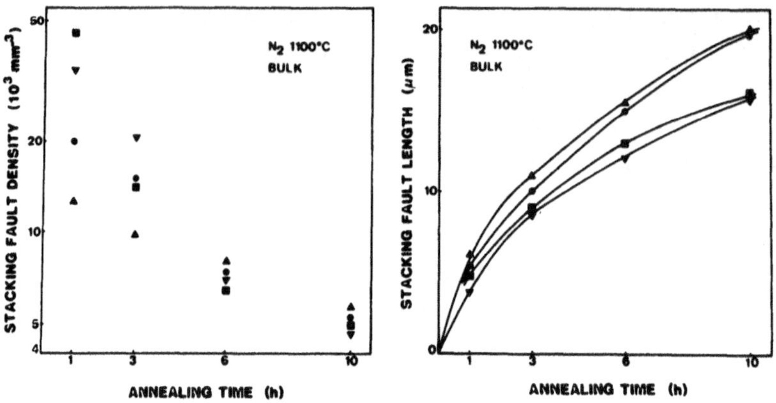

Fig. 1. The stacking fault density and the length distribution as a function of time for an N_2 annealing at 1100 °C. The same symbols correspond to measurements on a single wafer

[4]. The defect density variations in the first type of wafers are due to concentration differences of the grown-in nucleation centers. These can have been induced by differences in respectively the crystal growth conditions, pretreatment temperatures and times or interstitial oxygen concentration.

TEM-analysis of the wafers in the as-received condition does not reveal any contrast features that can be interpreted as grown-in nucleation centers, indicating that their sizes must be below the detection limits. However, already after a 5 minutes annealing, small SF's are present in the wafers. Their density does not increase during the further treatments. This indicates that the nucleation of the SF's is heterogeneous. They are generated by means of a Bardeen-Herring climb mechanism [5] at SiO_2 precipitates which are formed if an interstitial oxygen supersaturation exists. By the absorption of the silicon interstitials which are generated due to the concentration difference of the silicon atoms in the precipitate to that in the silicon matrix, the partial dislocation will climb further and close around the precipitate. These initial stacking faults have always a circular shape; they are of the Frank-type with $b = \frac{1}{3}\langle 111\rangle$ and are extrinsic in nature. They occur with equal density in the four {111}-planes. As the material is $\langle 111\rangle$-oriented, the growth and dislocation reactions can easily be observed on those defects lying in the basal plane, as these remain completely inside the thinned specimen.

As the anneal time is increased, a second SF can be generated at the same precipitate and will also grow out to a circular shape (Fig. 2a). These double Frank stacking faults are transformed into a polyhedral SF with the bordering segments in $\langle 110\rangle$ directions and surrounded by a

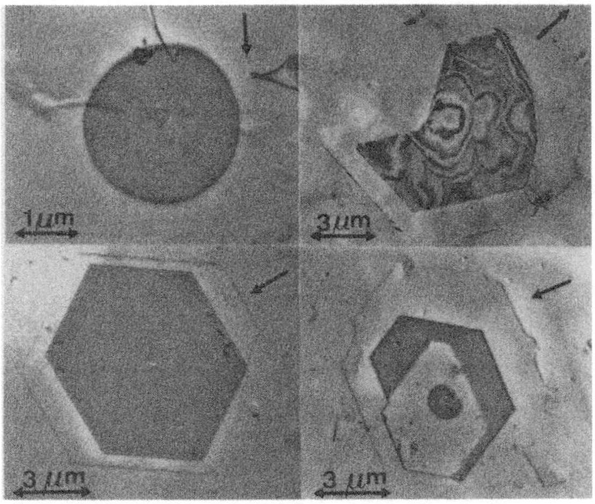

Fig. 2. The transition from Frank-type SF towards perfect loops: a) nucleation of double Frank loop; intermediate b) and final c) state of the transition towards a 1/6<114>-SF and perfect loop; d) further unfaulting (all diffraction vectors are of the <111>-type)

perfect dislocation loop. A transition and final state of this process are shown in respectively Fig. 2b and 2c. The diffraction contrast of these polyhedral SF's is quite different from that of the Frank type SF's. They have for only one of the three <220> diffraction vectors in the fault plane, perfect extinction on the sides parallel with the diffraction vector; whereas the Frank defects show perfect extinction parallel to all three of these diffraction vectors. In a previous report, we suggested a Shockley $\frac{1}{6}$<112> burgersvector [3]. However the strong residual diffraction contrast of the bordering dislocation sides non-parallel with the diffraction vector left us with some doubts concerning this burgersvector. The expected residual contrast is therefore computed with the programs published by Head et al. [6]. In Fig. 3 the

g [20$\bar{2}$] u [011]	Shockley	Frank	
b	$\frac{1}{6}$[121]	$\frac{1}{3}$[1$\bar{1}$1]	$\frac{1}{6}$[1$\bar{4}$1]
g·b	0	0	0
g·(b×u)	0	1,41	1,41
g·b_e	0,5	0	0,5

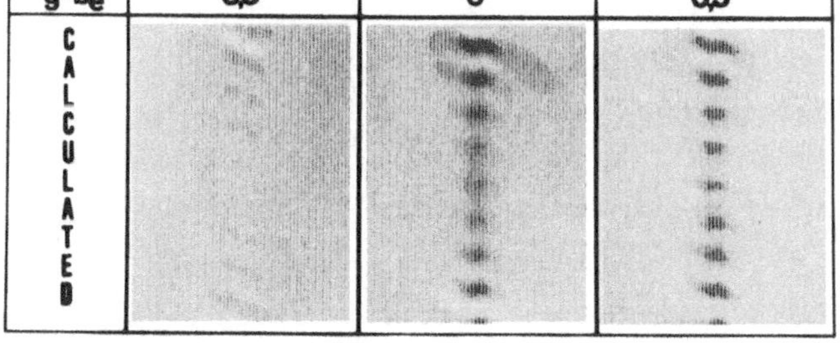

Fig. 3. Residual contrast image calculations of respectively a Shockley, a Frank and a 1/6<114> partial dislocation for g.b = 0 - conditions

results are shown for an inclined partial dislocation in the [011]-direction with the diffraction vector g = $[20\bar{2}]$ for the burgersvector b respectively $\frac{1}{6}[121]$, $\frac{1}{3}[1\bar{1}1]$ and $\frac{1}{6}[1\bar{4}1]$. The residual contrast for the Shockley dislocation is much too weak compared with the electron microscopical observations. Therefore it is concluded that the burgers-vector of the polyhedral SF's is given by $\frac{1}{6}$<114>, which is also in agreement with the higher described diffraction contrast. The generation of these stacking faults occurs by the reaction:

$$\frac{2}{3}[111] \rightarrow \frac{1}{6}[114] + \frac{1}{2}[110]$$

This is energetically favourable and is consistent with the observation that the polyhedral SF's are surrounded by a perfect loop which has also perfect extinction on the sides parallel to those on which the SF-loop has perfect extinction. The polyhedral $\frac{1}{6}$<114> SF's are intrinsic in nature, as was already suggested by Grilhé et al. [7]. This type of SF's is also reported in the case of the annealing of implantation damage [8-10]. Under these experimental conditions, they are always elongated into a <110> direction and are much smaller than in our cases. It was suggested [8] that these rod-like defects are generated from more complex defects such as 0.04 <116>. There are no indications from our experiments for this mechanism to occur. Neither do we believe in the generation of $\frac{1}{6}$<114> SF's from single Frank partial dislocations due to the reaction:

$$\frac{1}{3}[111] + \frac{1}{6}[\bar{1}\bar{1}2] \rightarrow \frac{1}{6}[114]$$

This reaction is not energetically favourable as there is some evidence about the intrinsic stacking fault energy to be larger than or equal to the extrinsic one [11], and furthermore, there is no reduction of the dislocation energy in this reaction.

The surrounding perfect loop moves out of the stacking fault plane. Therefore single polyhedral SF's can occur.

More complex defect configurations are observed in the long time annealed wafers (Fig. 2d). These defect structures are due to the further unfaulting of the stacking faults, possibly by a reaction of the type:

$$\frac{1}{6}[114] + \frac{1}{6}[\bar{1}2\bar{1}] \rightarrow \frac{1}{2}[011]$$

This unfaulting will result in a decreasing SF-density, as is shown in Fig. 1.

A conversion of Frank partial loops to perfect loops is also reported during successive thermal oxidations by Kawado [12]. It was explained in terms of a single or double step annihilation of single Frank loops due to the reaction with two Shockley dislocations. He found no evidence of the formation of $\frac{1}{6}$<114>-type stacking fault.

4. Conclusion

It is demonstrated that stacking faults can nucleate and grow during the inert annealing of high interstitial oxygen containing silicon wafers. The defect density decrease can be explained by a transition model of Frank type SF's towards $\frac{1}{6}$<114>-type SF's and further to perfect loops.

Acknowledgements

This work is performed with the financial support of the Belgian Science Foundation (I.I.K.W.). G. Declerck is indebted to the Nationaal Fonds voor Wetenschappelijk Onderzoek (N.F.W.O.) for his fellowship as Onderzoeksleider. The authors thank Dr. H. Dekkers of S.C.K. - Mol for helping with the image calculations.

References

1. D.M. Maher, A. Staudinger, J.R. Patel, J. Appl. Phys. 47, 3813 (1976)
2. H. Takaoka, J. Oosaka, N. Inoue, Jpn. J. Appl. Phys. 18, suppl. 18-1, 179 (1979)
3. H. Bender, J. Van Landuyt, S. Amelinckx, C. Claeys, G. Declerck, R. Van Overstraeten, Inst. Phys. Conf. Ser. n°60, 313 (1981)
4. C. Claeys, H. Bender, G. Declerck, J. Van Landuyt, R. Van Overstraeten, S. Amelinckx, Proc. of the 12th International Conference on Defects in Semiconductors, 31/8-3/9/82, Amsterdam, to be published in Physica B
5. K. Wada, H. Takaoka, N. Inoue, K. Kohra, Jpn. J. Appl. Phys. 18, 1629 (1979)
6. A.K. Head, P. Humble, L.M. Clarebrough, A.J. Morton, C.T. Forwood, in "Computer Electron Micrographs and Defect Identification", S. Amelinckx, R. Gevers, J. Nihoul, eds., North-Holland (1973)
7. J. Grilhé, K. Seshan, J. Washburn, Rad. Effects 27, 115 (1975)
8. J.A. Lambert, P.S. Dobson, Phil. Mag. A 37, 441 (1978)
9. J.A. Lambert, P.S. Dobson, Phil. Mag. A 44, 1043 (1981)
10. I.G. Salisbury, to be published in Acta Met.
11. H. Föll, C.B. Carter, Phil. Mag. A 40, 497 (1979)
12. S. Kawado, Jpn. J. Appl. Phys. 19, 815 (1980)

OXYGEN PRECIPITATION AND THE GENERATION OF SECONDARY DEFECTS IN OXYGEN-RICH SILICON

F.-G. Kirscht, P. Gaworzewski, K. Schmalz,
I. Babanskaja, P. Zaumseil, U. Winter
Akademie der Wissenschaften der DDR
Institut für Physik der Werkstoffbearbeitung Berlin
DDR-1166 Berlin-Rahnsdorf, Seestraße 82

1. Introduction

Oxygen in silicon has changed its role in silicon device processing from an unwanted to a desired impurity because it can be used to generate gettering centres in the bulk of Si wafers and reduce wafer warpage depending on the actual oxygen state [1]. On the other hand, oxygen precipitates as a prerequisite for gettering sites (intrinsic gettering) cause wafer warpage under certain processing conditions resulting in a tradeoff between internal gettering and mechanical wafer strength [2,3].

There are different possibilities for condensation of interstitially solved oxygen O_i in oxygen-rich silicon ($6 \cdot 10^{17}$ cm$^{-3} \leqq [O_i] \leqq 1.8 \cdot 10^{18}$ cm^{-3}) due to heat treatment (h.t.). In dependence on the temperature of h.t., initial oxygen concentration $[O_i]$, carbon concentration $[C_s]$, and the thermal history of the crystal the interstitially solved oxygen can condense according to following mechanisms [4,5,6]:

- further growth of grown-in precipitates
- homogeneous nucleation and growth of clusters and SiO_2 precipitates due to O_i supersaturation
- heterogeneous nuclation at growing and/or h.t. induced structural defects.

There is a simple model for the description of growth of SiO_x-clusters/precipitates [7]. According to this model, it can be written formally

$$O_i \rightarrow O_p + \lambda Si_i \qquad (1)$$

where O_p means an oxygen inside the cluster/precipitate, $\lambda \leqq 2$, and Si_i a silicon interstitial. It is known that the kinetics-limiting step of process (1) is the diffusion of O_i [8]. From (1) it would follow that cluster/precipitate growth is influenced by the concentration of vacancies [V] and interstitials $[Si_i]$, respectively. That means, there are some possibilities to control the growth of clusters/

/precipitates by external sources of vacancies or interstitials. An opportunity for this can be the use of different ambients at the h.t.

One approach for intrinsic-gettering includes 3 h.t. steps, the extraction of O_i near the surface by out diffusion as a h.t. ①, a nucleation h.t. ②, and a h.t. step ③ for the growth of precipitates including the formation of secondary defects. Heat treatment ①, moreover, has an essential impact on the formation of volume defects at subsequent h.t.'s. The purpose of this work is an investigation of the formation of different kinds of oxygen-related defects in the course of h.t. cycles in dependence on

(i) ambients at h.t. of type ①

(ii) different h.t. of type ②.

Moreover, the influence of different h.t. cycles on the macroscopic strength/wafer warpage is studied.

2. Experimental

As starting material dislocation free P- and B-doped Si-wafers were used (ρ: 2...10 Ωcm, 350 μm thick) with initial O_i- and C_s-concentrations in the range $7.5 \cdot 10^{17}$ cm^{-3} $\leq [O_i] \leq 9.5 \cdot 10^{17}$ cm^{-3} and $[C_s] \approx 6 \cdot 10^{16}$ cm^{-3}, respectively. The h.t.'s were performed under the following conditions:

- h.t. ① 1050 °C $\leq T \leq$ 1250 °C in ambients nitrogen, oxygen, oxygen + 2 % HCl
- h.t. ② 650 °C $\leq T \leq$ 850 °C as one or multi-step h.t. in nitrogen
- h.t. ③ 1000 °C $\leq T \leq$ 1100 °C in ambients nitrogen, oxygen

Heat treatment ① is used also as a means of the control of grown-in precipitates.

After the heat treatments, bevels were prepared and Wright etch has been used to investigate the depth distribution of oxygen related defects especially the density and length of stacking faults. Furthermore, HVTEM and X-ray topography were used for the investigation of the microscopic nature of the defects and of their macroscopic distribution, respectively. The wafer warpage was measured by using double crystal technique [9]. The $[O_i]$ after h.t. was determined by thermal donor formation due to subsequent 450 °C h.t.

3. Results

The determination of the interstitial oxygen concentration $[O_i]$ determined after different h.t. cycles shows remarkable differences depending mainly on temperature and ambient used at h.t. ① (Fig. 1). It is shown that $[O_i]$ decreases with increasing temperature of h.t. ① under nitrogen. In the contrary, there is no significant change in $[O_i]$ after h.t. ① in HCl-added oxygen. A comparison of $[O_i]$ with the corresponding precipitate density - determined from the hillock density after preferential etching - shows nearly reversed behaviour. Generally, $[O_i]$ decreases with increasing D_p modified by the conditions at h.t. ② (Fig. 2).

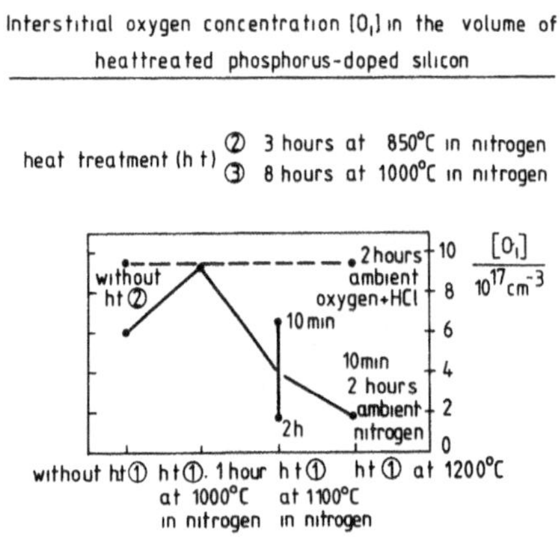

Fig. 1.

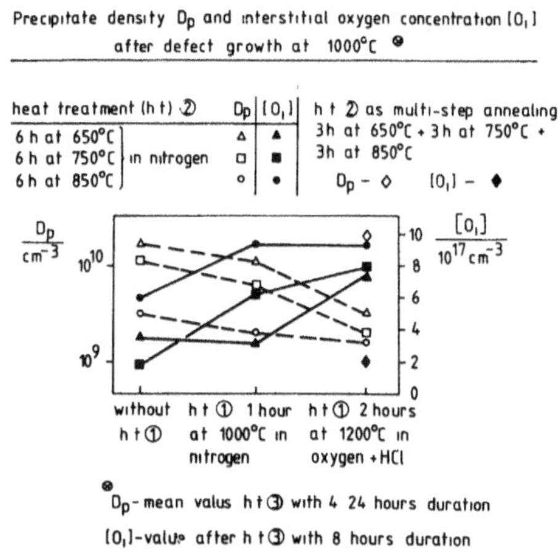

Fig. 2.

Especially at multi-step h.t. ② (in the concerning temperature range) a completely different behaviour is observed. Heat treatment ① under nitrogen, however, does not result in remarkably different defect generation and $[O_i]$-decrease after subsequent h.t.'s (Fig. 3a, 3b). In Fig. 3, also stacking fault density D_s is recorded. The slight difference in D_p and $[O_i]$ at Fig. 3a in relation to Fig. 2 is due to other silicon samples of the same crystal region.

From Fig. 3 it is evident that $[O_i]$-decrease is related rather to an increase in D_s than to D_p-increase. The special role of (particular) stacking faults is underlined by an other experiment with B-doped silicon (Fig. 4). It follows from Fig. 4 that only a relative high density of small stacking faults leads to a strong decrease of $[O_i]$.

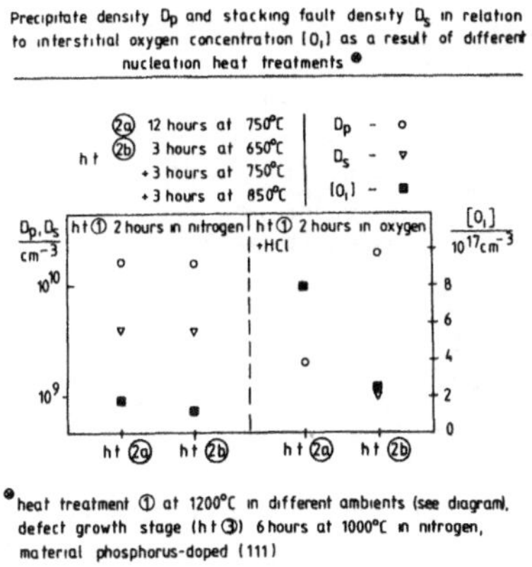

Fig. 3a.

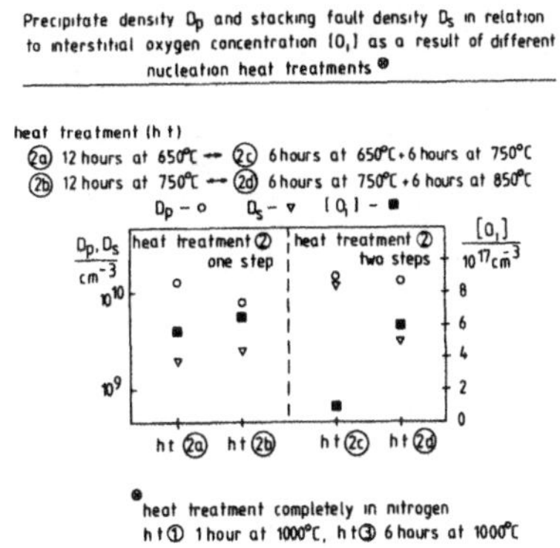

Fig. 3b.

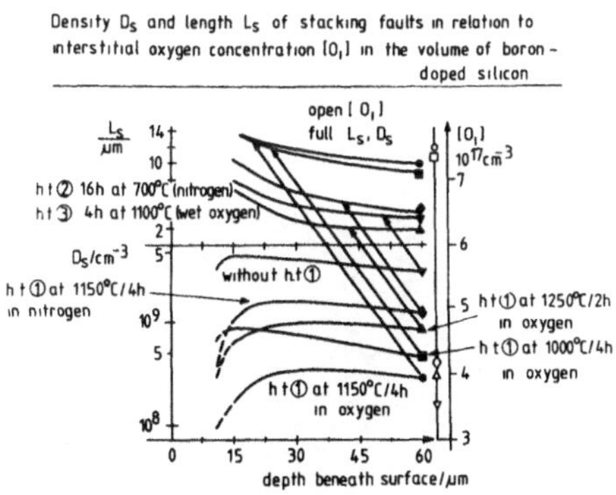

Fig. 4.

Evidently, such stacking faults can be generated using quite different h.t.'s including a nucleation h.t. without foregoing h.t. ① . On the other hand, a significant decrease of D_p only occurs after h.t. ① at 1250 °C (at a depth of 40 μm, volume density ≈ $2 \cdot 10^{10}$ cm^{-3}; corresponding value after the other h.t. cycles presented here ≈ $7 \cdot 10^{12}$ cm^{-3}). At the same time, only under these conditions a nearly defect-free zone beneath the surface is detectable.

Summarizing the observation of secondary defects we can state that completely different defect spectra are observed depending on h.t. conditions (as to stacking fault formation compare Fig. 3,4). Using HCl-added oxygen at h.t. ① , the by far dominating mechanism after h.t. ③ is the well-known punching of prismatic dislocation loops from plate-like SiO_2 precipitates (Fig. 5). Using nitrogen at h.t. ① , the dominating mechanism is the generation of stacking faults after h.t. ③ . Dry oxygen in h.t. ① leads to intermediate behaviour concerning defect spectrum observed after h.t. ③ .

After relatively long-extended h.t. ③ subsequent to h.t. ① in nitrogen, a characteristic defect transition stage can be observed indicating unfaulting reactions in connection with the formation of

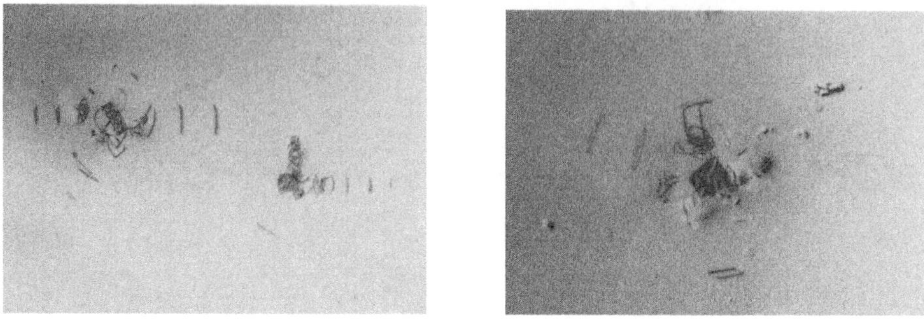

Fig. 5. HVTEM image (bright field, 1 MeV, (2$\bar{2}$0)-type of typical defects after h.t. ③ . (a) h.t. ① : 2 h at 1200 °C in HCl-added oxygen; h.t. ② : 6 h at 650 °C in nitrogen; h.t. ③ : 2 h at 1000 °C in nitrogen. (b) h.t. ① : 2 h at 1200 °C in HCl-added oxygen; h.t. ② : 6 h at 750 °C in nitrogen; h.t. ③ : 24 h at 1000 °C in nitrogen

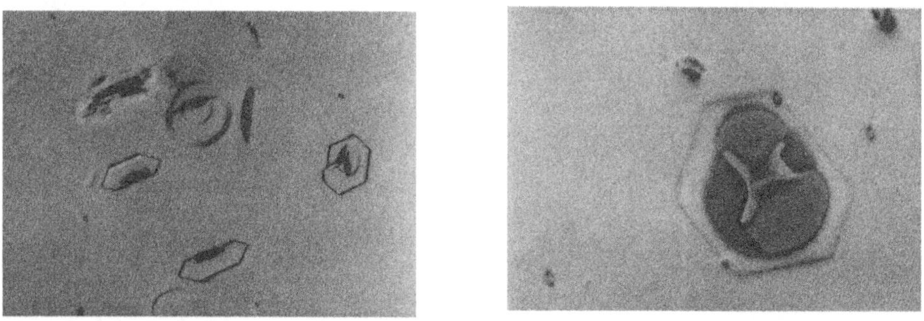

Fig. 6. HVTEM image (bright field, 1 MeV, (a) (2$\bar{2}$0)-type (b) (11$\bar{1}$)-type of typical defects after h.t. ③ . (a-b): h.t. ① : 2 h at 1200 °C in nitrogen; h.t. ② : 6 h at 750 °C in nitrogen, h.t. ③ : 24 h at 1000 °C in nitrogen

perfect hexagonal dislocation loops (Fig. 6).

Measurements of the radius of lattice curvature R reveal significant differences in wafer warpage in dependence on h.t. cycles used (Fig. 7a). Performing only h.t. ① or h.t. ③ , the smallest wafer warpage occurs. An additional h.t. ② leads to a decrease of R according to an increase of warpage. After a subsequent h.t. ③ ,

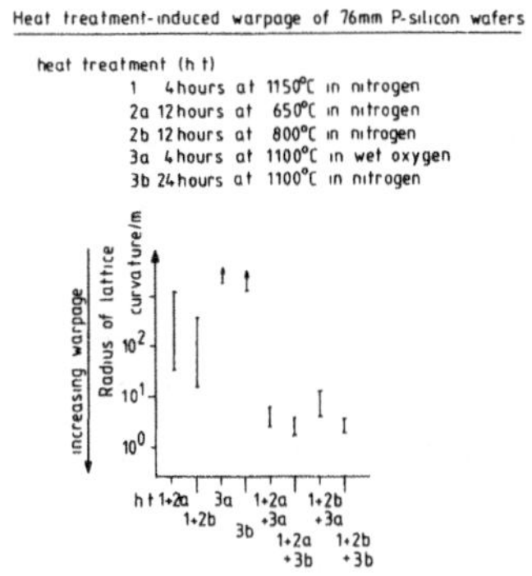

Fig. 7a.

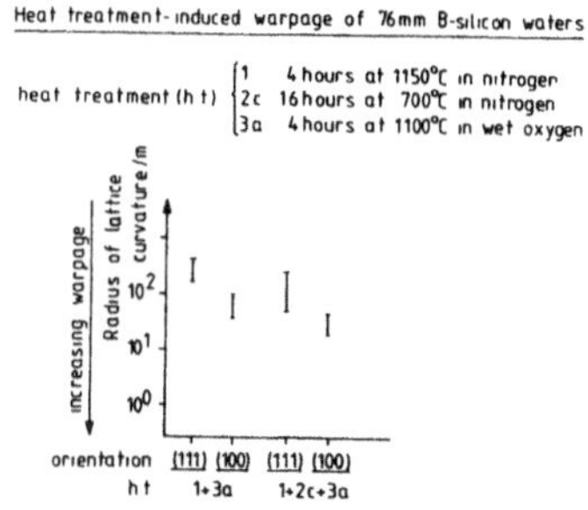

Fig. 7b.

further warpage increase is detected slightly depending on the duration of h.t. ③ .

X-ray topographs (Lang, section) at corresponding wafers reveal characteristic differences of the macroscopic defect distribution within the wafer volume (Fig. 8a, 8b). Especially, in connection with defect-rich central wafer regions, a locally enhanced warpage within these regions is detected. Figure 7b shows that warpage depends on wafer orientation and is indeed influenced by h.t. ② .

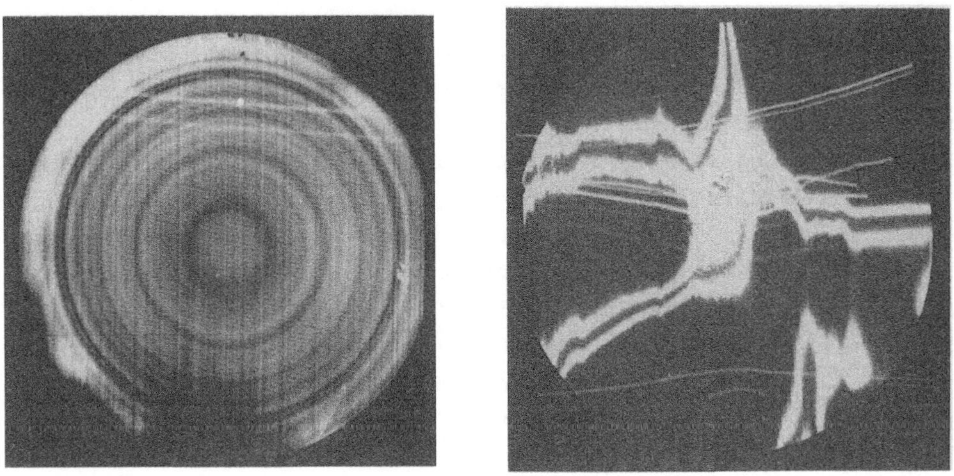

Fig. 8. Lang topographs $(2\bar{2}0)$-type of heat-treated Si wafers. (a) only h.t. of type ③ : 24 h at 1100 °C in nitrogen; (b) h.t. ① : 4 h at 1150 °C in nitrogen, h.t. ② : 12 h at 800 °C in nitrogen, h.t. ③ : 24 h at 1100 °C in nitrogen

4. Discussion

In the case of h.t. ① at 1000 °C in nitrogen, the $[O_i]$-decrease measured after h.t. ③ is governed by a homogeneous nucleation process [4] during h.t. ② as it can be deduced from Fig. 2 and Fig. 3. This is also noticed for the h.t. ① at 1200 °C under HCl-added oxygen. Furthermore, some possibilities of controlling the volume defect formation (in a wide range of density and type of defects generated) by appropriate varying the conditions of h.t. ② are demonstrated. At "gentle" stepwise h.t. cycles within particular temperature ranges, the stacking fault formation is strongly promoted (Fig. 3b). Such kind of h.t. ② can also result in completely different precipitation behaviour as compared to one-step h.t. ② (Fig. 3a). Obviously, a steadily progressing growth stabilizes the newly-generated precipitates. Similar results have been reported in the recent literature [10].

The strong $[O_i]$-dependence observed only after h.t. ① at $T \geq 1100$ °C under nitrogen as well as the time dependence of the $[O_i]$-decrease at this ambient (Fig. 1) suggest the indiffusion of a fast diffusing impurity lowering the activation energy of precipitate nucleation on heterogeneous centers already present in the as-received wafers. This follows from the fact of reaching a nearly constant saturation density as well as the in-depth distribution of volume defects after h.t. ③, for example, the depth of defect-denuded zones and the depth of corresponding defect saturation levels after h.t. ③ are strongly reduced in case of h.t. ① in nitrogen compared to h.t. ① in HCl-added oxygen if h.t. ① is performed at 1200 °C. Furthermore, the presented results (Fig. 4) indicate that (probable) in-diffusion of impurities is also inhibited under dry oxygen at h.t. ① (at least up to 1150 °C) but with raising temperature similar effects as after h.t. ① in nitrogen become dominating the further defect growth.

The well-established correlation between density of small stacking faults and $[O_i]$ (Fig. 4) would be in accordance with the simple growth model (1) but we can not decide whether oxygen is incorporated in the fault plane (instead of or together with Si) or it is limited only at central precipitates. The observed transition of stacking faults into perfect dislocation loops is assumed to be caused by shrinkage/dissolution of precipitates in connection with redistribution of oxygen. A model to explain the mechanism of transition has been recently discussed for similar defect configuration [11].

The density of volume defects observed after h.t. ③ subsequent to h.t. ① in different ambients let us conclude that the simple model (1) mentioned above does not hold consequently. For example, the expected promotion of precipitate growth by vacancy-supersaturation in case of HCl-added oxygen ambient at h.t. ① is not observed. Particularly, under certain conditions described, completely other precipitation governing processes become dominating resulting also in very different secondary defect generation. It is worthwhile to note that quite similar effects influence defect generation/transition and related intrinsic gettering phenomena as well as mechanical properties of wafers during certain actual high-temperature device processing steps.

References

1. P. Butler, Semicond. International, Febr. 1982, p. 95
2. L. Jastrzebski, Transactions on Electron Devices, Vol. ED-29, No. 4 (1982) p. 475
3. M. Mai, P. Gaworzewski, F.-G. Kirscht, H. Richter, poster at "12th International Conference on Defects in Semiconductors", Amsterdam, Sept. 1982
4. N. Inoue, K. Wada, J. Osaka, Semicond. Silicon, Minneapolis, USA, 1981, proceed. p. 282
5. S. Kishino, Y. Matsushita, M. Kanamori, T. Iizuka, Jap. J. Appl. Phys. 21, No. 1, 1 (1982)
6. H. Tsuya, F. Shimura, K. Ogawa, T. Kawamura, J. Electrochem. Soc. 129, No. 2, 374 (1982)
7. R.W. Series, K.G. Barraclough, W. Bardsley, Semicond. Silicon, Minneapolis, USA, 1981, proceed. p. 304
8. K. Wada, N. Inoue, K. Kohra, J. Cryst. Growth 49, No. 4, 749 (1980)
9. P. Zaumseeil, Cryst. Res. and Techn. 17, 639 (1982)
10. H. Tsuya, K. Ogawa, F. Shimura, Jap. J. Appl. Phys. 20, No.1, L31 (1981)
11. H. Bender, C. Claeys, J. Van Landuyt, G. Declerck, S. Smelinckx, R. Van Overstraeten, report at DCSS, Mátrafüred, 1982

ELECTRICAL AND OPTICAL PROPERTIES OF OXYGEN-RELATED DONORS IN SILICON FORMED AT TEMPERATURES FROM 600 TO 850 °C

F. Spiegelberg, D. Wruck, P. Gaworzewski* and K. Schmalz*

Academy of Sciences of GDR
Central Institute of Electron Physics
*Institute of Physics of Materials Processing

1. Introduction

The behaviour of oxygen in Czochralski-grown silicon due to heat treatment, though investigated over more than 25 years, has attracted again increasing interest in the latter years. This is due to the importance of the electrically active (the so-called "thermal donors") as well as of the electrically inactive defects (precipitates) formed, for the development of LSI and VLSI circuits.

As to the electrically active modifications of oxygen in silicon, the most work in the literature has been performed on the thermal donors (td) formed by a heat treatment at 450 °C. The main features of these donors obtained from electrical [1,2,3], photoluminescence [4], ESR [5], and infrared absorption [6] data are given in Table 1.

Moreover, in the last years, oxygen-related donors (od) of obviously different origin being formed by heat treatment at temperatures above 600 °C have been observed by electrical [10] and photoluminescence [4] investigations. The main properties of these new oxygen donors (od) are also summarized in Table 1.

In order to obtain more insight into the electronic structure of these new oxygen-related donors and its possible relations to the thermal donors formed at 450 °C, in the present investigation combined electrical and infrared absorption measurements have been carried out.

2. Experimental

The samples used were prepared from a dislocation-free Czochralski-grown silicon crystal, which before heat treatment was of p-type with the following concentrations: acceptor (netto) $4 \cdot 10^{13}$ cm^{-3}, interstitial oxygen: $8.2 \cdot 10^{17}$ cm^{-3}, and substitutional carbon: $1.2 \cdot 10^{16}$ cm^{-3}. For studying the thermal stability of the oxygen donors, samples of another part of this crystal with $[O_i] = 6.2 \cdot 10^{17}$ cm^{-3} have been used.

Table 1. Main properties of Si-O modifications formed in Cz silicon due to heat treatment (h.t.) at different temperatures

h.t. T[°C]	Formation kinetics and other properties	Proposed structural nature of the defect
400 600	1. $\frac{dN_{td}}{dt}\Big/_{t\to 0} \sim [O_i]^p$; $p \approx 4$ 2. $N_{td,max} \sim [O_i]^m$; $m \gtrsim 3$ (?) 3. inhibition of td formation by high carbon content 4. several species of double donors with $E_1 \approx 60$ meV and $E_2 \approx 135$ meV for $t \to 0$ [3,6]	SiO_4 unit [2], $[O\text{-}V_2]$ or $[O\text{-}V_3]$ complex [7], O_2 molecule [8]
800	1. $\frac{dN_{od}}{dt}\Big/_{t\to 0} \sim [O_i]^p$; $4 \lesssim p \lesssim 5$ [9] 2. $N_{od,max} = N_{od,max}$ $(T, [O_i], [C_s]$, thermal history of the crystal) 3. promotion of od formation by high carbon content [4]	SiO_x cluster of unknown structure [4]
1200	formation of clusters/precipitates by homogeneous or heterogeneous nucleation	amorphous or crystalline SiO_2 precipitates [11]

Details of heat treatment and electrical measurement are given in [3]. Infrared transmission spectra of samples after heat treatment at 600 and 800 °C were taken at liquid neon temperature.

3. Results

3.1. Electrical Measurements

The kinetics of formation of thermal donors at 600 and 800 °C shown in Fig. 1 are characterized by the following facts: The detectable start of the formation of 600 °C od's is retarded with regard to the formation of the 800 °C od's and the formation rate of the 600 °C od's is significantly lower than that of the 800 °C od's and than that of the 450 °C td's.

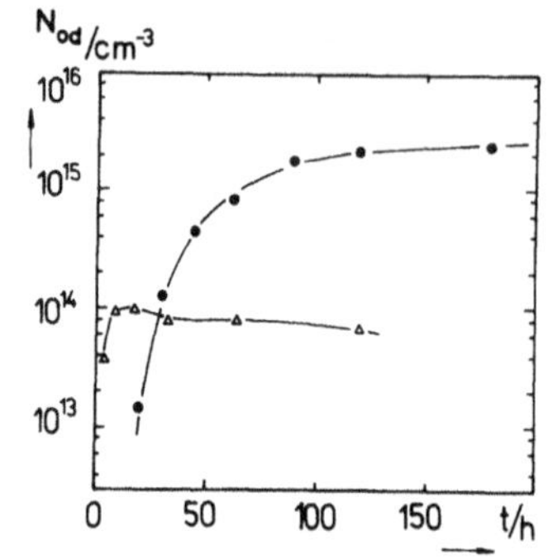

Fig. 1. Oxygen donor (od) concentration after heat treatment on duration of heat treatment (● 600 °C, Δ 800 °C)

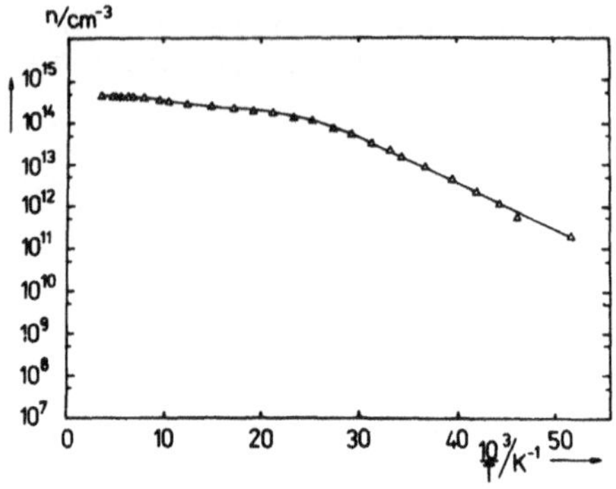

Fig. 2. Temperature dependence of electron concentration n(1/T) for heat treatment of 40 h at 600 °C

On the contrary, the maximum od concentration decreases with the increase of the temperature of heat treatment. This concentration is probably influenced by the simultaneously occurring precipitation process consuming a considerable amount of interstitial oxygen.

The time dependence of od concentration for different interstitial oxygen and substitutional carbon concentrations [9] suggest a possible connection between precipitation and od formation.

Electronic properties of the oxygen donors are studied by Hall effect measurements. From Fig. 2 it is seen that the temperature dependence of the electron concentration shows two regions of saturation suggesting a divalent donor; this character is less pronounced at long heat treatment. The analysis of n(1/T) gives an ionization energy for the shallow level E_I = 18 meV (for heat treatment of 100 h at 600 °C; E_I = 21 meV); in comparison, the thermal donors formed at 450 °C after short heat treatment are characterized by $E_I \approx$ 60 meV [3].

As manifested by Fig. 3 which shows the behaviour of 600 °C od's at heat treatment at higher temperatures, the oxygen donors exhibit high thermal stability at temperatures between 700 and 900 °C.

3.2. Infrared Absorption

600 °C Heat Treatment. In contrast to the narrow-like spectrum obtained after heat treatment at 450 °C shown in Fig. 4 for comparison (for details see [6]) the donors formed after heat treatment at 600 °C are characterized by broad-band spectra (Fig. 5), rather those of photo-ionization.

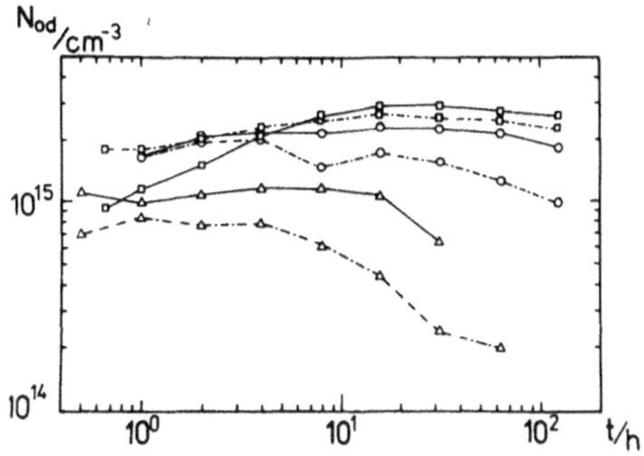

Fig. 3.
Oxygen donor concentration in dependence on heat treatment duration (□ 700 °C, o 800 °C, Δ 900 °C) after preceding heat treatment for 100 h at 600 °C (starting oxygen concentrations: —— $6.1 \cdot 10^{17}$ cm^{-3}, –·– $8.2 \cdot 10^{17}$ cm^{-3})

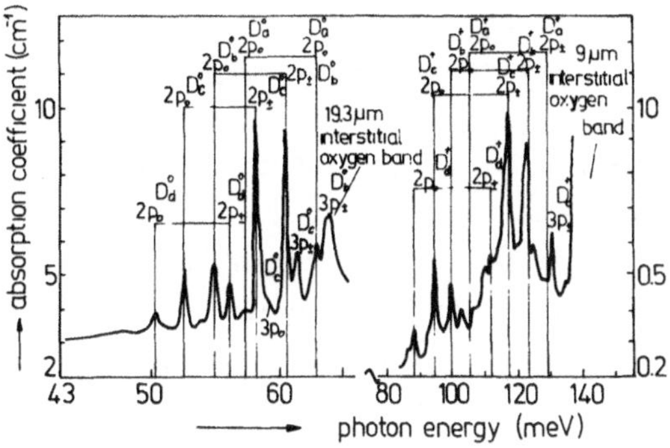

Fig. 4.
IR absorption spectra typical for 450 °C thermal donors, heat treatment duration 7.5 h, thermal donor concentration 10^{15} cm^{-3}

Fig. 5. IR absorption spectra corresponding to 600 °C heat treatment
(a) 30 h, $n_{od} = 1.9 \cdot 10^{14}$ cm^{-3};
(b) 40 h, $n_{od} = 4.6 \cdot 10^{14}$ cm^{-3};
(c) 100 h, $n_{od} = 2.1 \cdot 10^{15}$ cm^{-3}.
The two curves in (c) are taken with different illuminations of the sample

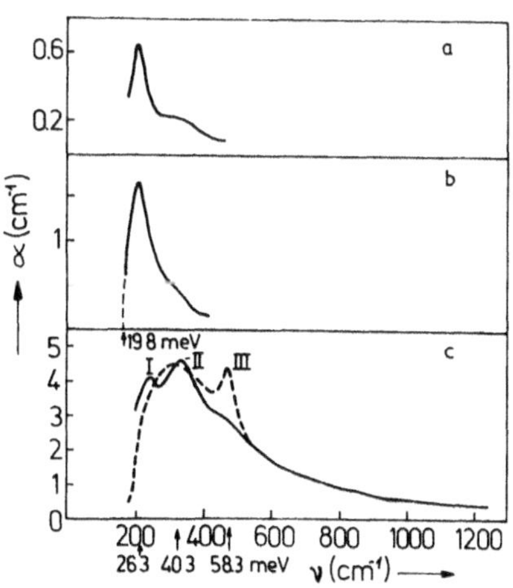

Assuming the spectra to be composed of the three photoionization bands I, II and III, ionization energies (low-energy thresholds) can be estimated of $E_I = 19.8$ meV, $E_{II} \approx 30$ meV and $E_{III} = 51.8$ meV, respectively.

Further, the mutual dependence on background illumination of the donor levels associated with bands I and III points to a double donor character, whereas band II dominating after prolonged heat treatment is obviously related to another type of oxygen-related donor.

800 °C heat treatment. Oxygen donors formed at ≈ 800 °C are more difficult to be detected by IR absorption because of both lower donor concentration and higher ionization energy [10], the latter being in a wave number range of high lattice absorption. The main features of the absorption spectra measured are silicon oxide precipitate bands already investigated and discussed [11].

Two-step heat treatment. A two-step heat treatment performed in connection with IR absorption measurement may give insight into the processes of formation and of remodelling of oxygen-related donors. Fig. 6 gives a first example for the observation of the co-existence of a narrow-line spectrum, being characteristic of 450 °C long-time heat treatment [6], and a developing broad band, typical for a 600 °C heat treatment. This fact leads to the suggestion that the 450 °C thermal donors may be precursors of the new oxygen donors formed at 600 °C. However, further investigations are needed to clarify this question.

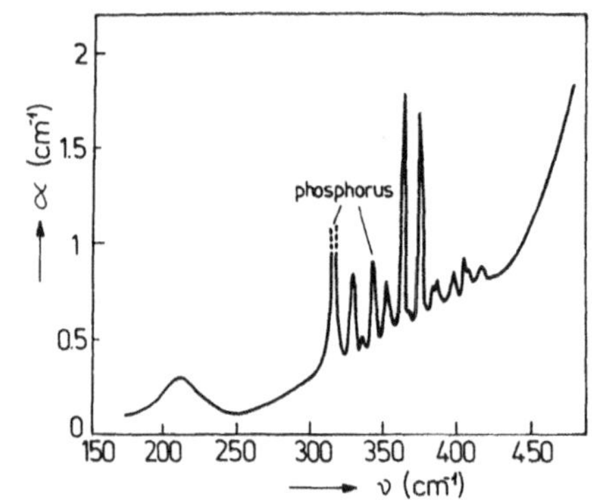

Fig. 6. IR absorption spectrum taken after a two-step heat treatment (I: 120 at 450 °C; II: 1 h at 650 °C; final donor concentration $3.3 \cdot 10^{14}$ cm^{-3})

4. Conclusions

(i) Combined electrical and IR spectroscopic investigations of oxygen-related donors formed due to heat treatment at temperatures ≤ 600 °C point to a different origin of these donors as compared to the "usual" 450 °C thermal donors.

(ii) The main feature concerning these new donors is their high thermal stability and different electronic properties (ionization energies).

(iii) The values for the ionization energies and the indications of a double-donor character especially for the 600 °C donor obtained from electrical and IR spectroscopic measurements are in good agreement.

(iv) The shape of the IR absorption spectra show a similar broad-band behaviour as found by photoluminescence techniques [4] indicating a rather complex nature of the defects associated with these donors.

(v) Infrared spectroscopic investigations after two-step heat treatment offer a promising tool for studying formation and remodelling of the "usual" 450 °C thermal donors and the "new" 600 °C oxygen donors.

(vi) Conclusions as to the microscopic structure of these new donors cannot be made examined at the present state of investigations and will be the aim of further work.

References

1. C.S. Fuller and R.A. Logan, J. Appl. Phys. 28, 1427 (1957)
2. W. Kaiser, H.L. Frisch and H. Reiss, Phys. Rev. 112, 1546 (1958)
3. P. Gaworzewski and K. Schmalz, phys.stat.sol.(a) 55, 699 (1979)
4. M. Tajima, T. Masui, T. Abe, T. Iizuka, in "Semiconductor Silicon 1981", ed. R. Huff, R.J. Kriegler, Y. Takeishi (The Electrochem. Soc., Pennington, 1981) p. 72;
   K. Yasutake, M. Umeno, H. Kawabe, H. Nakayama, T. Nishino and Y. Hamakawa, Japan J. Appl. Phys. 19, L544 (1980);
   H. Nakayama, J. Katsura, T. Nishino, Y. Hamakawa, ibid. L547
5. S.H. Muller, M. Sprenger, E.G. Sieverts and C.A.J. Ammerlaan, Solid State Comm. 25, 987 (1978)
6. D. Wruck and P. Gaworzewski, phys.stat.sol.(a) 56, 557 (1979)
7. O. Helmreich and E. Sirtl, in "Semiconductor Silicon 1977", ed. R. Huff and E. Sirtl (The Electrochem. Soc., Princeton, 1977) p. 626
8. U. Gösele, T.Y. Tan, Appl. Phys. A 28, 79 (1982)
9. P. Gaworzewski and K. Schmalz, phys.stat.sol.(a), to be published
10. K. Schmalz and P. Gaworzewski, ibid. 64, 151 (1981)
11. K. Tempelhoff, F. Spiegelberg, R. Gleichmann, D. Wruck, ibid. 56, 213 (1979)

ON THE FIELD DEPENDENCE OF CAPTURE AND EMISSION PROCESSES AT DEEP CENTRES

H. Klose, K. Irmscher and K. Maass
Sektion Physik, Humboldt-Universität zu Berlin
1040 Berlin, Invalidenstr. 42, GDR

1. Introduction

Deep impurity centres in space charge regions of semiconductor devices cause recombination and generation currents and change the characteristics accordingly. This fact can be used for special devices, for instance switching diodes and photodetectors. The improvement of the quantum efficiency of LED's is connected with the lowering of deep centre concentrations, too.

In the last years depleted layer spectroscopy was essentially improved and enables the determination of the recombination parameters of such levels with currently increased accuracy. By means of this new capacitance methods discrepancies were established in the well known data of e.g. gold in silicon [1].

In space charge regions of devices electric field strengths exist of 10^3 Vcm^{-1} up to some 10^5 Vcm^{-1}. Such fields effect the recombination and emission parameters of deep centres. In 1970 Tasch and Sah determined the field dependence of the thermal emission rate of gold acceptor level in silicon [2] for the first time. Recently experimental and theoretical results were published on Cr in GaAs, EL2 centre in GaAs and A-centre in Si [3-5].

Investigations on the influence of the electric field on the capture and emission rates can provide information about the nature underlying capture and emission processes.

For the experimental determination of the field dependence the exact knowledge of the spatial field distribution in the depletion layer and its position at the moment of the measurements is required. These conditions can be very well realized in DDLTS measurements [6].

In this paper we report on measurements of the field dependence of the electron emission rate in the case of gold acceptor and A-centre in silicon.

2. Experimental

2.1. Experimental Details

All samples under study were prepared from n-type silicon grown by the Czochralski technique. Gold was introduced in silicon wafers of various resistivities by vacuum evaporation on or ion implantation into the wafers and by a following diffusion to give homogeneous gold concentration. After the diffusion several microns of silicon were removed by chemical etching.

Ion implantation was used for the generation of the A-centre (oxygen-vacancy-complex). Chemically cleaned silicon wafers with nominal resistivities of 0.7 and 7 Ωcm were implanted with He (energy 20 keV, dose $2 \cdot 10^{10}$ cm^{-2}) and protons (300 keV, 10^{10} cm^{-2}), respectively.

For the formation of Ohmic back contacts AuSb was evaporated over the back of the wafers and then sintered at 450 °C for 15 min in N_2.

In the last preparation step Schottky diodes were fabricated on all wafers by vacuum evaporation of gold through circular masks with 0.25-1 mm diameter.

In Table I the net donor and the interesting deep level concentrations of the samples obtained from capacitance-voltage and DLTS measurements are summarized.

Table I.

Label	Au-acceptor			A-centre	
	A	B	C	D	E
net donors N_d/cm^{-3}	$3.1 \cdot 10^{17}$	$1.0 \cdot 10^{17}$	$5.8 \cdot 10^{14}$	$8.7 \cdot 10^{15}$	$7.0 \cdot 10^{14}$
deep centres N_t/cm^{-3}	$1 \cdot 10^{16}$	$1 \cdot 10^{15}$	$1 \cdot 10^{14}$	$3 \cdot 10^{15*}$	$1 \cdot 10^{14*}$

* These values correspond to the maximum of the implantation damage.

Only samples with a homogeneous net donor concentration were investigated allowing the use of the approach that the electric field strength in the space-charge layer is a linearly varying function of distance from the junction edge.

All measurements were carried out with a special DLTS set-up based on a modified lock-in filtering method working in double correlation mode as described in [6] for boxcar-systems and in [7] for various correlation functions. In the nomenclature of Crowell and Alipanahi [7] the type of measurement used can be characterized as DD^2LTS-filtering.

The width of the "spatial window" was choosen small enough to be sure that the electric field strength variation over the whole interval was less than 10 %.

The point where the charge state of a deep level changes immediately after a filling pulse can be determined from the width of the space charge layer for the given pulse voltage reduced by a distance which is a function of the free carrier concentration, the deep level parameters and the pulse duration [8]. Only for pulses longer than the inverse thermal emission rate this point coincides with the position where the deep level crosses the Fermi level. The latter was not fulfilled in our experiments and relations which took the pulse duration into account were used in connection with CV-measurements at the interesting temperatures. The average value of the "spatial window" was used for the evaluation of the electric field strength. There were two possibilities to change the electric field within the "spatial window":

(i) changing the pulse voltage U_1 and keeping U_0 and ΔU constant, i.e. a variation of the local position of the "spatial window" by a fixed charge layer;

(ii) changing the reverse bias U_0 while keeping the pulse voltage U_1 and $U_1 + \Delta U$ constant, i.e. fixed local position of the "spatial window".

The measurements were performed by choosing a constant "emission rate window" and registering the temperature shift of the DDLTS-peak maximum as a function of the electric field.

2.2. Experimental Results

In Figure 1 examples of DDLTS-spectra are shown demonstrating the effect of the electric field strength on the temperature position of the peak maximum for a given "emission rate window". The first of the two explained possibilities of field strength variation is presented for the Au-acceptor level (Fig. 1a) and the second for the A-centre (Fig. 1b). The apparent decrease in the peak heights with increasing electric field is mainly due to the spatial variation of the sensitivity in capacitance measurements which depends on the trap location or in our case on the location of the "spatial window".

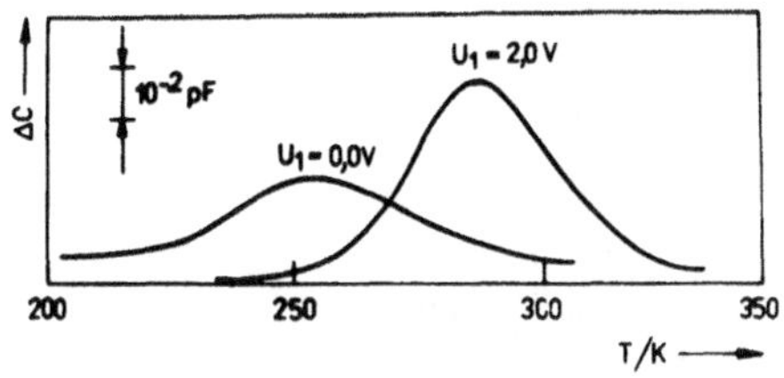

Fig. 1a

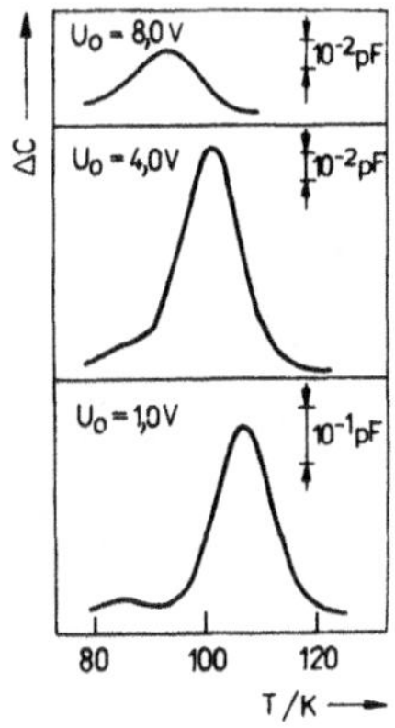

Fig. 1b

Fig. 1: DDLTS spectra showing the temperature shift of a peak maximum to a given "emission rate window" caused by the electric field.
(a) Au-acceptor level in sample A: The rate window was 675 s^{-1}. For a constant reverse bias U_o = 3.0 V and a pulse voltage difference ΔU = 0.2 V the electric field strength was fixed by the pulse voltages U_1 of 2.0 V and 0.0 V to $2.3 \cdot 10^5$ and $4.7 \cdot 10^5$ Vcm^{-1}, respectively.
(b) A-centre in sample D: The rate window was 6750 s^{-1}. The pulse voltages U_1 = -0.2 V and $U_1 + \Delta U$ = 0.0 V were kept constant and the reverse biases U_o of 1.0 V, 4.0 V and 8.0 V corresponded to electric field strengths of $0.32 \cdot 10^5$, $0.77 \cdot 10^5$ and $1.18 \cdot 10^5$ Vcm^{-1}

A lot of measurements done in the described way provide information as depicted in Fig. 2. Whilst for the Au-acceptor data an Arrhenius plot is used (Fig. 2a) the A-centre data are shown in a graph (Fig. 2b) reflecting our measurements directly.

These graphs contain the desired information about the emission rate dependence on the electric field. Using the temperature as parameter this function is presented for the Au-acceptor in Fig. 3 and for the A-centre in Fig. 4.

The experimental results of [5] for the emission rate dependence of the A-centre on the electric field strength at 77 K varying the field from 4 to $7 \cdot 10^4$ Vcm^{-1} are in good agreement with our measurements concerning the A-centre.

3. Discussion

On the basis of the Poole-Frenkel effect used in [2,5] and by the pure tunneling emission the observed strong electric field dependence of the emission rates cannot be explained. Currently Makram-Ebeid and Lannoo [3] have developed a theory for phonon-assited tunnel ionization

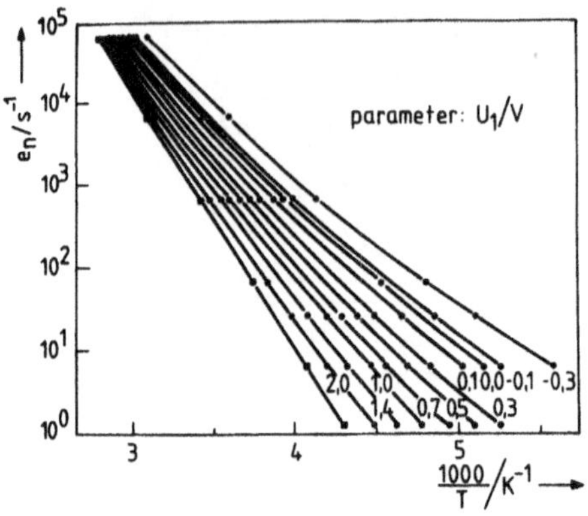

Fig. 2a

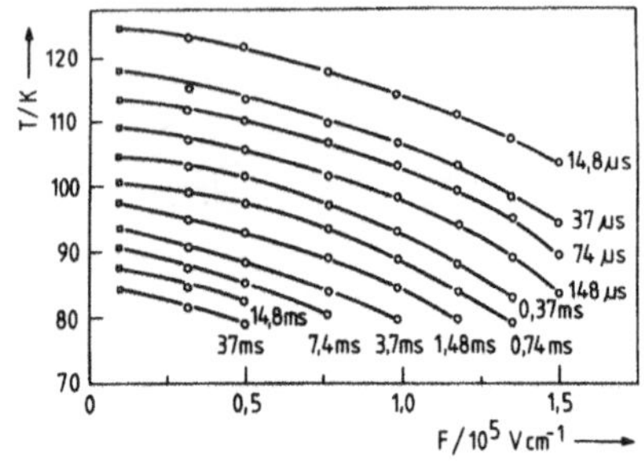

Fig. 2b

Fig. 2. Presentation of the DDLTS peak maximum temperature shift caused by the electric field in an Arrhenius plot for the Au-acceptor level (a) with the pulse voltage as parameter and in a direct manner for the A-centre (b). (a) - ●●● sample A with $U_o = 3.0$ V and $\Delta U = 0.2$ V kept constant while the pulse voltage U_1 was varied from 2.0 to -0.3 V corresponding to an electric field strength variation from about $2 \cdot 10^5$ to $5.4 \cdot 10^5$ Vcm^{-1},■■■ sample C with F between $1 \cdot 10^5$ and $2 \cdot 10^5$ Vcm^{-1}. (b) - The parameter was the inverse rate window. ooo sample D, □□□ sample E

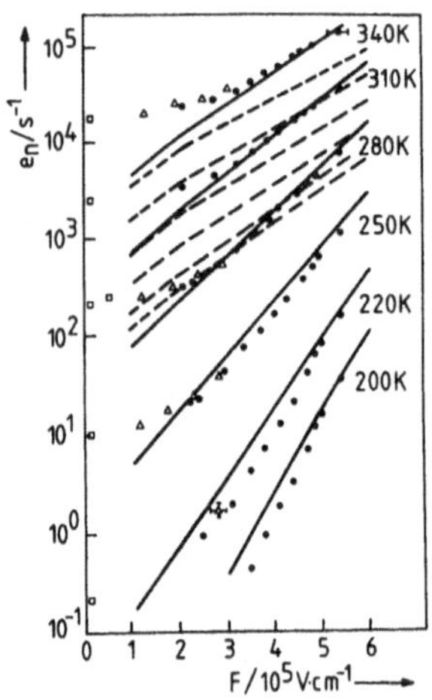

Fig. 3. Electric field dependence of the electron emission rate at different temperatures for the Au-acceptor level in Si (●●● sample A, ΔΔΔ sample B,□□□ sample C). Calculations of the phonon assisted tunneling rate R using an analitic approximation of the model in [3] with the following parameter sets are also shown: $S = 15.0$, $\hbar\omega = 10$ meV, $\Delta_T = 0.44$ eV (full lines); $S = 2.2$, $\hbar\omega = 68$ meV, $\Delta_T = 0.55$ eV (dashed lines). (Note that the thermal emission rate e_{no} was omitted for the calculations.)

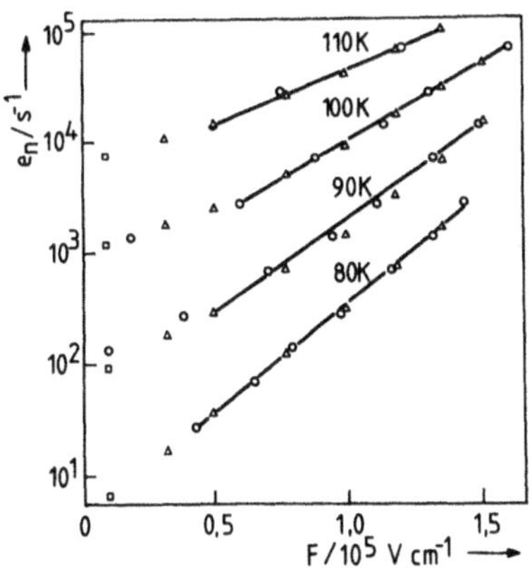

Fig. 4. Electric field dependence of the e_n at different temperatures for the A-centre in Si. ooo sample D, points gained from Fig. 2b; ΔΔΔ sample D, points gained from Arrhenius plot; □□□ sample E

and have explained experimental results in GaAs. For the discussion of our experimental data the equation

$$e_n = e_{no} + R(F) \qquad (1)$$

was used for the emission rate, where R(F) represents the phonon-assisted tunneling rate.

The relative good agreement between experimental data and the given theory was only achieved by a parameter set ($S = 15.0$, $\hbar\omega = 10$ meV, $\Delta_T = 0.44$ eV) which seems to be not characteristic for gold in silicon. Using the parameter set given by [9] a very strong deviation is obtained from experimental results.

The parameter set was found to yield a good fitting of the theoretical to the experimental curves for the temperature range of the experiment in the case of A-centre in silicon.

In the future the differencies between theoretical and experimental results for deep centres in silicon has to be explained. A detailed treatment is in preparation.

The authors would like to thank Dr. D.Suisky for helpful discussion.

References

1. D.V. Lang, H.G. Grimmeiss, E. Meijer, M. Jaros, Phys. Rev. B 22, 3917 (1980)
2. A.F. Tasch, C.T. Sah, Phys. Rev. B 1, 800 (1970)
3. S. Makram-Ebeid, M. Lannoo, Phys. Rev. B 25, 6406 (1982)
4. G. Vincent, A. Chantre, D. Bois, Journ. Appl. Phys. 50, 5484 (1979)
5. B.A. Komarov, V.I. Sopryakov, phys. stat. sol. (a) 66, 139 (1981); Fiz. Techn. Pol. 16, 191 (1982)
6. E. Lefévre, M. Schulz, Appl. Phys. 12, 45 (1977)
7. C.R. Crowell, S. Alipanahi, Solid State El. 24, 25 (1981)
8. A. Zylberstejn, Appl. Phys. Letters 33, 200 (1978)
9. J.R. Morante, J.E. Carceller, P. Cartujo, J.J. Barbolla, phys. stat. sol. (b) 111, 375 (1982)

PART IV

DEFECTS IN COMPOUND SEMICONDUCTORS

LATTICE MATCHED HETEROLAYERS

E. Lendvay
Research Institute for Technical Physics of the HAS
H-1325 Budapest, Újpest 1, P.O.Box 76, Hungary

1. Introduction

In epitaxial growth of semiconductors both the substrate and epitaxial layer have a cubic (sphalerite) structure. Various examinations of epitaxy show that generally discrete islands form first and later coalasce into a homogeneous film. During coalescence the islands do not recrystallize and the results are less complicated: the principal defects are inverted truncated stacking fault tetrahedra [1], caused either by the collapse of vacancy discs [2] or by faulted deposited layers [3]. These faults are characteristic eg. for epitaxial silicon layers.

Heterojunctions with different chemical composition are often used in microelectronics to form optoelectronic or microwave devices. These devices, however, require low interfacial state density and long minority carrier life time which demands good lattice matching between the materials. Using of different composition materials generally the difference between the lattice constants (a_o) introduces misfit and strains. Lattice misfit can also be generated in growth with impurity doping or diffusion [4,5]. Misfit dislocations form when the strain exceeds a critical value, so the heteroepitaxial systems contain not only the mentioned stacking faults, but a misfit dislocation system, too.

Among the binary compounds only the GaAs and AlAs have nearly the same lattice constants. All of the other binaries suffer from severe problems associated with lattice mismatch, when grown onto other binary semiconductor. In ternaries the lattice parameter can be adjusted between the limiting values (the lattice parameters of binary components) but the band gap E_g is also changing with composition. The advantage of using a quaternary material is that independent control of E_g over a wide range is possible for a given lattice constant and vice versa, so the quaternaries are the most important materials for lattice matched heteroepitaxial systems.

2. Misfit and Lattice Matching

In optoelectronics there are a lot of heterojunction applications; LED's, lasers, detectors, etc. are based on heteroepitaxial systems. In these applications generally two main classes are distinguished. The first is related to the growth of epitaxial layers with a given forbidden band gap (E_g) starting from substrate crystals with differing lattice parameters using either continuous or step grading of the lattice parameter (composition). The second possibility relates to the growth of lattice matched interfaces between layers with different physical (optical, electrical) parameters.

In semiconductor devices particularly in optoelectronics, structures all rely on obtaining long carrier life-times (τ) and diffusion lengths (L). Since lattice defects in general reduce these parameters, it is extremely important to obtain low lattice defect densities which form as a result of lattice mismatch.

The influence of lattice mismatch in a particular device is established but not well understood in detail: generally mismatch results in dislocation generation and propagation influencing not only the carrier lifetime but they may increase the interfacial recombination rate, too.

The realisation of a given band gap that is optimum for a particular device application (laser, LED, photodetector, etc.) requires constant composition material in the device active region. Moreover, to achieve lattice matched layers a precise control of a_o lattice parameter and hence the composition have to be reached.

Generally the device design needs a detailed knowledge of the system, and one has

i) to plan the composition (a_o)
ii) to study the stability of phases involved in the system and
iii) to determine the experimental circumstances for heterojunction preparation.

The experimental data for most of ternaries and quanternaries are poor, thermodynamical considerations have to be applied. In these respect many works have been published about lattice matched layers mostly grown by liquid phase epitaxy (LPE), and considerable amount of phase diagram information pertinent to such growth has been reported. However, the complexity of phase diagrams of the ternary or quaternary systems means that they are much more difficult to determine either experimentally or theoretically. The main problem is that in the strict thermodynamical sense only a solid quaternary can be in equilibrium with a quaternary solution. This means that we can use these data only in cases where (generally using a relatively high supercooling) on the very moment of the substrate and solution contact a thin layer of the solid alloy is formed on the seed crystal preventing it from further dissolution. Using this technique one can prepare layers of high quality.

The thermodynamical processes determining the phase relations can be described with the aid of various models. The important points regarding the phase diagrams are the following [6,7]:

The equilibrium conditions are determined by the solidus equation

$$\ln(\gamma_{ij}^{S} N_{ij}^{S}) = \frac{\Delta S_{ij}^{F}(T_{ij}^{F}-T)}{kT+\ln \dfrac{4\gamma_i^1 \gamma_j^1 N_i^1 N_j^1}{\gamma_i^{s1}\gamma_j^{s1}}} \qquad (1)$$

where γ_{ij}^{S}, N_{ij}^{S}, S_{ij}^{F} and T_{ij}^{F} are the activity coefficients, and the mole fractions in the solid, the entropy of fusion and the temperature of fusion for the binary constituents and s1 subscript denotes the stoichiometric composition. For ideal solid solution $\gamma_{ij}^{S} = 1$, and the activity coefficient of the i-th component in the melt is

$$RT\ln\gamma_i^1 = \sum_{\substack{j=1\\ j\neq 4}} \alpha_{ij} N_j^1 + \frac{1}{2} \sum_{\substack{k=1\\ k<j}} \sum_{\substack{j=1\\ i\neq k\\ i\neq j}} N_k^1 J_n^1 (\alpha_{ij} + \alpha_{ik} - \alpha_{kj}) \qquad (2)$$

where α is the interaction parameter. From the knowledge of the interaction parameters the solidus composition N_{ij}^{S} at a given T can be predicted for a liquidus composition allowing to determine the conditions for lattice matching.

In particular systems different results are obtained for lattice matched growth on polar (III) and (100) substrates. For the most important quaternary system at present, - for GaInAsP - the composition of alloys obtained by LEP at a given T and liquid composition strongly depends on orientation [8], too. Consequently not only the knowledge of the thermodynamical data, but orientation data are also required to lattice matching. Similarly, for GaInAs the Ga distribution coefficient is higher for (100) growth than for (III) one [9], and as a consequence, there is a difference (Δa_o) between the lattice constant. This difference usually decreases with increasing growth temperature. Assuming Vegard's law the alloys lattice parameter "a" depends on x and y composition parameters (in $A_xB_{1-x}C_yD_{1-y}$). If the lattice constant of the constituent binaries is a, then [10]

$$a(x,y) \sim xya_{AC} + x(1-y)a_{AD} + (1+x)ya_{BC} + (1-x)(1-y)a_{BD} \quad (3)$$

and from the mentioned Δa value this orientation effects correspond to differences of 2-3 °C in the liquidus temperature.

If no complete lattice matching is achieved, large interfacial state densities and lattice misfit occur. The usual definition of misfit is

$$f = \frac{\Delta a}{a_o} = \frac{a-a_o}{a_o}(\%) \quad (4)$$

where a is the lattice constant of the epitaxial layer and a_o is the lattice parameter of the substrate. Another possible definition of misfit uses the data of strained layers [11]. According this

$$f = \frac{\Delta a_f^{\perp}}{a_o} = \frac{1 - \nu}{1 - \nu}\frac{\Delta a}{a_o} \quad (5)$$

where $\Delta a_f^{\perp} = a^{\perp} - a_o$, $a_f^{\perp}$ is the lattice constant of strained hetero-epitaxial layer perpendicularly to the heterojunction and ν is the Poisson rate (~1/3). In Eq.(5) $\Delta a = a_f - a_o$ where a_f is the lattice constant of the stress-free heteroepitaxial material. If the f value surpass a particular value, the consequences are:

i) elastic stresses are formed
ii) dislocations generate and propagate forming nonradiative recombination centres and increasing the interface recombination velocity
iii) "cross-hatched" or microcrystalline epitaxial layer form
iv) microcracks appear.

The mentioned morphological variations are shown in Fig. 1.

If we want to avoid these unfavourable phenomena, the lattice constants of all adjoining layers should be matched as closely as possible preparing coherent phase boundaries; this is the lattice matching.

3. Dislocations in Heteroepitaxial Systems

As it is clearly shown from theoretical investigations in heteroepitaxy, if the epitaxial layer is very thin, the layer is expected to strain elastically, and this reduces the misfit between the different lattices. This is equivalent with the statement that the separation of dislocations in misfit-network increases. If this thickness tends toward 0,

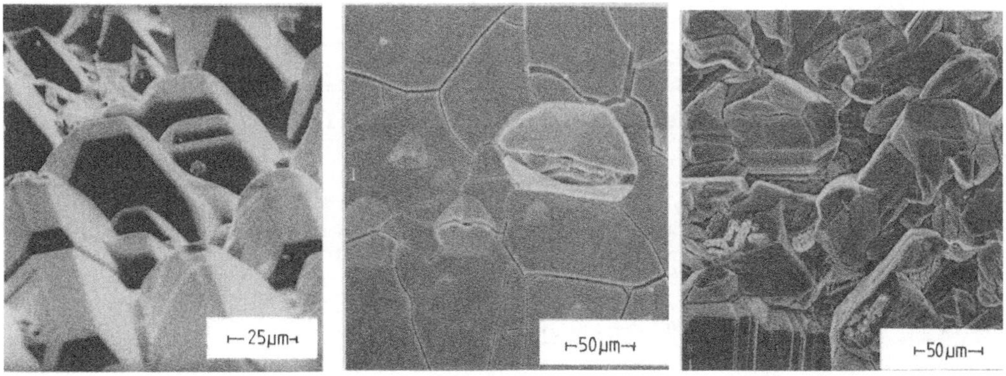

Fig. 1. Characteristic surface morphologies of heteroepitaxial layers with misfit above the critical value. (a) Surface morphology of $GaSb/GaAs_xSb_{1-x}$ heterolayer (f=7.7 %). The cross-hatched pattern consists of <110> aligned cubic elements. (b) Microcracks in GaSb/ $/Al_xGa_yIn_{1-x-y}Sb$ heterolayers (f=0.8 %). (c) Characteristic microcrystalline overgrowth of $Al_xGa_yIn_{1-x-y}Sb$ on GaSb grown by near-equilibrium LEP

the separation becomes infinite and all misfit is accomodated by elastic strain [12,13].

In a square network which is the most common misfit network configuration, the energy of an edge dislocation is [14]

$$E_d = \frac{Gb^2}{4\pi(1-\nu)}(\ln \frac{h}{b} + 1) \tag{6}$$

where b is the Burgers vector of the dislocation, h is the separation from the surface, G is the shear modulus and ν is the Poisson ratio. If we denote the elastic strain with ε, then the separation of adjacent misfit dislocations is

$$\frac{b}{f-\varepsilon}$$

and the energy of misfit grid consists of two orthogonal and noninteracting arrays of edge dislocations

$$E_g = (f-\varepsilon)\frac{Gb}{2\pi(1-\nu)}(\ln \frac{h}{b} + 1) \tag{7}$$

The elastic energy in the layer can be expressed as

$$E_\varepsilon = \frac{2G(1 + \nu)}{1 - \nu} e^2 h \tag{8}$$

Minimizing $(E_\varepsilon + E_g)$ the corresponding ε value is

$$\varepsilon^{*} = \frac{b}{8\pi h(1+\nu)} \left(\ln \frac{h}{b} + 1\right) \qquad (9)$$

If $\varepsilon^{*} > f$, the film will be strained to match the substrate ($\varepsilon^{*} = f$). If $\varepsilon^{*} < f$, then the $f - \varepsilon^{*} = \alpha$ part of misfit will be accomodated by misfit dislocations.

From these above follows that there is a critical layer thickness, where $\varepsilon^{*} = f$ and where the generation of misfit dislocations begins. This value is

$$h^{*} = \frac{b(1+\nu)}{8\pi} \left(\ln \frac{h}{b} + 1\right) \qquad (10)$$

This equation was expereimentally proven for different heteroepitaxial systems [15].

In heteroepitaxial systems there are three different kinds of dislocation

i) threading dislocations

ii) dislocations that accomodate misfit between the heterolayer and its substrate

iii) dislocations that accomodate misfit between individual layers in multilayer structures.

Dislocations that accomodate misfit between individual layers can be avoided by choosing layer thickness below the critical layer thickness, h^{*}. Below this value energetically the situation is not favourable for misfit dislocations to be made (~0.25 nm).

If there is a misfit between the substrate and the epitaxial layer, dislocation pile ups, threading dislocations, slip lines and cracks generate as a result.

One way avoiding these defects is to match the a_o values forming a coherent phase boundary between the substrate and the layer. The condition for this is that the lattice parameter of the usually used binary

$$a_o = a\left[1 + \frac{G_1^e h_1^e f}{G_1^e h_1^e + G_2^2 h_2^2 + \ldots}\right] \qquad (11)$$

where G_i^e are the shear moduli of the films in a multilayer and h_i is their thickness. f is the misfit between stress-free lattice parameters of the component.

Using a substrate with 10^4-10^5 cm^{-2} dislocation density the epitaxial layer generally contains about ~10^8 cm^{-2} threading dislocations [16].

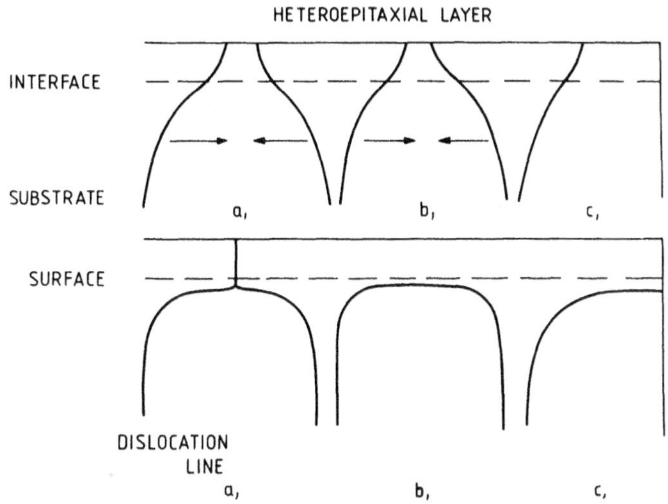

Fig. 2. Reduction of threading dislocations in heteroepitaxial systems. (a) bowing of dislocations and dislocation reaction; (b) annihilation of dislocation segments with opposite signs; (c) escaping of dislocation segment near to the crystal surface

This 10^4 fold increase is probably caused by the nucleation of half-loops during the very first moment of heteroepitaxy. To reduce their density there are different possibilites which are seen in Fig.2. As a result of coherent strain, these dislocations are bowed. Bowing increases with growing film thicknesses. These dislocations can move laterally and leave misfit dislocations in the interface. This motion also may lead to the combination of a and b dislocation forming a third one, e.g.:

$$\frac{1}{2}a[01\bar{1}] + \frac{1}{2}a[101] = \frac{1}{2}a[110]$$

or if the components have opposite signs, the lateral motion can result in the annihilation of dislocations. The escape of the dislocation segment at the crystal edge is also possible.

If we have no possibility to match the adjoining layers, grading techniques must be used. Applying a continuous composition change the importance of low compositional grading rate ($G = \Delta C/\Delta x$) has been determined by many authors [17-21]. However, even for low G values the dislocation density in the constant composition layer is high ($\sim 10^5 - 10^6\ cm^{-2}$) and generally strain in present [22]. According to the measurements made on $GaAs_{1-x}P_x$ grown onto GaAs and GaP the dislocation density (ρ_D) linearly depends on G, giving the lowest value of ρ_D using a very low grading rate [18,19]. During the experiments the 1 mole% μm^{-1} value was found as sufficient. In practice the lowest limit of ρ_D is determined not by G, but by the substrate dislocation density. This is the reason e.g. why the low dislocation density in substrate crystals is necessary for semiconductor lasers.

Alternatively, the use of step grading [23] is also often applied enabling the necessary final composition. The composition steps are small ones (2-3 μm) corresponding typically to 3-5 % change in composition. The principal advantage of this method is that lower values of ρ_D in the final composition range can be gained comparing with equivalent continuous grading rate [17]. Using this method the formed dislocations are confined to the interface bending over the compositinal step. The step-grading is generally used for increasing lattice constants producing compression in the epitaxial layer which is necessary to bend the dislocations. At decreasing a_o values tension forms and it results in microcrack formation.

From these mentioned above follows that misfit dislocations in the epitaxial layers generate more easily, if tensile stresses are present at growth temperature.

For different heterosystems the tolerable amount of misfit is different. Binaries are the most senitive semiconductor compounds, and the power for generation of misfit decreases as

binaries > ternaries > quaternaries.

This is a very important factor of designing multilayer heterostructures. In Fig. 3 misfit free regions are indicated for InP, InGaAs and GaInAsP, illustrating the mentioned prediction.

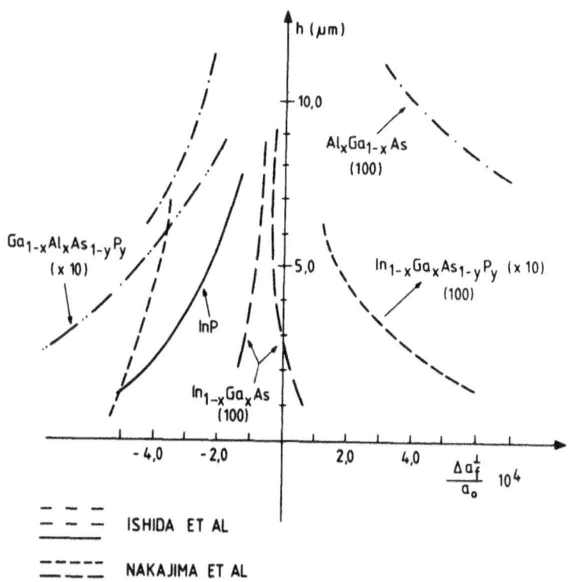

Fig. 3. Formation of misfit dislocation-free heteroepitaxial layers at different stresses. Above the detected h^*-f lines misfit dislocations are formed. Below the curves the misfit is accomodated by elastic strain

3. Binary/Ternary of Quaternary Systems

Among the binary III-V compounds only the AlP-GaP and AlAs-GaAs pair have nearly the same lattice constants, these systems are able to form matched heterolayers. In practice only the binary-ternary pairs have importance, because the binary Al compounds are very reactive ones. They react vigorously with O_2 and H_2O traces, as well as with the quartz which is the usual component of the growth systems. AlSb reacts e.g. with water:

$$AlSb + 3H_2O \rightarrow Al(OH)_3 + 3/2H_2 + Sb$$

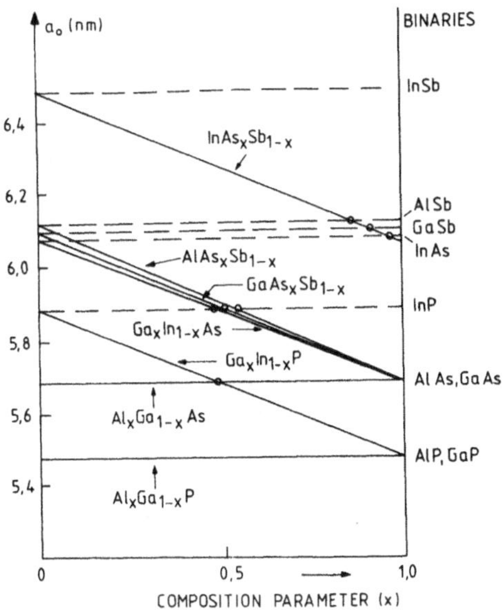

Fig. 4. Lattice constants of III-V binary and ternary compounds

In Fig. 4 the lattice constant of the binaries and of the corresponding ternary alloys are illustrated assuming that the Vegard's law holds. As it can be seen in the figure, lattice matched epitaxial layers of $Al_xGa_{1-x}P$ and $Al_xGa_{1-x}As$ can be grown onto the particular binary (GaP or GaAs) in the whole composition range (for $Al_xGa_{1-x}As$ e.g.

$$2.16\ eV \leq E_g(x) \leq 1.42\ eV \quad or$$
$$0.57\ \mu m \leq \lambda(x) \leq 0.87\ \mu m).$$

For the other compounds there are either only a particular composition parameter (x) value where the lattice matching can be achieved (for $Ga_xIn_{1-x}P$ at x=0.51 where λ=0.65 µm) or composition grading must be applied. In the first group there are important systems, as $In_{0.53}Ga_{0.47}As/InP$; $GaAs_{0.57}Sb_{0.43}/InP$; $GaAs_{0.91}Sb_{0.09}/InAs$ and $InAs_{0.91}Sb_{0.09}/GaSb$. The second group contains e.g. $GaAs_xSb_{1-x}/GaAs$, $GaAs_xP_{1-x}/GaAs$ and $InAs_xP_{1-x}/InAs$. This means that the binary/ternary systems have severe limitations:

i) the heteroepitaxy needs close compositional control , because the x value is fixed

ii) generally grading technique is used

iii) the physical parameters (E_g, n, a_o, etc.) of the lattice matched layers is determined by the substrate material.

Practically the only exception is the GaAs-$Al_xGa_{1-x}As$ system, where the lattice constants are nearly the same. It was a very fortunate moment when this system was discovered and applied as a heterosystem, because it made possible to develop the DH lasers with no remarkable mismatch forming at the interface. The difference between the lattice constants small comparing the pure binaries and this difference is even smaller when we take into account a ternary with a composition of $Al_xGa_{1-x}As$ or if the temperature is high enough (see Fig. 5). At 920 °C both binaries (GaAs and AlAs) have the same a_o value, while at room temperature Δa_o is about only 0.2 %. In the commonly used laser diodes e.g. this value is only 0.06 % [24,25]. Adding some P into the system the matching would even be better [26], because the effective radius of P is less than the radius of As in the solid.

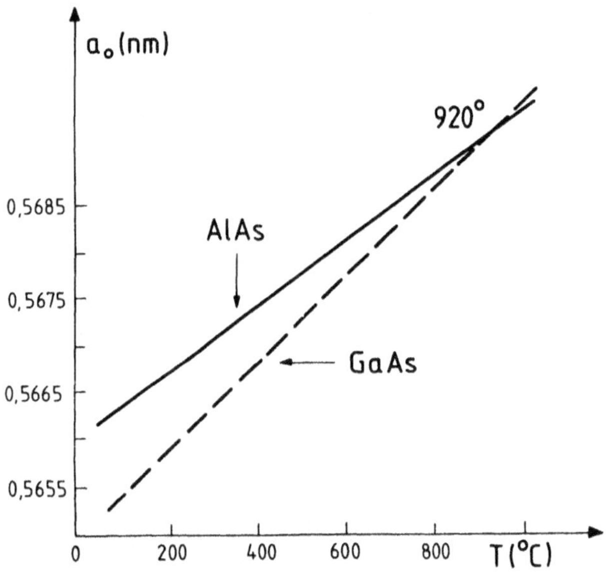

Fig. 5. Temperature dependence of GaAs and AlAs lattice constants

Similar matching can only be achieved using quaternaries. In the GaAs/GaAlAs and GaP/GaAlP systems the good matching is caused by the similarity of Ga and Al atoms. The substitution of one of them by the other does not disturb the lattice periodicity too much. None of the ternaries similar to these ones can be found.

The limitations of ternary semiconductors can be circumvented applying quaternaries. Quaternaries have one degree of freedom more than ternaries (two compositional parameters x is y, T and P) and owing to this fact they have whole composition ranges, where lattice matching can be achieved. However, only limited number of quaternaries are known in details. There are only three compounds, two so called symmetric (2:2) quaternaries ($Ga_xIn_{1-x}As_{1-y}P_y$ and $Al_xGa_{1-x}As_ySb_{1-y}$) and one pseudo-ternary ($Al_xGa_yIn_{1-x-y}Sb$) which were prepared as hetero-epitaxial layers so far.

One can found some application of $GaP_xAs_ySb_{1-x-y}$ pseudoternary, too [26]. Adding some P to the melt during LPE growth of GaAsSb, the above mentioned quaternary forms. Owing to the strong segregation of P, the melt is depleted in P and grading composition layer grows, in which x tends rapidly to 0. This graded layer is matched to GaAs substrate decreasing the misfit dislocation density in GaAsSb layer.

For a quaternary the a_o lattice constant can be approximated supposing Vegard's law (see (3)). In these alloys, e.g. in InGaAsP the ionic radii differ and if we want to match the quaternary to the substrate (to InP e.g.) we must add 2.2 As for all Ga atoms. This means that in $Ga_xIn_{1-x}As_yP_{1-y}$ the x/y = 2.2, but this is a wide range (1.3-0.75 eV) and contains as a limiting case the $Ga_{0.47}In_{0.53}As$, too.

The usual representation of the E_g-a_o diagrams gives help to determine the lattice matching conditions. In Figs 6 and 7 the important 2:2 quaternary diagrams are shown.

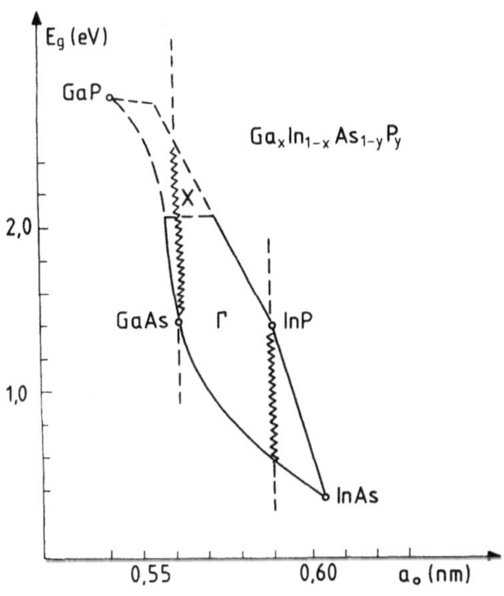

Fig. 6. Forbidden band gap and lattice constant relation for $In_{1-x}Ga_xAs_{1-y}P_y$

Looking at Fig. 6 one can observe that the InGaAsP can be matched with two binaries, with GaAs and InP.

The matching conditions are

$$y_{InP} = \frac{1-x}{0.4837-0.0327x} \qquad (12)$$
$$y_{GaAs} = \frac{1-2.0672x}{1-0.0677x}$$

The lattice matched regions cover wide E_g ranges. (On InP substrate 0.75 eV ≤ $E_g(x,y)$ ≤ 1.35 eV which corresponds to the 1.65-0.92 μm spectral range. On GaAs the band gap of matched layers can change between 1.32-1.9 eV, that is between 0.87 μm and 0.65 μm.) For InP, e.g. the GaInAsP compositions that are lattice matched to InP, decrease in P content from InP itself to the limiting P free ternary (see Fig. 6).

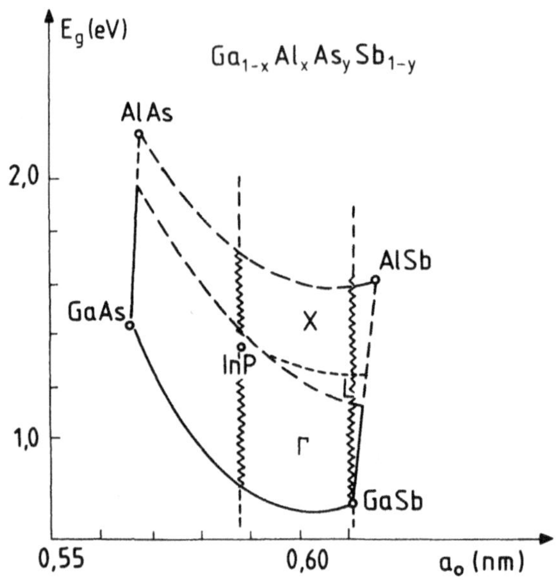

Fig. 7. E_g-a_o curves for $Ga_{1-y}Al_xAs_ySb_{1-y}$ quaternary alloy

It is very important that using the same melt at a given temperature the lattice matching condition depends on the orientation of the substrate, too. Namely, the alloy composition obtained by heteroepitaxy from a given melt at a definite temperature depends on substrate orientation. Consequently, different melt compositions and growth temperatures are required to obtain layers of the same composition by growth on differently oriented substrates. This was recognised for GaInAsP first, but it was observed for other quaternaries, too. Similar effect was found for the threshold of the misfit dislocation generation. In Fig. 3 the strong asymmetry and the different limitations for dislocation free zones is presented [11,27] for (111) and (100) InGaAsP, as well as for InGaAs. Corresponding to those mentioned the ternary zone is narrow comparing with the quaternary. If there is a misfit, the experiments show that at $f > 0.3$ % cross-hatched InGaAsP surface is formed. If $f > \pm 0.1$ %, a strong increase is deep levels in InGaAsP can be detected.

A novel class of quaternaries is the Sb - containing pseudo-ternaries. In these alloys it is characteristic that in their cationic or anionic sublattice there are three different kinds of atom; either Al, Ga and In, or P, As and Sb. The formation of graded hetero-anionic $GaP_yAs_ySb_{1-x-y}$ was already observed earlier [26], but the production of homogeneous phases has been described only at the present time [28]. In Fig. 8 the theoretically predicted E_g-a_o curve is presented for this compound. It can be seen that this alloy has lattice matched binary substrates (GaAs and InP) and this makes it very promising for the further applications as heterosystems. Microprobe analysis showed that the single crystalline phase prepared has a composition of $x = 0.366$ and $y = 0.619$ [29]. According to theoretical approximations

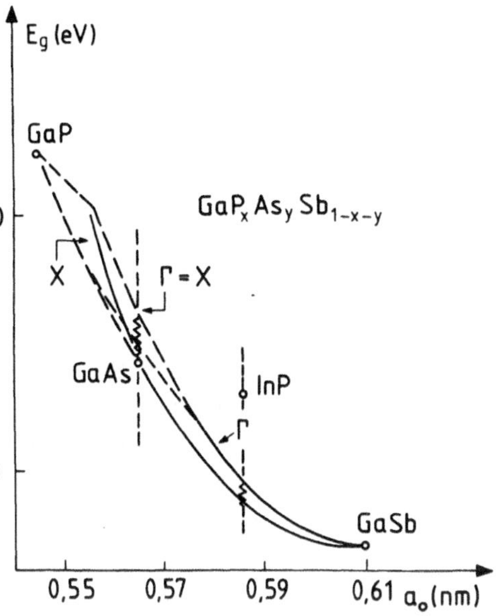

<u>Fig. 8.</u> E_g–a_o curves for pseudoternary $GaP_xAs_xSb_{1-x-y}$ heteroanionic alloy

this composition falls into the direct bandgap range of the quaternary, as it is seen in Fig. 9. Preliminary optical experiments support this assumption, as the crystals show highly efficient brillian red photo- and cathodoluminescence. This phase does not match to binary substrates, but modifying the composition one can hope to achieve the proper conditions for the preparation of lattice matched $GaP_xAs_ySb_{1-x-y}$/GaAs or InP heterojunctions.

The other interesting pseudoternary compound is $Al_xGa_yIn_{1-x-y}Sb$. In contrast with GaPAsSb, in this alloy the cationic sublattice consists of different kinds of atom. This alloy was prepared recently by Zbitnew and Woolley [30] using directional freezing, but their bulk samples were inhomogeneous ones. However, homogeneous epitaxial layers have been prepared by LPE methods [30]. Investigating the lattice matching conditions in Fig. 10, one can observe that only AlSb can be lattice-matched to the quaternary. Owing to the reactivity of this binary this is not a useful solution. GaSb has a similar lattice constant, therefore for LPE this substrate material was used as a nearest attainable material of slightly different

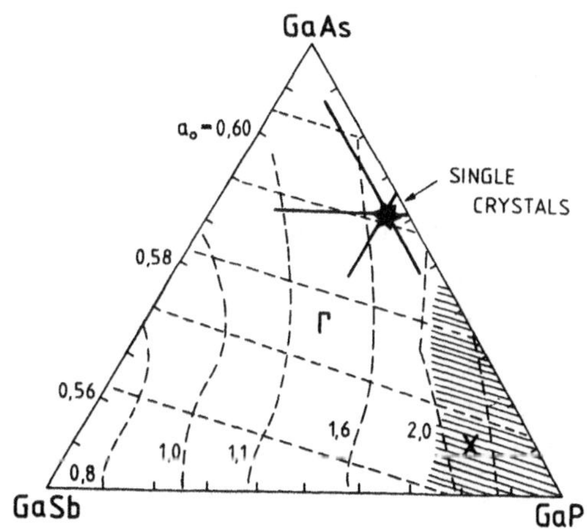

<u>Fig. 9.</u> Triangular diagram of $Ga_xAs_ySb_{1-x-y}$ pseudo-ternary. In the figure the composition of single crystals prepared is labelled by star

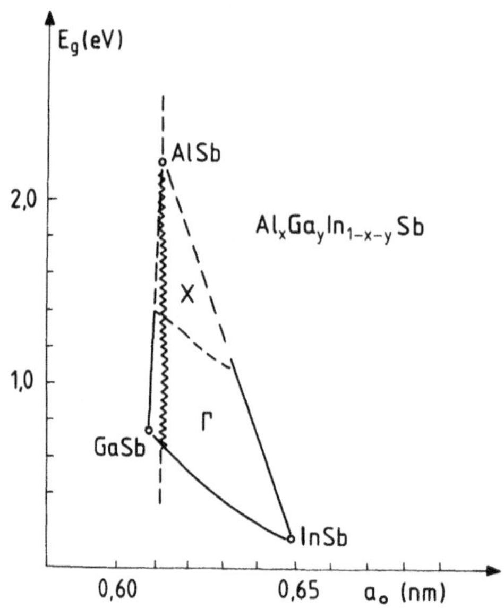

Fig. 10. E_g-a_o curves for $Al_xGa_yIn_{1-x-y}Sb$ quaternary alloy system

a_o value. Quaternary layers were grown on both (111) single crystalline and polycrystalline GaSb substrates to study the formation of misfit. Epitaxial layers grown on single crystals have exhibited rocking curves shown in Fig. 11. The rocking curve consists of peaks of the substrate, as well as peaks belonging to the heteroepitaxial layer. These latter ones are doublets, with slightly different lattice constants and their half-width is narrower than those ones which are usual among the similar heteroepitaxial systems. Taking into account the measured lattice constants [32] and the composition determined by microprobe-analysis, the heteroepitaxially grown pseudo-ternary is in the indirect bandgap region of the alloy (see Fig. 12).

On polycrystalline GaSb substrate samples a strong growth orientation dependence have been found. The substrates used were prepared by a modified Bridgman-method [33] and consist of grains with the size of some cm in dia. The SEM electron channelling experiments, X-ray investigations, selective chemical etching showed that most of the grains are in twin po-

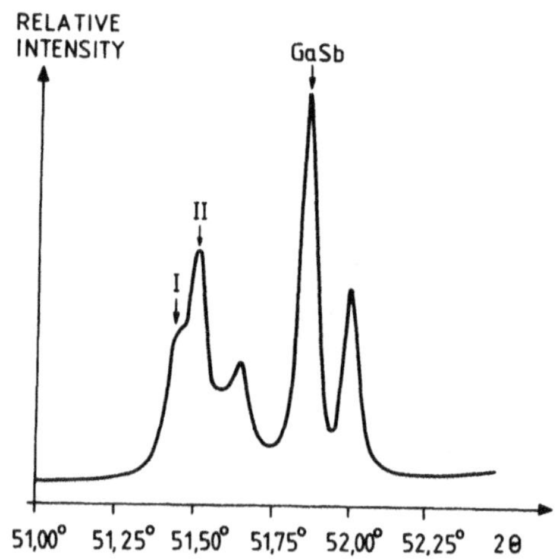

Fig. 11. Rocking curve of $Al_xGa_yIn_{1-x-y}Sb$/GaSb system. The substrate orientation is (111)

sition. On adjoining twins one can found either a thick heteroepitaxial overgrowth, or no layer at all. Figure 12 shows layerfree regions on one orientation, while on the other adjoining twin thick epitaxial layer is shown. Exact reasons for such substrate orientation dependence of the hetero-epitaxial growth are not known yet, and it is completely unexpected. It means that for lattice matching not only the lattice constant values and melt compositions, but the substrate orientation have to be taken into consideration. Some substrate orientation dependence of the solid composition (and hence lattice matching) has already been observed in the InP-InGaAsP system [5], but the AlGaInSb is the first system, where extrem selectivity in LPE growth was found.

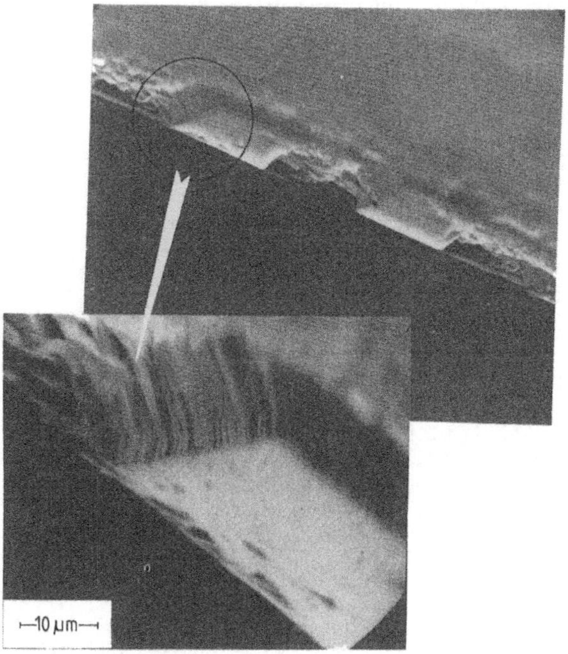

Fig. 12. Selective heteroepitaxial -rowth of $Al_xGa_yIn_{1-x-y}Sb$ onto polycrystalline GaSb sbustrate. The layers form only particular orientations, in figure regions are shown where no overgrowth can be found

If we have no sufficient substrate, lattice mismatch can seriously degrade the surface morphology. It is interesting that for a given mismatch surface morphology is poorer for the narrower band gap materials [34]. In GaInAsP system good quality crystals can only be grown when $\Delta a/a_o \sim 0.03$ %. In InGaAs at -0.22 % mismatch cracking of the layer occurs [35].

Since the growth of these heteroepitaxial systems involves large temperature regions the thermal expansion differences are also very important. Generally substantial differences are between the substrate and layer thermal expansions and the temperature dependence of these functions are also different. This means that a layer exactly lattice matched at a particular T temperature will be mismatched at a T+ΔT. The study of misfit dislocations show that faults are more easily generated at high temperatures, so the lattice matching can be achieved at the growth temperature. The existence of mismatch affects the chemical composition width at the heterojunction; mismatches of 0.03 % give sharp composition profile whereas grading occur at mismatches of -0.11 % and 0.2 % [36].

From the mentioned facts it is clear that good quality devices can be prepared with care taking into account many factors. Lattice matched heteroepitaxial systems can only be realised under special conditions.

References

1. S. Mendelson, Single Crystal Films, ed. M.H. Francombe and H. Sato (Pergamon Press, N.Y., 1964)
2. J. Jacodine, Appl. Phys. Lett. 2, 201 (1963)
3. G.R. Booker and R. Stickler, J. Appl. Phys. 33, 3281 (1962)
4. B.G. Cohen, Solid State Electron. 10, 33 (1967)
5. Y. Sugita, M. Tamura and K. Sugawara, J. Appl. Phys. 40, 3089 (1969)
6. G.A. Antypas and J. Edgecumbe, H. Cryst. Growth 34, 132 (1976)
7. A.S. Jordan, J. Electrochem. Soc. 123 (1972)
8. J.J. Hsieh, IEEE J. Quant. Electron. QE-17, 118 (1981)
9. T.P. Pearsall, M. Quillec and M.A. Pollack, Appl. Phys. Lett. 35, 342 (1979)
10. P.A. Houston, J. Mater Sci. 16, 2935 (1981)
11. K. Ishida, Y. Matsumoto and K. Taguchi, Phys. Stat. Sol. (a) 70, 277 (1982)
12. F.C. Frank and J.H. van der Merwe, Proc. Royal Soc. 189, 205 (1949)
13. C.A.B. Ball and J.H. van der Merwe, Phys. Stat. Sol. 38, 355 (197)
14. S.W. Matthews, J. Vac. Sci. Technol. 12, 126 (1975)
15. V. Gradmann, Ann. Phys. 17, 91 (1966)
16. J.W. Mathews and A.E. Bleakeslee, J. Cryst. Growth 32, 265 (1976)
17. C. Schiller and J.P. Gowers, Acta Electron. 21, 167 (1978)
18. M.S. Abrahams, L.R. Weisberg, C.J. Buiocchi and J. Bland, J. Mater. Sci. 4, 223 (1969)
19. L. Block, J. Cryst. Growth 31, 250 (1975)
20. R.H. Saul, J. Appl. Phys. 40, 3273 (1969)
21. M. Dupny and D. Lafenike, J. Cryst. Growth 31, 244 (1975)
22. W.A. Jessev and D. Kuhlmann-Wilsdorf, Phys. Stat. Sol. 19 (1967)
23. M. Ettenberg, C.J. Nuese, J.R. Appert, J.J. Gannon and R.E. Enstrom, J. Electron. Mater 4, 37 (1975)
24. G.H.B. Thompson, "Physics of Semiconductor laser Devices" (John Wiley and Sons, Chicester, 1980)
25. H. Kressel and H. Nelson, "Physics of Thin Films" (Acad. Press, N.Y., 1973) Vol. 7.
26. Y. Nishitani, K. Akita, A. Yamaguchi and T. Kotani, J. Electrochem. Soc. 127, 949 (1980)
27. K. Nahajima, S. Yamazaki, S. Komiya and K. Akita, J. Appl. Phys. 52, 4575 (1981)
28. E. Lendvay, J. Cryst. Growth (to be published)
29. E. Lendvay and I. Pozsgai, to be published
30. K. Zbitnew and J.C. Wooley, J. Appl. Phys. 52, 6611 (1981)
31. E. Lendvay, Electronic Letters 18, 407 (1982)
32. L. Petrás, M. Farkas-Jakube, E. Ledvay and J. Kardos, Oric, Ubtk, Doffiraction Conference, Györ, 1982
33. M. Hársy, T. Görög, E. Lendvay and F. Koltai, J. Cryst. Growth 53, 234 (1981)
34. M. Feng, M.M. Tashima, T.H. Windhorn and G.E. Stillman, Appl. Phys. Lett. 33, 533 (1978)

35. S.B. Hyder, G.A. Antypas, J.S. Escher and P.E. Gregory, Appl. Phys. Lett. 31, 551 (1977)
36. M. Feng, L.W. Cook, M.M. Tashima, G.E. Stillmann and R.J. Blattner, Appl. Phys. Lett. 34, 697 (1979)

COMPOSITIONAL TRANSITION LAYERS IN HETEROSTRUCTURE

Jun-ichi Nishizawa and Masayoshi Koike
Research Institute of Electrical Communication
Tohoku University, Sendai, Japan

1. Introduction

The heterojunction is a very important structure for light emitting diodes, photo-voltage devices, laser diodes, and so on. Because of the optical and carrier confinement in an active layer, a window effect or wider bandgap occures. The heterojunctions show some very difficult problems of lattice mismatch at the interface of each layer, because, according to earlier reports [1], the lattice mismatch causes deep levels and dislocations which are non-radiative centers increasing the threshold current for lasing, and reduce the lifetime of operation.

Another important problem for compound semiconductors containing high vapor pressure elements is the the way of controlling the stoichiometric composition, because the deviation from the stoichiometric composition, for example Ga:As, Al:As in the case of $Ga_{1-x}Al_xAs$, has also a great influence on the lattice constant [2].

2. Misfit Dislocation

The conventional way of lattice matching is the adjustment of composition in mixed crystals and the addition of composing atoms, i.e. producing of a quanternary mixed crystal and so on. But, even if at growth temperature, the lattice constant of each layer can considerably be fitted, misfit dislocations arise during decreasing the temperature due to the differences of the thermal expansions or contractions in each layer respectively [3].

2.1. Epitaxial Growth of Heterostructure

The lattice constants of $In_{1-x}Ga_xP$ considerably alter in the region of 5.45117 - 5.86875 Å with composition X. Epitaxial growths of X = 0.61,

0.80, 0.86 and 0.90 are possible on GaP substrate, but the surface morphology of the epitaxial layer appears to be drastically different as shown in Fig. 1(a), (b), (c) and (d) [4,5]. In the case of X = 0,61, the lattice mismatch between epitaxial layer and GaP substrate ($|\Delta a|/a_{GaP}$; $|\Delta a| = |a_{GaP}-a_{InGaP}|$) is $3.05\ 10^{-2}$. Such a large mismatch in $In_{0.39}Ga_{0.61}P$ epitaxial layer is rather detrimental and island like formations build up as shown in Fig. 1(a). As the composition increases to 0.80, 0.86, the quality of the epitaxial layer becomes relatively better. The triangular patterns appear at X = 0.80, which means that the lattice mismatch dislocations seem to be formed in the direction of <110>.

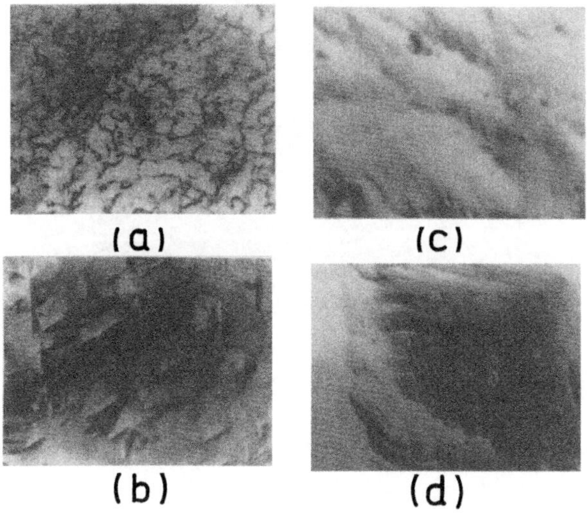

Fig. 1. Optical photographs of $In_{1-x}Ga_x$ epitaxial surface grown on GaP substrate (a:x = 0.61, b:x = 0.8, c:x = 0.86, d:x = 0.9)

2.2. "Cross-Hatch" Pattern

Various combinations of hetero-epitaxial growth have been tested, for example GaAlAsP (epi)/GaAs (sub), GaAsP/GaAs, and InGaP/GaP. The "cross-hatch" patterns due to the lattice misfit dislocations appear in the X-ray topographs as shown in Fig. 2. The "cross-hatch" patterns intersect at normal angles on an (001) substrate and at an angle of 60° on a (111) substrate. These patterns also correspond to the <110> direction and imply that the lattice mismatch dislocations are edge-type dislocations which have the components of strain parallel or perpendicular to the interface between epitaxial layer and substrate. Figure 3 shows a misfit dislocation in the heterojunction of GaAlAsP (epi)/GaAs (sub) observed by an electron microscope.

The mismatch dislocations are introduced in the direction of <110> and in two other diections of <110>,<1$\bar{1}$0> [6,7]. When the lattice spacing in the GaAlAsP epitaxial layer is larger than that in the GaAs substrate, i.e. $\Delta\Theta_{p-p}$ is positive ($\Delta\Theta_{p-p}$ is a difference between the

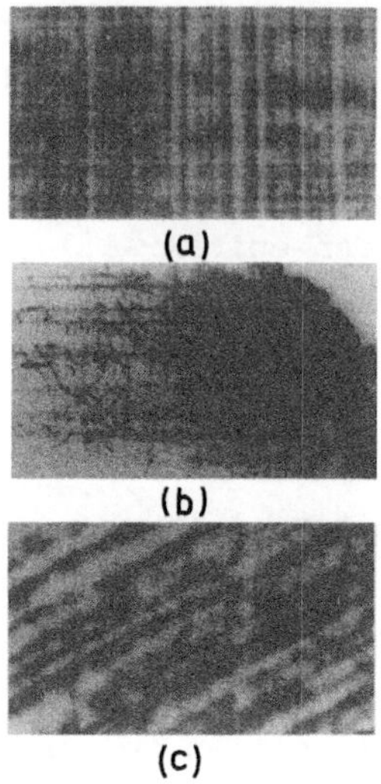

Fig. 2. X-ray reflection topographs of heteroepitaxial surface (a) GaAlAsP/GaAs (001), (b) GaAsP/GaAs (001) and (c) inGaP/GaP (111)

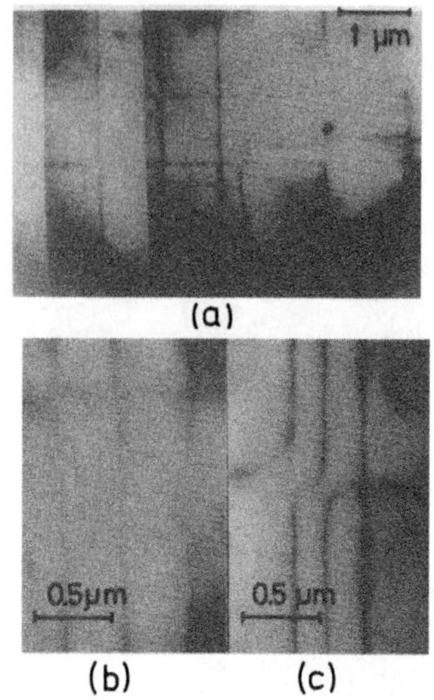

Fig. 3. Electron microscope topographs of misfit dislocations of GaAlAsP/GaAs heterojunction

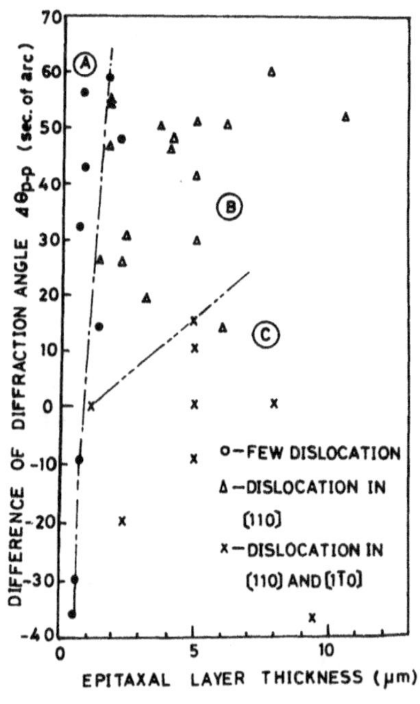

Fig. 4. Dependence of misfit dislocations of GaAlAsP/GaAs heterojunction on the lattice strain and the thickness of the epitaxial layer

diffraction angle observed in the epitaxial layer and the substrate), the mismatch dislocations are inclined to be incorporated in one direction as shown in Fig. 4, B . When $\Delta\Theta_{p-p}$ is small or negative, two directional hatch patterns appear as shown in Fig. 4, C . As shown in Fig. 4, the very thin epitaxial layer; A region has only a few dislocations. The density of hatch-pattern is inclined to increase according to the decrease of the difference $\Delta a/a_o$ (Δa is the difference of lattice constant between the epitaxial layer and the substrate, a_o is the lattice constant of the substrate). This tendency is shown in Fig. 5, and these "cross-hatch" pattern almost disappears below a certain limit of lattice mismatch $\Delta a/a_o$. Actually, the lattice

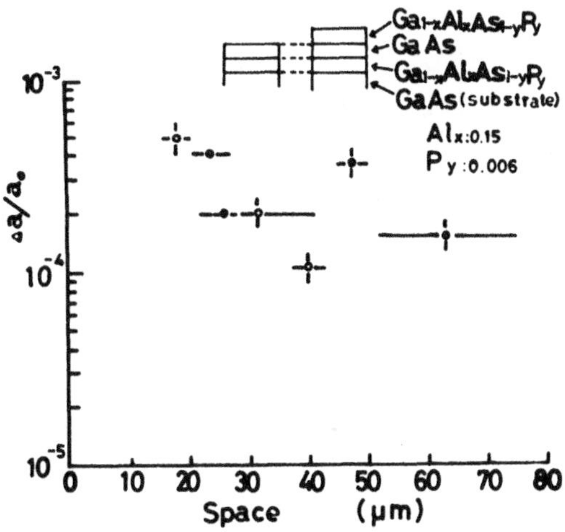

Fig. 5. Relation between the distance of "cross-hatch" pattern and $\Delta a/a_o$

mismatch $\Delta\Theta_{p-p}$ can be minimized by an optimum dope of phosphorus in $Ga_{1-x}Al_xAs$.

One of the conventional reduction methods of lattice mismatch dislocations is the incorporation of a so-called "Buffer layer" on the substrate as follows (a), grading composition changes (b) [8-10], abrupt composition changes. These methods, however, are not always the best ones to solve the problems of dislocation-free heterojunctions.

2.3. Numerical Calculation of Lattice Strain in Heterojunction

When the lattice constants of the epitaxial layer and substrate are different, the crystal bends and the lattice strain in induced. The conventional numerical analysis of crystal bending was mainly reported with regard to a stress without boundary conditions at the crystal edges. A two-dimensional mathematical model of a simple cubic lattice have been numerically analized [11]. When lattice spacing changes, the stress influences both lattices, but the sum of the stress forces acting on both must be zero. Figure 6 shows the distribution of the differential strain which corresponds to the relative differences of lattice position at each layer. The change in the differential strain is large at the interface and becomes maximum at the edge of the crystal. The critical limit of differential strain introducing dislocation is about several Ångstroms.

3. Stoichiometry Control

3.1. Crystal Growth-Temperature Difference Method (TDM)

The conventional method of liquid phase epitaxial growth is mainly the "cooling method". In this method, the composition of mixed crystal changes in the direction of the epitaxial growth thickness, because

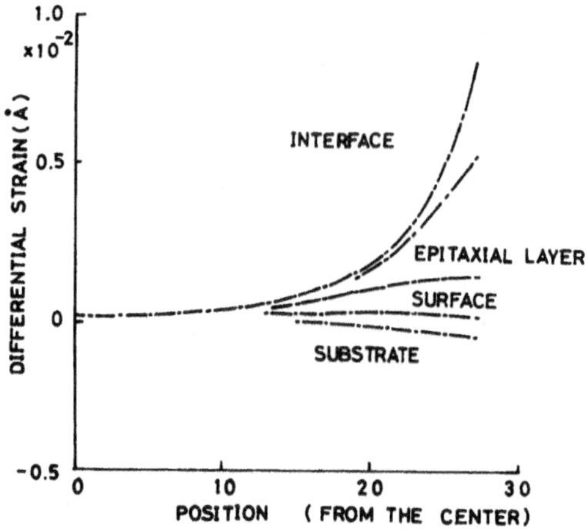

Fig. 6. Distributions of differential strain

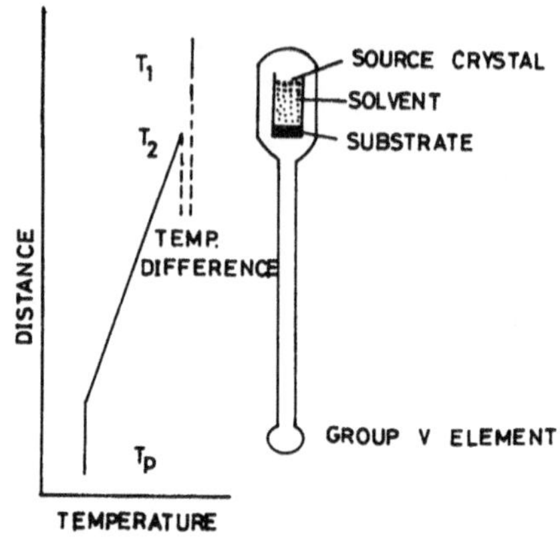

Fig. 7. Schematic diagram for temperature difference method under lattice misfit ($|\Delta a/a_o|$) dependences of $Ga_{1-x}Al_xS_{1-y}P_y$

the growth temperature decreases monotonically [12]. On the contrary, the temperature difference method can produce a uniform compositional epitaxial layer due to the constant growth temperature [13-16]. The principle of this method is explained by the schematic diagram described in Fig. 7. The source material is carried to the substrate through a solvent by diffusion due to concentration difference as well as by thermal diffusion. High vapor pressure elements must be controlled as shown in Fig. 7, and will be mentioned later.

3.2. Effects of Controlled Vapor Pressure (CVP) on $Ga_{1-x}Al_xAs_{1-y}P_y$ and $In_{1-x}Ga_xP$ Epitaxial Layer

The most important problem of compound semiconductors containing high vapor pressure elements is how to control the stoichiometric composition. The deviation from the stoichiometric composition, for example that of Ga:As, Al:As in the case of $Ga_{1-x}Al_xAs$, also influences the lattice constant [2,13,14]. Another typical important feature is the lattice misfit dependence of $Ga_{1-x}Al_xAs_{1-y}P_y$ epitaxial layer on GaAs substrate at the applied arsenic pressure during epitaxial growth. In the case of x = 0,3 and y = 0.006, 0.009, 0.016 and 0.02, the lattice misfit is defined as $|\Delta a/a|$, Δa is the difference of the lattice constant between the $Ga_{1-x}Al_xAs_{1-y}P_y$ epitaxial layer and the GaAs substrate, a is the lattice constant of the GaAs substrate. For each constant composition y (|Δa/a|) varies in the range of 5×10^{-5} - 1×10^{-3} for the applied arsenic pressure of 0.2 - 300 Torr at growth temperature 800 °C. At a given pressure of 60 Torr, |Δa/a| becomes a minimum of 5×10^{-5} and both for lower and higher arsenic pressures the value of |Δa/a| is always larger than that according to Fig. 8.

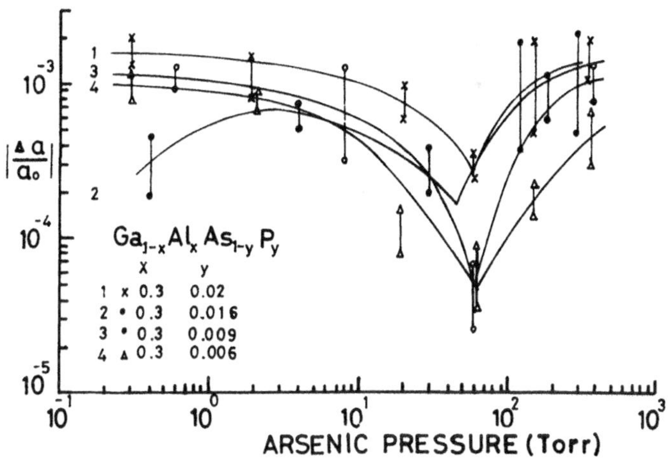

Fig. 8. Lattice misfit ($|\Delta a/a_o|$) dependences of $Ga_{1-x}Al_xAs_{1-y}P_y$ epitaxial layer grown on GaAs substrate at the applied arsenic pressure.

From these experimental results it is obvious that the lattice constants of mixed crystals are defined by not only the compositional ratio but also by the deviation from the stoichiometric composition, that means by the vacancies or by the interstitial atoms [16-20], too .

From these results one can deduce that high quality of heteroepitaxy can be produced by the compositional matching of the lattice constant and by the control of non-stoichiometric defects for deep levels, because the large number of phosphorus vacancies changes the lattice constant of $In_{1-x}Ga_xP$, but does not change directly the compositional ratio of Ga with respect to In.

4. Homo-Epitaxial Growth of Mixed Crystal on Mixed Crystal Substrate and More Suitable Combinations in Many Kinds of Mixed Crystals

In order to achieve high quality epitaxial layer of $In_{1-x}Ga_xP$ without misfit dislocation, an $In_{1-x}Ga_xP$ substrate itself, the composition of which is the same as that of the epitaxial layer, is more ideal than a graded compositional layer. The growth of bulk $In_{1-x}Ga_xP$ single crystal is in the practice very difficult,because composition of $In_{1-x}Ga_xP$ must be kept constant, therefore an appropriate technology of crystal growth is required. The temperature difference method under controlled vapor pressure (TDM-CVP) is the best way for the crystal growth.

A wide band gap mixed crystal is usually constructed by active layers on both sides, in order to confine carrier in active layer. We had proposed the following three conditions for high efficient heterojunction lasers [21],

(1) a larger refractive index in the active layer compared to that in the adjoined non-active layers;

(2) the lattice constants matching at heterojunctions;

(3) direct transition type active layer and indirect transition type non-active layers adjoining the active layer.

From these three conditions, we have chosen good combinations for high efficient heterojunction optical devices. One of them is the combination of $In_{1-x}Ga_xP$ (sub)/$GaP_{1-y}Sb_y$ (indirect)/$In_{1-x}Ga_xP$(direct)/ /$GaP_{1-y}Sb$(indirect). Other good combinations are listed in Table 1.

Table 1. Combinations of direct and indirect transition types

	Indirect transition	ε_∞	a (A)	E(eV)	Direct transition	ε_∞	a(A)	E(eV)
1	$Ga_{0.18}Al_{0.82}Sb$	11.0	6.13	1.43	GaAs	11.1	5.65	1.43
2	AlSb	8.5	6.14	1.60	$In_{0.3}Ga_{0.7}P$	9.0	5.58	2.2
3	$GaP_{0.88}Sb_{0.12}$	8.8	5.53	2.14	$In_{0.35}Ga_{0.65}P$	9.2	5.61	2.18
4	$GaP_{0.65}Sb_{0.35}$	10.2	5.73	1.84	$Ga_{0.67}Al_{0.33}As$	10.3	5.65	1.85
5	AlN	4.7	a=3.104 c=4.965	3.8	GaN	5.8	a=3.18 c=5.16	3.39

5. Conclusion

The transitional layer in heterostructure is associated with the very serious problem of the lattice mismatch dislocation. A conventional method to overcome this problem is the application of different kinds of buffer layers between the epitaxial layer and the substrate, and on the other hand, the use of substrates for the epitaxy with nearly the same lattice constant and thermal expansion coefficient. Another important question is the deviation of the composition of the compound crystal from the stoichiometric one, because these deviations correspond to different kinds of defects and reduce the radiative efficiency, promote the propagation of dislocations and change the lattice constant even if the composition throughout the mixed crystal is the same. Therefore the lattice mismatch dislocation can be minimized by both, adjustment of the composition of the mixed crystal and controlling the deviation from the stoichiometric composition.

References

1. Y.T. Lee, N. Miyamoto, J. Nishizawa, J. Electrochem. Soc. 122, 530 (1975)
2. H. Ohtsuka, K. Ishida and J. Nishizawa, Japan J. Appl. Phys. 8, 632 (1969)
3. I. Kudman and R.J. Paff, J. Appl. Phys. 43, 3760 (1972)
4. O.N. Ermakov, L.S. Garbe, Y.A. Golovanov, V.P. Sushkov and M.V. Chekichev, IEEE ED-26, 1190 (1979)
5. G.B. Stringfellow, J. Appl. Phys. 43, 3455 (1972)
6. G.A. Rozgonyi, P.M. Petroff and M.B. Panish, Appl. Phys. Lett. 24, 251 (1974)
7. M.S. Abrahams, J. Blanc and C.J. Buiocchi, J. Appl. Phys. 45, 3277 (1974)
8. G.B. Stringfellow and P.E. Greene, J. Appl. Phys. 40, 502 (1969)
9. M.S. Abrahams, L.R. Weisberg, C.J. Buiocchi and J. Blanc, J. Master, J. Mater. Sci. 4, 223 (1969)
10. G.H. Olsen, M.S. Abrahams, C.J. Buiocchi and T.J. Zamerowski, J. Appl. Phys. 46, 1643 (1975)
11. J. Nishizawa, Y. Okuno, M. Fukase and H. Tadano, J. Crystal Growth 52, 929 (1981)
12. H. Mariette, V. Thierry-Mieg, A. Etcheberry, J.C. Guillaume, A. Marbeuf and M. Rommeluere, J. Crystal Growth 53, 413 (1981)
13. J. Nishizawa, Y. Okuno, IEEE Trans. Electron Devices ED-22, 716 (1975)
14. J. Nishizawa, Y. Okuno, H. Tadano, J. Crystal Growh 31, 215 (1975)
15. W.R. Hitchens, N. Holonyak, Jr., M.H. Lee, J.C. Campbell, J.J. Coleman, W.O. Groves and D.L. Keume, Appl. Phys. Lett. 25, 352 (1974)
16. H.M. Macksey, J.C. Campbell, G.W. Zack and N. Holonyak, Jr., J. Appl. Phys. 43, 3533 (1972)
17. J. Nishizawa, M. Koike, K. Miura and Y. Okuno, Japan J. Appl. Phys. 19, 25 (1980)

18. J. Nishizawa, Y.J. Shi, K. Suto and M. Koike, J. Appl. Phys. 53, 3878 (1982)
19. V.G. Kermidas, H. Temkin and W.A. Bonner, Appl. Phys. Lett. 40, 731 (1982)
20. J.M. Parsey, Y. Nanishi, J. Lagowski and H.C. Gatos, J. Electrochem. Soc. 128, 937 (1981)
21. J. Nishizawa, OYO BUTURI 41, 912 (1972) (A monthly publication of the Japan Society of Applied Physics)

DEFECT COMPLEXES IN III-V COMPOUNDS

Lars-Åke Ledebo

Innovance AB
Magle Stora Kyrkogata 8, 223 50 Lund, Sweden

ABSTRACT

Some technically important examples of defect complexes in III-V semiconductors will be briefly discussed in this paper. It is now clear that the midgap donor level in GaAs, often referred to as EL2, is not as previously assumed due to substitutional oxygen on an As site, but is a defect complex involving a Ga vacancy. This defect makes it possible to grow undoped semi-insulating GaAs for the production of integrated circuits. Defect reactions can be monitored by studying the luminescence from ZnO pairs, which are used to obtain red light from GaP LED's. The A and B levels often present in liquid phase GaAs are identified as the Ga_{As} antisite. A normally unwanted but frequent contaminant in III-V semiconductors is Cu. We now know that the two copper-related levels which may appear in all large bandgap III-V's are not two levels of the same defect. At least one of the defects therefore has to be a complex. Studies in series of alloys demonstrate that the copper-related levels have the same defect origin in GaP, InP, GaAs and AlAs.

Much knowledge has been collected about the electrical and optical properties of various energy levels in the bandgap of semiconductors. Deliberate doping with deep level impurities can be made for example by the diffusion of transition metals or by addition of dopants during epitaxial growth. However, in spite of many years of work, little is known about the chemical identity of the defects causing the levels. This is at the moment the limiting factor for progress within "defect science". There is no known general method for making the identification, and the cases where convincing proofs of identity have been given are few. Instead we have to resort to indications of identity and to use as many different methods of characterization as possible to get complementary pieces of information. By doing so, a picture is slowly emerging at least for the larger bandgap III-V semiconductors, where simple substitutional defects are much less common than what has often been assumed. Many of the well-known impurity levels in GaAs and GaP have turned out to be due to defect complexes, even though their exact compositions and configurations are not yet known.

Defect complexes may dissociate or re-form depending on the temperature and the concentration of the constituents. The chemistry of defect reactions is similar to the chemistry of dilute solutions. The crystal, which is the matrix in which the various defects are embedded, can normally be regarded as a homogeneous, smeared-out background. The defects that react with each other may be single substitutional atoms, interstitial atoms, vacancies or different combinations of such building-blocks. The symmetries of a defect may deviate considerably from the normal symmetries in the matrix. The chemistry of defect reactions was early treated in works by Reiss [1] and by Kröger [2]. The experimental data that initiated these studies were mainly from germanium and alkali halides. The basic theory is of course also valid for Si and III-V semiconductors.

The equations describing the equilibria between the various types of defects are of the type

$$X + Y = XY + \text{thermal energy} + \text{electronic energy} \tag{1}$$

The energies may be positive or negative. The concentrations of the constituents at thermal equilibrium are determined by a normal chemical equation of the type

$$\frac{[XY]}{[X] * [Y]} = K(T) \tag{2}$$

The equilibrium constant K is strongly temperature-dependent and also depends on the charge state of the reactants. K(T) can be expressed in terms of energy as

$$K(T) = \frac{\exp(\Delta E/kT)}{ZN} \tag{3}$$

where ΔE is the energy gain in creating the complex XY, k the Boltzmann constant, T the absolute temperature, N the concentration of lattice sites in the crystal and Z the coordination number (number of possible equivalent orientations of the defect) [1].

At low temperatures, non-equilibrium situations may be frozen in because of slow diffusion of the constituents X and Y, or because of energy barriers for the reaction. At higher temperatures when diffusions starts, complex formation may go to equilibrium according to the above equations within experimentally feasible times. At still higher temperatures, where kT is no longer small compared to the energy gain at the formation of the complex, dissociation will start. This may lead to the formation of new types of complexes that are often larger. The defect reactions in irradiated silicon are a good example of this.

Pair Spectra

An instructive example of defect reactions in GaP is the behaviour of the isoelectronic ZnO centre responsible for the wellknown red electroluminescence in GaP LED's Isolated Zn is a single substitutional atom on a Ga site, and isolated O sits on a P site.At low temperatures a number of lines appear which are characteristic for pair-spectra [3]

$$h\nu(R) = h\nu_{\infty} + \frac{e^2}{4\pi\varepsilon\varepsilon_o R} \qquad (4)$$

where R is the distance between the Zn-atom and the O-atom in the pair for which recombination occurs. When Zn and O are allowed to diffuse, they would be randomly distributed if no energy gain was obtained by approach. For random distribution the dependence of the luminescence intensity I on the distance R can be approximated by [4]

$$I(R) \approx R^6 \exp(-\frac{4}{3}\pi N_{Zn} R^3) \qquad (5)$$

where the capture cross section for ionized pairs is supposed to follow $\sigma(R) = CR^2$ [5], and the exponential term is the probability that there is no zinc atom closer to the oxygen atom than R. This equation fits the experimentally observed intensities very well with one important exception; a much larger peak is seen for nearest neighbour Zn and O than predicted. Obviously the forces involved favour the formation of Zn-O near-neighbour pairs. By heat treatment it is possible to change the equilibrium between randomly distributed, isolated Zn and O and near-neighbour complexes. Furthermore, the nearest-neighbour pair Zn-O can be made to dissociate by laser excitation. This has recently been demonstrated by Feenstra and McGill [6] in one of the few investigations in III-V's where identified defect reactions have been monitored. A substantial part of the end products is next-nearest and third-nearest Zn-O pairs, which normally are not formed. It is suggested [6] that the defect reactions is due to excitation of local phonon modes by non-radiative electron-hole recombination as a consequence of the laser excitation. This is thus an example of a *photoinduced* defect reaction.

Antistructure and Antisite

In the III-V compounds several different types of native defects may occur. The simplest defect complex is the antistructure pair. Taking GaAs as an example, the antistructure is a reversed Ga-As pair, with the Ga sitting on an As site and the As sitting on a Ga site. Because of the similarity between the Ga and the As atoms, it has been calculated that in stochiometric material the concentrations of such defects should be high [7]. The defect however has probably not been noticed since it is not expected to be electrically active, nor to produce a local phonon mode. It could however add to the scattering of free carriers [7].

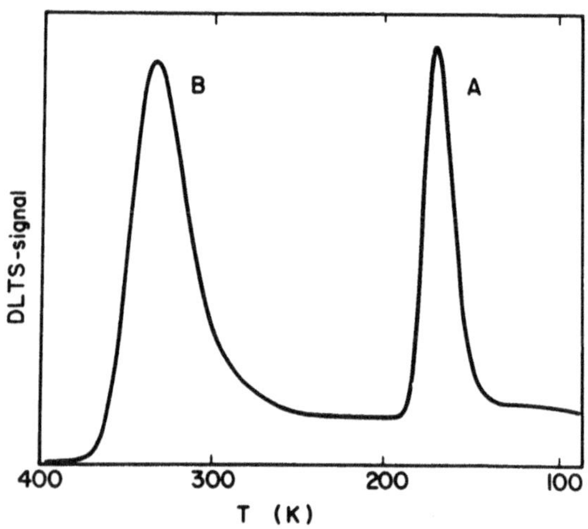

Fig. 1. A DLTS-spectrum for the so-called A and B levels in a liquid phase epitaxial p^+n-junction of GaAs. Note the similarity in amplitude, indicating equivalent concentrations. The apparant thermal activation energies are E_V+0.40 eV and E_V+0.70 eV respectively [25]

More important are the antisite defects, which in GaAs are Ga on an As site (Ga_{As}, the Ga antisite) or As on a Ga site (As_{Ga}, the As antisite). From the simple electron counting, Ga_{As} is a double acceptor and As_{Ga} a double donor. The Ga antisite should exist mainly in material grown under Ga-rich conditions, such as liquid phase material. Indeed, the so-called levels A and B often found in LPE GaAs have recently been demonstrated to be double acceptor, most likely the Ga_{As} native defect (Fig. 1-3) [8]. The double acceptor is present in GaAs and AlGaAs grown by liquid phase epitaxy (LPE) in many laboratories. In the liquid phase process that we use, the acceptors do not appear. To produce samples that had the levels, the process had to be manipulated either by rapid cooling after the epitaxial growth, or by the addition of small amounts of Ni to the gallium melts. The defect obviously anneals during slow cooling. The nickel impurity (which itself produces no deep level in GaAs [9]) probably occupies Ga sites, thereby promoting the availability

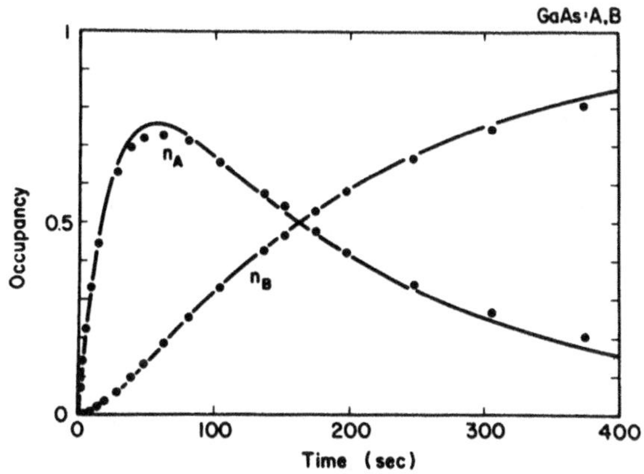

Fig. 2. Time-dependence of the occupancy for the A and B-levels during optical excitation with a photon energy large enough to induce hole transitions from both levels to the valence band. A special capacitance method was used to monitor the occupancy [8]. The full line is the theoretical time-dependence with the assumption that the two levels are coupled. In the uncoupled case this dependence is significantly different

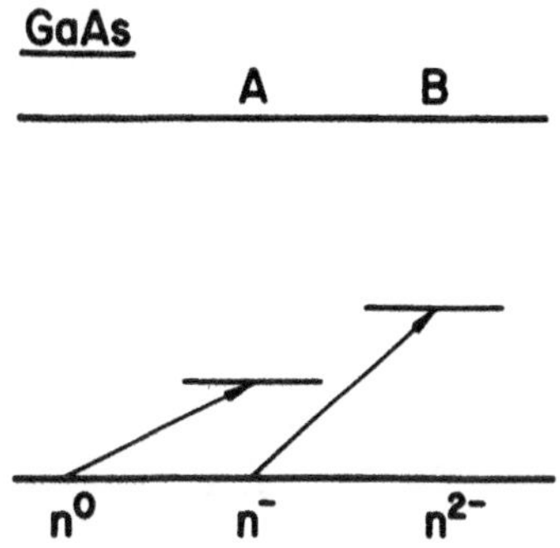

Fig. 3. Level scheme of the A and B levels, identified as the levels of the Ga_{As} antisite double acceptor

of interstitial Ga which is needed in the formation of the Ga antisite.

The As antisite is expected to be found in material grown under As-rich conditions, which is the case of LEC (Liquid Encapsulation Czochralsky) and VPE (Vapour Phase Epitaxy) material, and often for Bridgman material. A ubiquitos level in such material is the technically important EL2 or "O"-level with an optical threshold for electron excitation of E_c-0.79 eV at 77 K (Fig. 4). By studying the concentration-dependence for this level on the Ga/As ratio in a MOV-PE system it has been possible to conclude that one part of the impurity complex is a gallium vacancy [10]. It has been suggested from the study of EPR in plastically deformed samples that EL2 precisely is the As_{Ga} antisite defect [11].

A peculiar property of the EL2 centre is the possibility to optically remove the midgap donor level from a depletion region or from a high-resistivity material. The spectral distribution for the cross-section for removal is shown in Fig. 5 as measured from the photoconductivity quenching in a high-resistivity sample at low temperature. After removal, the level reappears if the sample is heated above about 120 K, or if it is flooded with electrons. It has been suggested that the anomalous behaviour is due to a metastable state of the donor level [12], and a configurational coordinate diagram has been suggested. Such a diagram however does not help towards an understanding of the

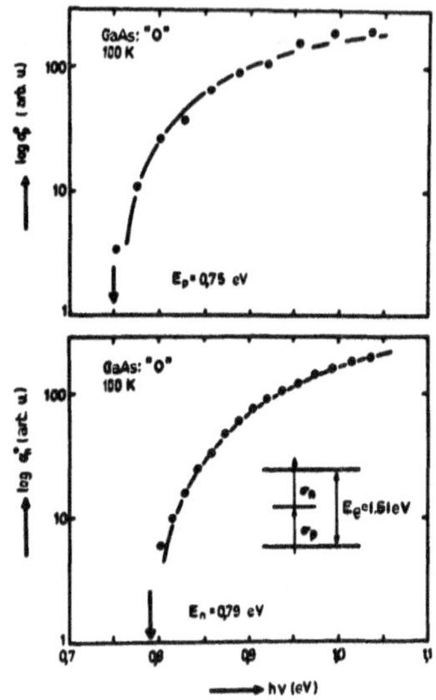

Fig. 4. Spectral distribution of the photo-ionization cross-sections for the defect EL2 in GaAs. The cross-section for holes can not be obtained at low temperatures by the normally used capacitance methods. To obtain the data a transient current method was used in a p^+n-junction. The result illustrates that the lattice relaxation is smaller than sometimes suggested [18]

defect on the atomic level, but merely is a pictorial summary of the experimental parameters obtained.

Another unusual property of GaAs:EL2 is that the defect only appears in n-type material. In addition, if a pn-junction is made by diffusing zinc into n-type material, the concentration of EL2 is substantially reduced also in the n-type region of the pn-junction. The level, however is somehow dormant, because a subsequent anneal again increases the concentration of the defect in the n-type region. This is in contrast to samples that have not previously contained EL2, where an anneal does not create the defect.

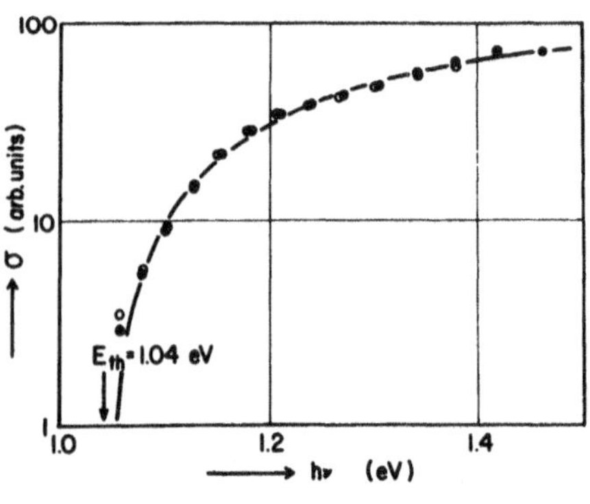

Fig. 5. Spectral distribution of the optical cross-section for removal of the EL2 midgap donor level. The shape of the cross-section indicates a connection with a bound-to-free electronic transition. The method used is constant quenching in a photoconductor with two different quenching values as indicated with the different symbols. The solid line is a theoretical fit [24]

By considering the above-mentioned properties of EL2, it is tempting to suggest a photo-induced defect reaction as the mechanism for removing the level, with subsequent reformation of the complex upon heating or in the presence of electrons. The dormant state may be caused by Zn-atoms temporarily occupying Ga-vacancies which are one of the essential ingredients in the defect. Some details in this interpretation are however still difficult to understand, and at the time of writing this paper the detailed identity of the EL2 level still remains unknown.

Cu-related Levels

In GaAs a number of vacancy-impurity pairs have been throughly studied by luminescence [13]. It seems that those defects normally do not give rise to impurity levels in the bandgap, and they are technically of little importance. Levels resulting from the diffusion of transition metals into III-V compounds are well-known. Whereas a number of arguments exist that Cr, Mn and Fe are simple substitutional impurities [16], particularly Cu but also Ni [17] readily form complexes in GaAs and GaP. Luminescence from Cu-related centres has been investigated mainly in GaP. The best understood emissions so far are the so-called COL (Copper Orange Luminescence) and the "1.91 eV" emission. The COL is suggested to arise from a substitutional Cu on a Ga site surrounded by two interstitial copper atoms, creating an isoelectronic centre [14]. The "1.91 eV" emission could possibly originate from a linear (100) oriented Cu-Ga split interstitial [15]. Both the COL and the "1.91" are readily quenched by the diffusion of Li, upon which other types of luminescence are seen. This can be interpreted in terms of various types of complexes of increasing complexity involving vacancies, copper and lithium. In similarity to GaAs the luminescent centers probably do not have energy levels in the bandgap.

Some of the Cu-related defects have no strong luminescence but instead manifest themselves by producing levels in the bandgap. By diffusion of Cu into GaAs, GaP or InP, two clearly dominant levels are observed in a DLTS measurement [19]. By following the levels through the composition range from pure GaAs to pure GaP it has been possible to establish that the levels have the same origin in both semiconductors. The two levels can also be traced from GaAs

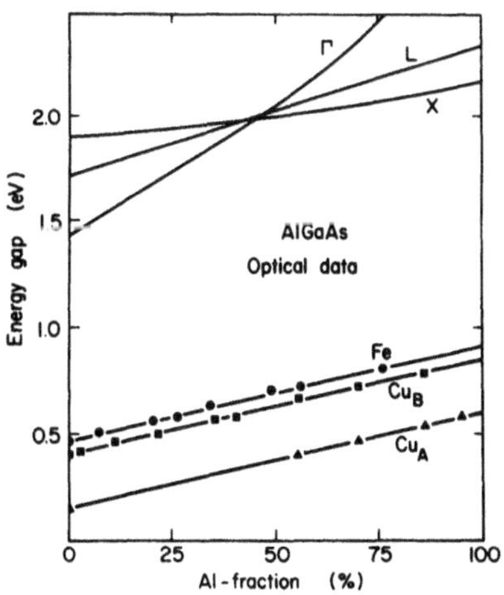

Fig. 6. Optical threshold energies for Cu_A [20], Cu_B [20] and Fe [25] in AlGaAs as a function of alloy composition. The data demonstrate a smooth transition of the level positions from GaAs to AlAs, indicating a similar defect structure in the entire composition range

towards AlAs, and as seen in Fig. 6 they have a continuous dependence of binding energy on composition up to at least 80 % Al [20]. Evidence can be given that the two levels in InP also have the same origin [19]. The variation of the binding energy with matrix composition is in good agreement with the assumption that the level position is constant with respect to the vacuum level (Fig. 7) [21]. Contrary to what has sometimes been suggested, it is now clear that in each semiconductor the two levels are not connected to the same defect i.e. they are not coupled. Therefore at least one of them has to be a complex. Several suggestions for a possible identification can be found in the literature [22,23], but a convincing argument leading to a positive identification of the copper-related defects still remains to be presented.

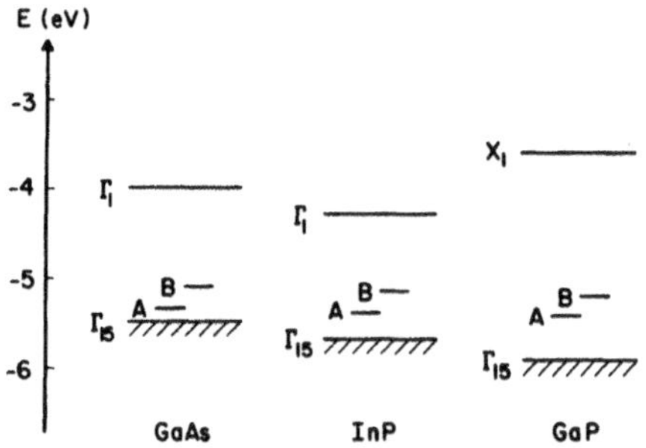

Fig. 7. Absolute energy positions (with reference to the vacuum level) for the two Cu-related levels in GaAs, InP and GaP [21, 19]

Much of the work needed to understand more about defect complexes involves expensive and fascinating instruments such as ESR, SIMS, ODMR, dye-lasers and equipment for photocapacitance. It is, however, important to point out that also the normally tedious sample preparation is very crucial, and many mistakes have been made because of contamination in the processing. Once published, erroneous results may float around in the literature for a long time, confusing the discussion. If we combine the use of good equipment with careful and critical sample preparation it is likely that we will soon know more about defect complexes in semiconductors as long as we decide to put our minds and money to it.

Discussions with H. Gislason and L. Samuelsson are gratefully acknowledged. Some of the original results presented in this paper were obtained while the author was with the Department for Solid State Physics, Lund University.

References

1. H. Reiss, C.S. Fuller and F.J. Morin, Bell. Syste. Techn. J. 35, 535 (1956)
2. F.A. Kröger, The Chemistry of Imperfect Crystals, North-Holland, 1964
3. J.J. Hopfield, D.G. Thomas and M. Gershenzon, Phys. Rev. Letters 10, 162 (1963)
4. R. Bindermann and K. Unger, phys. stat. sol. (b) 66, 133 (1974)
5. D.G. Thomas, J.J. Hopfield and W.M. Augustyniak, Phys. Rev. A 140, 202 (1965)
6. R.M. Feenstra and T.C. McGill, Phys. Rev. Letters 47, 925 (1981)
7. J.A. van Vechten, J. Electrochem. Soc. 122, 423 (1975)
8. L-Å. Ledebo and Zhan-Guo Wang, submitted to J. Appl. Phys.
9. V. Kumar and L-Å. Ledebo, J. Appl. Phys. 52, 4866 (1981)
10. L. Samuelson, P. Omling, H. Titze and H.G. Grimmeiss, J. Cryst. Growth 55, 164 (1981)
11. E.R. Weber, H. Ennen, U. Kaufmann, J. Windscheif, J. Schneider and T. Wosinski, submitted to Appl. Phys. Lett. (1982)
12. G. Vincent and D. Bois, Solid State Commun. 27, 431 (1978)
13. E.W. Williams and H.B. Bebb, in: "Semiconductors and Semimetals" ed. R.K. Willardson and A.C. Beer, ch 5 (1972)
14. B. Monemar, H.P. Gislason, P.J. Dean and D.C. Herbert, Phys. Rev., to be published 1982
15. H.P. Gislason, B. Monemar, P.J. Dean, D.C. Herbert, S. Depinna, B.C. Cavenett and N. Killoran, Phys. Rev., to be published 1982
16. U. Kaufmann and J. Schneider, Festkörperprobleme XX, p 87 (1980)
17. H. Ennen, U. Kaufmann and J. Schneider, Appl. Phys. Lett. 38, 355 (1981)
18. A. Chantre, G. Vincent and D. Bois, Phys. Rev. B 23, 5335 (1981)
19. N. Kullendorff, L. Jansson and L-Å. Ledebo, submitted to J. Appl. Phys. 1982
20. L. Jansson, L-Å. Ledebo and H.G. Grimmeiss, unpublished
21. L-Å. Ledebo and B.K. Ridley, to be published in J. Phys. C 15 (1982)
22. P.O. Fagerström, H.G. Grimmeiss and H. Titze, J. Appl. Phys. 49, 3341 (1978)
23. P.J. Dean, B. Monemar, H.P. Gislason and D.C. Herbert, J. Luminescence 24/25, 401 (1981)
24. G. Lucovsky, Solid State Commun. 3, 299 (1965)
25. Zhan-Guo Wang, L-Å. Ledebo and H.G. Grimmeiss, unpublished

LOW FREQUENCY CURRENT OSCILLATIONS DUE TO ELECTRON RETRAPPING BY THE As_{Ga} ANTISITE DEFECT IN GaAs

M. Kamińska*, J. Lagowski, J.M. Parsey and H.C. Gatos

Massachusetts Institute of Technology
Cambridge, Mass. 02139, USA
*On leave from Institute of Experimental Physics
Warsaw University, Warsaw, Poland

ABSTRACT

Large amplitude current oscillations with a thermally activated frequency and thermally - stimulated current oscillations were observed in semi-insulating GaAs for the first time. These oscillations were found at DC electric fields above 500 Vcm^{-1}, about one order of magnitude below the electric-field threshold for Gunn oscillations. The activation energy of the oscillation frequency was found to be 0.8 eV corresponding to the dominant electron trap energy in semi-insulating GaAs (the so-called EL2 level recently attributed to the antisite arsenic [1]). A model proposed to explain the phenomena of voltage - controlled negative resistance in semi-insulating GaAs exposes the role of EL2 defect in the trapping process. The capture of electrons from the conduction band by the EL2 defect has a configurational barrier of approximately 70 meV. The presence of an electric field enhances this electron capture process which makes possible the migration of slow domains.

In semiconductor materials two types of current oscillations have been observed upon application of a high DC electric field. The so-called Gunn oscillations exhibiting microwave frequency [2] have been found in n-type GaAs and InP. The other kind with much lower frequencies has been observed in many semiconductors (e.g. Ge, GaAs, CdS) exposed to incident light [3-5]. Both types of oscillations have been attributed to negative resistance characteristics of semiconductors. In the case of Gunn effect the transfer of heated electrons by high DC electric field from a high mobility sub-band to a higher energy low mobility sub-band has been used for an explanation of negative resistance characteristics [6,7]. The mechanism responsible for a negative resistance leading to low frequency oscillations has not been entirely understood. Most of the explanations, however, relate this phenomenon to an enhancement of recombination of excess carriers by electric field [8-10].

The new type of low frequency current oscillations activated by the thermal release of electrons from deep levels in semiinsulating (SI) GaAs was found. The samples used in the experiment were liquid - encapsulated Czochralski (LEC) - grown undoped SI GaAs. The 300 K resistivity was greater than 10^7 ohm/cm and the Hall effect mobility was of about 4500 cm^2/Vs. The free electron concentration exhibited a thermal activation energy of about 0.8 eV, which is typical for high quality undoped SI GaAs and indicate a dominant role of the deep donor [11] related to the antiside As_{Ga} defect [1], which compensates the net shallow ionized acceptors. The concentration of EL2 in our samples was estimated from optical absorption [12] to be of about 10^{16} cm^{-3}. The other electron traps of the concentration of about 10^{15} cm^{-3} were found using thermally stimulated current (TSC) measurements in which a double-peak structure between 180 K and 220 K was seen, which corresponds to thermal emission of electrons from the deep traps at E_c - 0.34 eV and E_c - 0.4 eV commonly present in bulk GaAs [13].

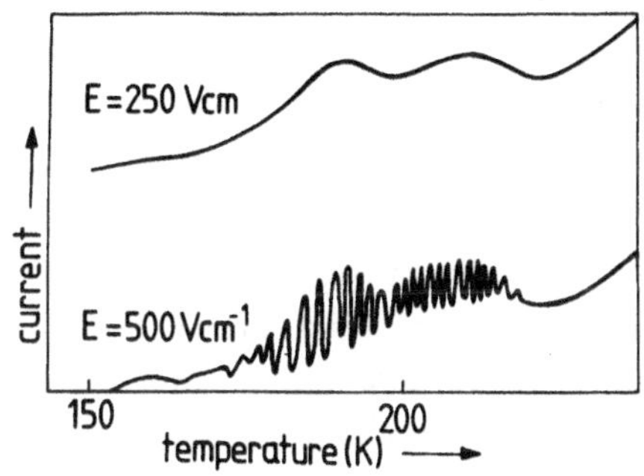

Fig. 1. Thermally-stimulated current in SI GaAs under different electric field. TSC oscillations were observed at high DC field conditions, E = 500 Vcm^{-1}

At high DC electric field (greater than 500 Vcm^{-1}) large amplitude current oscillations were found in two temperature regions: 180-220 K (Fig. 1) and above 300 K (Fig. 2). The current oscillations occurring in temperature region 180-220 K correspond to TSC oscillations and decayed with time for a given (fixed) temperature. For creating them it was necessary to illuminate the sample in low temperature to load the traps responsible for TSC. After that the sample kept under high DC bias was heated in darkness. At a constant heating rate the oscillation frequency exhibited a maximum value of about 2 Hz around each of the two TSC peaks.

The oscillations found above 300 K were thermal equilibrium current (TEC) oscillations which did not decay in time (Fig. 2) and they did not require any prior illumination of the sample. TEC oscillation frequency increased strongly with increasing temperature changing from ~10 Hz to above 2000 Hz in the temperature region 330-410 K. The oscillation frequency was an exponential function of inverse temperature yielding an activation energy of ≃0.8 eV (Fig. 2).

The model proposed to explain thermally activated oscillations, i.e. to realize a voltage controlled negative resistance characteristic in SI GaAs, relies on an electric field, enhanced capture of electrons

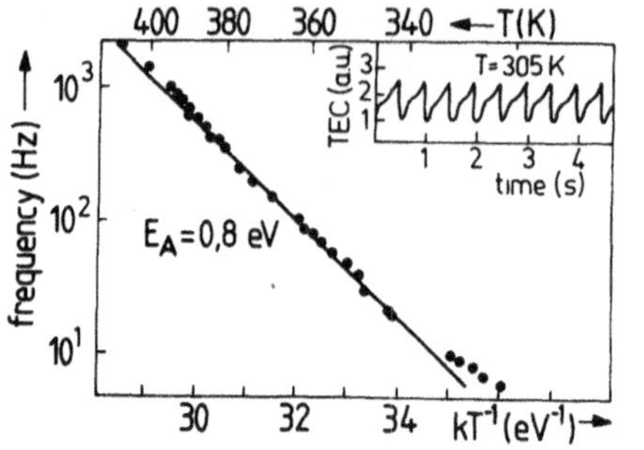

Fig. 2. Temperature dependence of the thermal equilibrium current oscillation frequency. Inset: TEC oscillations at T = 305 K, DC electric field E = 500 Vcm^{-1}

by the major deep donor in GaAs EL2. The negative resistance due to the field - enhanced recombination of photogenerated carriers has been originally proposed in ref. [8] in order to explain low frequency oscillations observed under illumination. It has been pointed out that the rate of capture of electrons by a centre with repulsive potential barrier should increase as the field rises. However, the EL2, a dominant deep level in GaAs, is a donor and in its neutral or ionized state it is not surrounded by a potential repulsive for electrons. Accordingly, the enhanced trapping of electrons by this level requires a different explanation, such as configurational barrier between the conduction band and EL2 level. In this model some excited electron-phonon states of the conduction band can act as an efficient channel of electron transfer from conducting to trapped level. These states can be reached by hot electrons, so field - enhanced recombination takes place. The resulting voltage - controlled negative resistance is a sufficient condition for current oscillations [6].

Assuming a linear dependence of electron energy on the applied electric field, it is possible to estimate (from Gunn oscillation electric field threshold, reported to be ∿3000 Vcm^{-1} [2]) the energy position of the states in the conduction band responsible for the enhanced trapping process. It should be ∿6 times smaller than the energy distance between Γ and L minima of GaAs conduction band, which gives ∿60 meV above the bottom of the conduction band. A configurational representation of EL2 proposed in ref. [14] assumes a barrier for electron capturing of about 70 meV, i.e. very close to a value estimated above. With the aid of a configurational model of Fig. 3 the quasi-steady state oscillations are explained as a dynamic balance between the thermal emission and electron capture process of the EL2. The thermal emission requires overcoming of the activation energy E_A and the emission rate is unaffected by an electric field of 10^3 V/cm. (High field effects /tunneling and Frenkel- Poole effect/ on emission rate become noticeable for fields about 2 orders of magnitude larger). The trapping process of low energy conduction electrons (minimum of the upper pa-

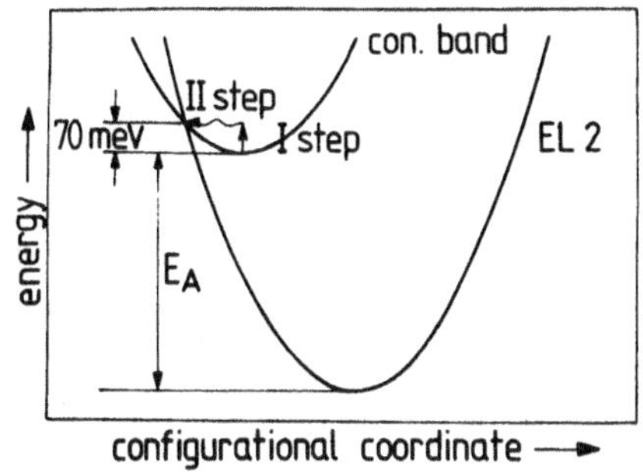

Fig. 3. The proposed model for voltage-controlled negative resistance in SI GaAs. Electron-trapping on the EL2 level is enhanced by electric field "heating" of electrons step I, which can then overcome the configurational energy barrier by releasing energy to the lattice step II

rabole in Fig. 3) proceeds via very inefficient radiative recombination process characterized by the capture cross section of about 10^{-17} cm^2 [12]. Electrons heated by the electric field (Step I, Fig. 3) can overcome the configurational barrier by releasing energy to the lattice (Step II, Fig. 3). This leads to the field - enhanced trapping by the EL2 what recombined with unchanged emission rate results in a decrease in the concentration of conduction band electrons in the electric field and thus in a negative resistance.

From the treatment of ref. [7] it follows that the velocity of high field domain movement and the oscillation frequency should be proportional to μn, where μ is the electron mobility and n is the concentration of free electrons. In the temperature range 300-410 K the electron mobility in SI GaAs varies only slightly. The frequency of current oscillations should increase with temperature similar to the increase of electron concentration, so with an activation energy of about 0.8 eV. The oscillation frequency dependence on invert temperature shown in Fig. 2 proofs that it is the case, so TEC oscillations occur under quasi-steady state conditions where a dynamic balance exists between the thermal ionization and the electron capture process of EL2.

The TSC oscillations are explained in a similar way to that of TEC. Thus, electrons thermally released from the E_c - 0.34 eV and E_c - 0.4 eV traps are capture by the EL2 level and because of field enhanced capture the TSC oscillations occur.

In summary, we have shown that the activated capture properties of the deep donor level (EL2) are responsible for a negative resistance characteristic of undoped SI GaAs leading to low frequency current oscillations controlled by thermal emission of electrons from deep levels. It is probable that the EL2 level, almost always present in melt-grown GaAs, is also responsible for the optically activated low-frequency current oscillations previously attributed to repulsive potential centers.

References

1. J. Lagowski, H.C. Gatos, J.M. Parsey, K. Wada, M. Kamińska and W. Walukiewicz, Appl. Phys. Lett. 40, 342 (1982)
2. J.B. Gunn, Solid State Commun. 1, 88 (1963)
3. A. Barraud, C.R. Acad. Sci. Paris 256, 3632 (1963)
4. K.W. Boer, Proc. 7th Int. Conf. on Phys. of Semicond., Paris 1964 (Paris: Dunod) p. 987
5. B.K. Ridley and R.G. Pratt, J. Phys. Chem. Solids 26, 21 (1965)
6. B.K. Ridley and T.B. Watkins, Proc. Phys. Soc. (London) 78, 293 (1961)
7. B.K. Ridley, Proc. Phys. Soc. (London) 82, 954 (1963)
8. B.K. Ridley and T.B. Watkins, J. Phys. Chem. Solids 22, 155 (1961); B.K. Ridley and T.B. Watkins, Proc. Phys. Soc. (London) 78, 710 (1961)
9. B.K. Ridley, Brit. J. Appl. Phys. 17, 595 (1966)
10. B.K. Ridley, J. Phys. C: Solid St. Phys. 7, 1169 (1974)
11. G.M. Martin, J.P. Farges, G. Jacob, J.P. Hallais and G. Poiblaud, J. Appl. Phys. 51, 2840 (1980)
12. G.M. Martin, Appl. Phys. Lett. 39, 747 (1981)
13. G.M. Martin, A. Mitonneau and A. Mircea, Electron. Lett. 13, 191 (1977)
14. G. Vincent and D. Bois, Solid State Commun. 27, 431 (1978)

MAIN ELECTRON TRAPS IN GaAs: AGGREGATES OF ANTISITE DEFECTS

T. Wosiński, A. Morawski and T. Figielski
Institute of Physics
Polish Academy of Sciences
Al. Lotników 32, 02-668 Warszawa, Poland

The origin of the main electron trap (commonly labelled EL 2) in GaAs crystals is still not definitely established, despite of the importance of this trap in forming the properties of GaAs. Several kinds of point defects were invoked in the past to account for the nature of EL 2 trap (among them gallium vacancy and oxygen impurity have been considered as the most serious candidates), but no strong arguments were found in favour of any kind of considered defects; moreover, the oxygen impurity has been definitely excluded by separate experiments [1].

The last experimental results suggest that arsenic antisite defect, As_{Ga}, may be responsible for occurring of the EL 2 trap [2,3,4]. The As_{Ga} is expected to behave as a double donor and, when singly ionized, to exhibit a paramagnetic behaviour. The As_{Ga}^{4+} paramagnetic center has been recently identified by the EPR due to its characteristic hyperfine structure [5]. The concentration of those centers has been found to increase considerably after plastic deformation of the crystal [6]. It has been also found that the EPR signal, originating from As_{Ga}^{4+} centers is quenched when the sample is irradiated with photons of energies exceeding two threshold values: $h\nu_1$ = 0.75 eV and $h\nu_2$ = 1.05 eV [7]. The first threshold energy can be attributed to the electron transition from the valence band edge to the upper donor level, E_2: $h\nu_1 = E_2 - E_c$. Consequently, the ionization energy of the second electron from the center, $E_c - E_2$, would be equal to $E_{gap} - h\nu_1 \approx 0.8$ eV. The latter value is in a good agreement with the activation energies of the EL 2 trap reported in literature.

In the studies being carried out in our lab we look for further correlations between the As_{Ga}^{4+} paramagnetic centers, detected by the EPR, and the main electron traps in GaAs, observed in the DLTS spectrum. We utilize the fact that the concentration of As_{Ga}^{4+} centers can be increased by the plastic deformation of the crystal. The crystals investigated were deformed by uniaxial compression at the temperature of 400 °C in the forming gas atmosphere.

The DLTS measurements have been done with LEC grown crystals having the initial electron concentration of $5 \cdot 10^{16}$ cm^{-3} and subjected to different plastic strains. It has been noticed, first of all, that EL 2 traps are hardly saturable while increasing the filling-pulse duration time, t_p, within a wide range of variation of t_p. Such a dependence, reported previously by Ferenczi [8] for GaP diodes, indicates to a barrier-limited capture rate at the traps. It is just the feature predicted for linear aggregates of centers, when the charge of electrons trapped at the centers gives rise to a macroscopic electrostatic potential linearly dependent on the charge density [4]. A slope of the amplitude-versus-ln(t_p) line should be proportional to the trap density and this relation brings independent information about the centers. An example of the behaviour described above is shown in Fig. 1 for two samples deformed to different strains.

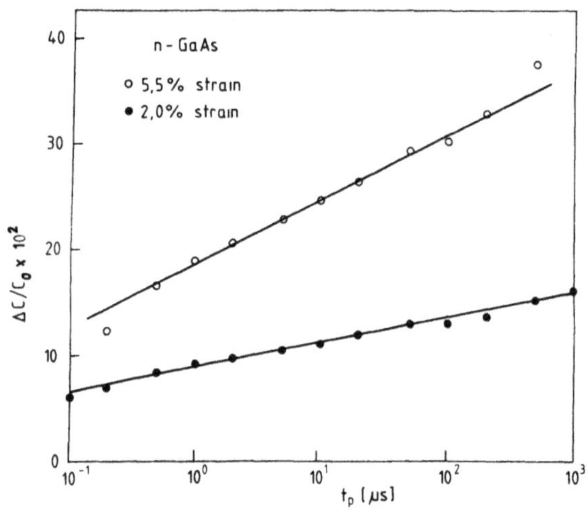

Fig. 1. Dependence of the DLTS peak amplitude for EL 2 trap versus logarithm of the filling-pulse duration time, t_p; $\tilde{e} = 1.4\ s^{-1}$, $V_R = -2.5$ V, pulse amplitude $V_p = 3$ V

Some quantitative results obtained by the DLTS measurements, as well as those obtained by the EPR, are shown in the tables. It is seen that the concentration of EL 2 traps grows up with the strain in a similar way as the concentration of As_{Ga}^{4+} centers, and that the values of both concentrations are comparable for comparable strains. The above correlation offers a strong argument that the EL 2 trap and As_{Ga}^{4+} center are due to the same defect.

The logarithmic-type dependence of the DLTS amplitude on the filling-pulse duration time suggests that the traps are not distributed randomly throughout the crystal but are rather arranged in rows. This feature, together with the pronounced effect of deformation on the trap concentration, points out a relation of the defects responsible for the traps to dislocations. On the other hand the antisite defects are expected to be produced in a considerable amount during the glide motion of dislocations [9].

Table 1

Concentration of EL 2 types (measured with t_p=200 μs)	
sample	N $[cm^{-3}]$
as-grown	$7 \cdot 10^{15}$
2 % strain	$1.3 \cdot 10^{16}$
5.5 % strain	$3.4 \cdot 10^{16}$

Table 2

Concentration of As_{Ga}^{4+} centers	
sample	N $[cm^{-3}]$
as-grown	$3 \cdot 10^{15}$
4 % strain	$2.0 \cdot 10^{16}$

References

1. A.M. Huber, N.T. Linh, M. Valladon, L.J. Debrun, G.M. Martin, A. Mtonneau, A. Mircea, J. Appl. Phys. 50, 4022 (1979)
2. J. Lagowski, H.C. Gatos, J.M. Parsey, K. Wada, M. Kamińska, W. Walukiewicz, Appl. Phys. Lett. 40, 342 (1982)
3. E.R. Weber, H. Ennen, U. Kaufmann, J. Windschief, J. Schneider, T. Wosiński, Appl. Phys. Lett. (in press)
4. T. Wosiński, A. Morawski, T. Figielski, Appl. Phys. (in press)
5. R.J. Wagner, J.J. Krebs, G.H. Stauss, A.H. White, Solid State Commun. 36, 15 (1980)
6. T. Wosiński, phys. stat. sol. (a) 60, K149 (1980)
7. T. Wosiński, Crystal Research and Technology 16, 217 (1981)
8. G. Ferenczi, I. Dózsa, Crystal Research and Technology 16, 203 (1981)
9. T. Figielski, Appl. Phys. (in press)

DEFECT REACTIONS IN GaP CAUSED BY ZINC DIFFUSION

P. Krispin and J. Maege

Central Institute of Electron Physics
of the Academy of Sciences of the GDR
1086 Berlin, Mohrenstr. 40/41, GDR

ABSTRACT

Shallow diffusions of zinc in GaP epitaxial layers were carried out varying the positions on the Ga-P-Zn phase diagram. Under special diffusion conditions zinc profiles were obtained as predicted by the subsitutional-interstitial theory in which the gallium vacancy equilibrium at the diffusion front is maintained. Otherwise, a tail was found in zinc profiles at concentrations below $5 \cdot 10^{16}$ cm^{-3} not reported till now in GaP epitaxial layers. It is demonstrated that the degradation behaviour of green-emitting diodes is determined by the occurence of zinc diffusion tails in the active zone.

1. Introduction

Zinc is the most common acceptor used in fabricating devices on the basis of III-V compounds. Nevertheless, there is no information concerning the role which the zinc profile plays for the long-term stability of diffused light-emitting diodes.

It is well known [1] that the zinc diffusion mechanism in GaP can be understood on the basis of the substitutional-interstitial model in which the equilibrium concentration of gallium vacancies is the controlling factor. The following defect reaction holds at equilibrium:

$$Zn_i^+ + V_{Ga} \leftrightarrow Zn_{Ga}^- + 2h^+$$

It is seen that gallium vacancies are used up, so the breakdown of vacancy equilibrium determines the actual shape of the zinc profile. In previous papers [2] it was argued that in the case of deep diffusions (depth above 10 μm) local equilibrium can be achieved by a kick-out mechanism or by climbing of dislocations. If the defect equilibrium at the diffusion front is maintained the profile is characterized by an abrupt fall of the zinc concentration down to zero at the diffusion front. Such profiles should yield abrupt p^+-n junctions as practical applications demand. But instead of a slope of greater than 10^{23} cm^{-4}

extracted from the calculated diffusion profile, values below $2.5 \cdot 10^{22}$ cm^{-4} have been measured by C-V technique [3].

Here, defects near the diffusion front have been detected. It will be shown that these defects cause an increased degradation rate of the quantum efficiency as well as a yellow shift of the emission spectra of green-emitting diodes.

2. Experimental Results and Discussion

Shallow diffusions of zinc in GaP were carried out in different regions of the gallium-phosphorus-zinc phase diagram [4] (Fig. 1). Without going into detail the regions of interest are those labelled 1, 2, and 3. Especially, regions 1 and 2 are monovariant regions, the diffusion conditions become fixed as soon as the temperature is fixed. The profile of the electrically active zinc atoms was monitored by combining differential Hall measurements and the C-V technique which stands for the measurement of the capacitance in dependence on the applied dc bias.

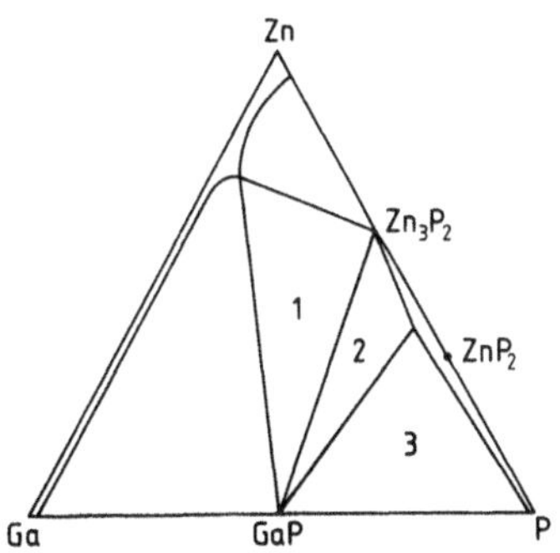

Fig. 1. Ga-P-Zn phase diagram [4]

Irrespective of the diffusion condition, we found that the zinc profile at high concentration can be explained by the substitutional-interstitial model as Fig. 2 shows. The measuring points obtained by differential Hall analysis of the p-side of the junction fit the theoretical curve very well. Therefore, it can be concluded that in the case of shallow diffusions where the junction depth is below 10 µm the equilibrium of gallium vacancies is maintained, probably due to the supply of vacancies from the surface.

In Table I data are collected concerning the high-temperature diffusion labelled A. A series of LPE samples was studied having net donor concentrations N_D^B in the bulk between 10^{16} and 10^{18} cm^{-3}. In all cases, a linear relationship between $1/C^2$ and the applied voltage was found. This behaviour is marked in Table I. Figure 3 demonstrates a typical curve. There is no deviation from the straight line and the extrapolated intercept on the voltage axis at about 2.0 V in agreement with the expected value. From this, the presence of an abrupt

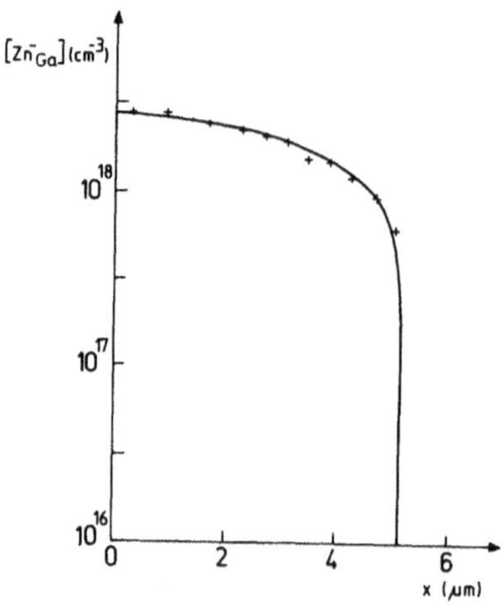

Fig. 2. Zinc acceptor profile (diffusion D): The theoretical curve was obtained with $D_S = 4.2 \cdot 10^{-12}$ cm²s⁻¹ and $C^S_{Zn} = 2.47 \cdot 10^{18}$ cm⁻³

p^+-n junction is indicated in each case. So, a sharp drop of the zinc profile follows from $5 \cdot 10^{18}$ down to 10^{15} cm^{-3} in agreement with the diffusion model. That is to say, neither donor- nor acceptor-like defects are additionally introduced into the n-side by the diffusion process A.

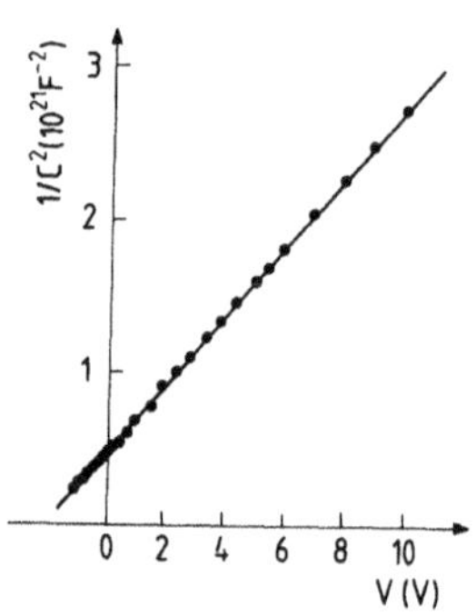

Fig. 3. C^{-2} vs. V plot (diffusion A)

Table II is a compilation of the results obtained by using the low-temperature diffusion processes labelled B and D. In region 2, again, plots of $1/C^2$ versus the bias V yield are straight lines, indicating abrupt junctions. On the other hand, in region 1 C^{-2} vs. V plots change slope as zero bias is approached indicating a change in doping level near the junction. In Fig. 4 the effective donor concentration N_{eff} of the n-side is plotted versus the position x. Two junctions are compared made on the same epitaxial wafer. It is clearly seen that the diffusion D leads to a change of the concentration near the junction whereas deeper inside the material the levels are equal. From Fig. 2 it follows that the p-n junction is located at the sharp drop of the zinc acceptor concentration. The change of the donor concentration is connected with an acceptor-like defect introduced by the diffusion process D. The defect intensity exponentially depends on the position (Fig. 5). In the case of diffusion D a p^+n^- junction arises. The main point is the much higher phosphorus vapour pressure in region 2 of the ternary phase diagram compared with region 1. Therefore, additional phosphorus vacancies are created during the diffusion process D, whereas in the other case no phosphorus vacancies exist near the diffusion front. Then, a satisfactory explanation of this tail would seen to be the formation of a nearest-neighbour complex involving a zinc atom on a gallium site and a phosphorus vacancy. This defect should be able to diffuse beyond the zinc acceptor front by a different diffusion mechanism.

Table I. High-temperature diffusion of Zn in GaP (A). Source ZnP_2; T_D = 855±5 °C; [Zn] = 1.2 mg/cm³ AV. Subsequent annealing procedure: 750 °C; 8h

Sample No.	Material	p-n junction depth [μm]	N_D^B [cm^{-3}]	$C^{-2} \sim V$	V_D [V]
1	LPE	5	$1.1 \cdot 10^{16}$	+	2.04
2	LPE	7	$1.3 \cdot 10^{17}$	+	2.05
3	LPE	5	$1.8 \cdot 10^{17}$	+	2.02
4	LPE	10	$2.4 \cdot 10^{17}$	+	2.00
5	LPE	5	$3.4 \cdot 10^{17}$	+	1.92
6	LPE	5	$5.7 \cdot 10^{17}$	+	2.00
7	LPE	7	$1.1 \cdot 10^{18}$	+	2.05
8	VPE	5	$2.0 \cdot 10^{16}$	+	2.00
9	VPE	5	$3.3 \cdot 10^{16}$	+	1.71
10	VPE	5	$6.4 \cdot 10^{16}$	+	1.95
11	LEC	5	$1.6 \cdot 10^{16}$	+	1.98

Table II. Low-temperature diffusion of Zn in GaP (B,D). Sources - Zn,P; Zn,GaP; Zn; Zn,P,GaP. T_D = 660 10 °C; [Zn] = 60 μg/cm³ AV

Sample	Material	Region of phase diagram	p-n junction depth [μm]	N_D^B [cm^{-3}]	$C^{-2} \sim V$	V_D [V]
12	VPE	2 (B)	4.9	$7.0 \cdot 10^{16}$	+	2.00
13	VPE	2 (B)	3.5	$5.4 \cdot 10^{16}$	+	2.36
14	VPE	2 (B)	5.9	$6.4 \cdot 10^{16}$	+	1.93
15	VPE	1 (D)	5.5	$7.0 \cdot 10^{16}$	-	-
16	VPE	1 (D)	5.5	$5.0 \cdot 10^{16}$	-	-
17	VPE	1 (D)	5.2	$2.0 \cdot 10^{17}$	-	-
18	VPE	1 (D)	5.0	$4.7 \cdot 10^{16}$	-	-
19	VPE	1 (D)	5.5	$3.5 \cdot 10^{17}$	-	-

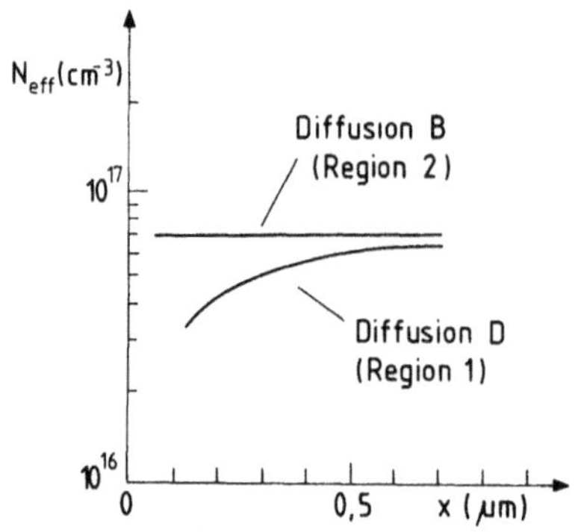

Fig. 4. N_{eff} vs. x plots of two diodes made on the same VPE wafer

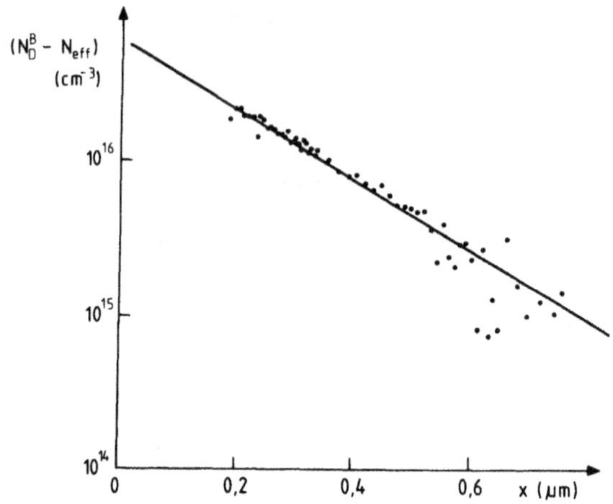

Fig. 5. $(N_D^B - N_{eff})$ vs. x plot (diffusion D)

Table III. Comparison of LED data. (GaP diodes fabricated on the same wafer by low-temperature diffusion)

	Diffusion B (Region 2) p^+-n junction		Diffusion D (Region 1) p^+n^- n junction	
	virgin	degraded (2500 h; 50 Acm^{-2})	virgin	degraded (2000 h; 50 Acm^{-2})
η/η_o	1.00	0.81	1.00	0.43
EL	(Fig. 6)	no yellow shift	(Fig. 7)	yellow shift
C-V	constant		constant	
defects	defect-"free"		V_P, $(Zn_{Ga}V_P)$	

Finally, results are reported concerning the influence of phosphorus vacancies on the properties of green emitting diodes. Again, diodes fabricated on the same wafer were selected in order to compare the emission spectra and the external quantum efficiency during operation. Typical data are given in the Table III. The mormalized efficiencies η/η_o after degradation differ by a factor of 2. The behaviour of the electroluminescence spectra of both types of diodes is also different. p^+-n junctions show constant electroluminescence spectra during operation (Fig. 6). Diodes having a tail within the zinc profile exhibit a yellow shift of the spectra as usually found (Fig. 7). Such

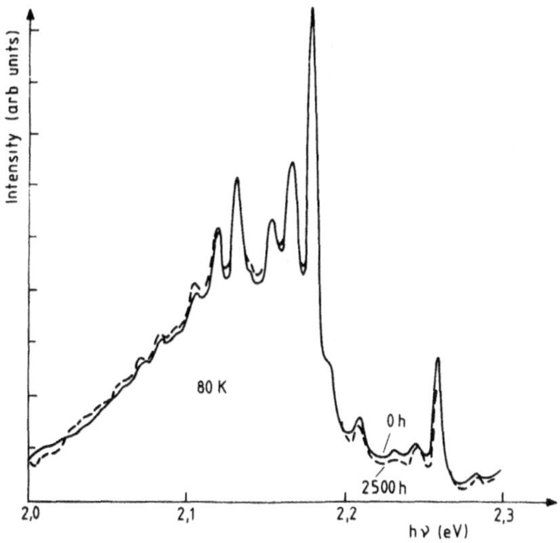

Fig. 6. EL spectra of a p^+-n junction (diffusion B) during operation

spectra are typical of highly nitrogen-doped GaP [5]. It is known [6] that during operation nitrogen atoms change their positions from non-radiative configurations in the lattice to neighbouring phosphorus sites generating new radiative channels. This change of position is only possible if phosphorus vacancies are present as in the case of p^+n^-n junctions.

The experimental results indicate that the degradation behaviour of green-emitting diodes is partly determined by the occurrence of phosphorus vacancies at the diffusion front. During operation additional defect reactions take place if phosphorus vacancies were introduced into the n-side of the junction by the zinc diffusion process.

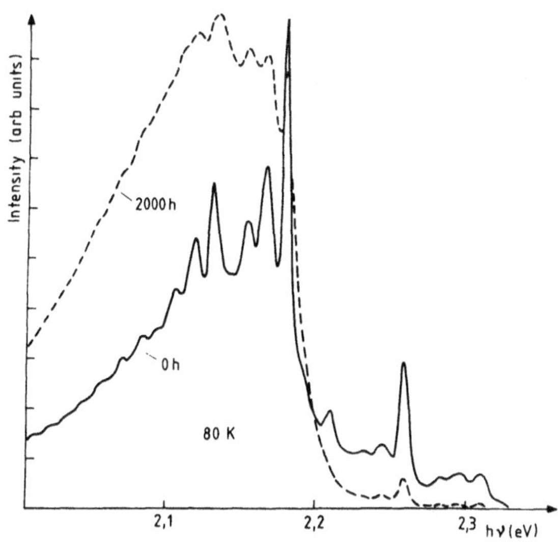

Fig. 7. EL spectra of a p^+n^-n junction (diffusion D) during operation

Acknowledgements

The authors acknowledge the contributions of E. Nebauer and H. Treptow. This work was partially supported by the Werk für Fernsehelektronic - Berlin (GDR).

References

1. See, e.g. H.C. Casey and G.O. Pearson, in "Point Defects in Solids" (Ed. J.H. Crawford and L.M. Slifkin) Vol. 2, 163 (1975)
2. R.K. Ball, P.W. Hutchinson and P.S. Dobson, Phil. Mag. A 43, 1299 (1981);
   U. Gösele and F. Morehead, J. Appl. Phys. 52, 4617 (1981)
3. J.C. Verplanke, J. Electrochem. Soc. 124, 469 (1977);
   B.W. Thomas, G.J. Rees and D.R. Wright, Solid State Electronics 23, 611 (1980)
4. M.B. Panish, J. Electrochem. Soc. 113, 224 (1966)
5. G.B. Stringfellow, M.E. Weiner and R.A. Burmeister, J. Electronic Materials 4, 363 (1975)
6. G. Ferenczi, P. Krispin and M. Somogyi, to be published

NONSTATISTICAL DEFECT SURROUNDINGS IN MIXED CRYSTALS - THE SELFACTIVATED LUMINESCENCE CENTRE IN ZnS_xSe_{1-x}

R. Mach and G. O. Müller

Zentralinstitut für Elektronenphysik
der Akademie der Wissenschaften der DDR
1086 Berlin, Mohrenstraße 40/41, GDR

Mixed crystals are special cases of disordered systems. While the lattice is periodically preserved, the composition in at least one sublattice - in ZnS_xSe_{1-x} in anion one - is thought to be randomly mixed. This is meant in the sense that by definition a "mixed crystal" is formed if the free enthalpy of a system with a statistically occupied sublattice is minimum. An alternative is the non-miscibility of certain systems which shows up in phase separation or segregations.

One obvious question arises if a miscible system is perturbed by doping: Does the dopant atom tend to organize its microsurrounding? A completely understood thermodynamics of doping in the two binary compounds would allow to answer the question whether the system could gain energy from the preference of one surrounding component.

The selfactivated centre in ZnS_xSe_{1-x} is one of the best investigated deep luminescence centres, giving rise to broad yellow luminescence (2.0 eV) in ZnSe and a blue emission (2.6 eV) in ZnS [1]. At low temperatures the luminescence has been shown to be donor-acceptor pair emission [2]. The chemical and crystallographic nature of the SA-centre had been identified using paramagnetic resonance by Schneider et al. [3]. The model has been confirmed recently by ODMR-investigations [4,5].

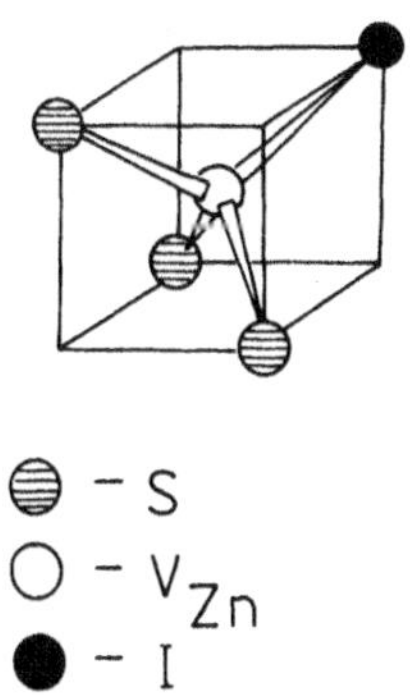

Fig. 1. The selfactivated centre in ZnS doped with elements from group VII

Figure 1 shows, as an example, the SA-centre in ZnS doped with elements from group VII. This centre consists of a Zn^{2+} vacancy with an isolated substitutional donor on one of the nearest neighbour anion sites and it is principally impossible to avoid this centre if there is a high concentration from group VII. Fonger [6] postulated a long time ago from photoluminescence investigations on ZnS_xSe_{1-x} powders the existence of four different SA-centres with one iodine on a V_{Zn} nearest site due to nearest neighbour splitting (Fig. 2). These are all possibilities of constituting nearest V_{Zn} surroundings from the available chalcogenes. In any further approximation the question arises over what distance the statistical composition is re-achieved around the centre which is far from x. Schneider et al. [7] investigated ZnS_xSe_{1-x} (x = 0.07) doped with Cl or Br by EPR and found evidence for the existence of $V_{Zn}S_3Cl$, $V_{Zn}S_2SeCl$, and $V_{Zn}SSe_2Cl$ centres (the same for Br). More recently Murcia et al. [8] investigated the SA-luminescence of $Zn_{0.52}Se_{0.48}$. The authors concluded that the SA-emission consists only of a single band.

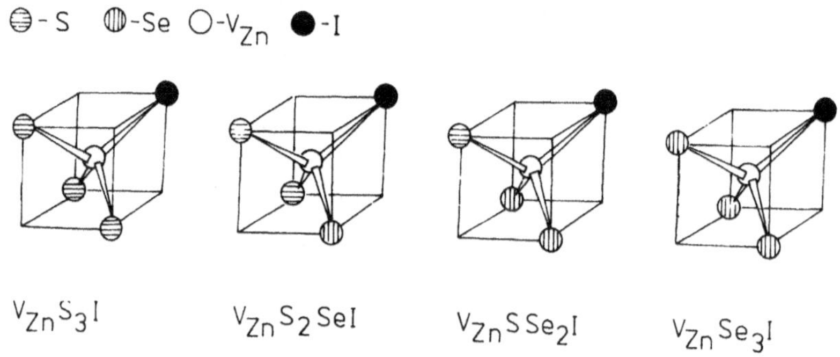

Fig. 2. The zinc vacancy (V_{Zn}) in ZnS_xSe_{1-x} mixed crystals can be surrounded by 3S + 1I, 2S + 1Se + 1I, 2Se + 1S + 1I, 3Se + 1I atoms

Our experimental results obtained on ZnS_xSe_{1-x} single crystals highly doped with iodine on the other hand clearly evidence the occurance of different subbands for each composition (Fig. 3), which can be selekted by the choice of the excitacion wavelength. This method has been widely used in various inhomogeneously broadened emissions and is usually termed "site selective spectroscopy" or "laser narrowing". In our case "defect or centre selective spectroscopy" would be adequate, as in the range of the excitacion wavelengths and/or x-values the same emission components are observed [9]. In Fig. 3 tentative assignements of the emission bands to the various SA centre are given (compare to Fig. 2). Assuming now that a random distribution of Se and S atoms in the relative concentrations 1-x/x is obeyed even in the anion configuration shell next to an iodine paired zinc vacancy, the relative concentrations of the four different SA centres must be the ones given in Fig.4.

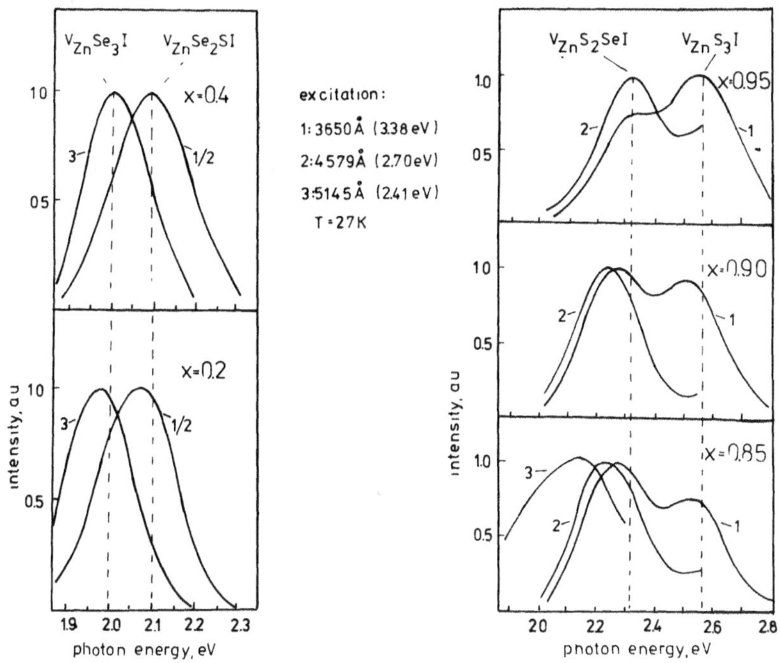

Fig. 3. Spectral distributions of photoluminescence of ZnS_xSe_{1-x} single crystals excited at different wavelength

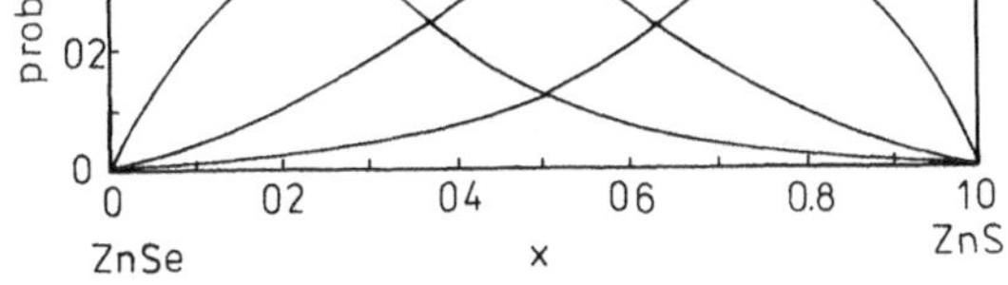

Fig. 4. Statistical probabilities for finding in the first anion shell around the fourfold coordinated zinc vacancy 3S + 1I, 2S + 1Se + + 1I, 2Se + 1S + 1I, 3Se + 1I atoms

Assuming for a moment furthermore that the emission probabilities of all SA centres are the same, the emission spectra resulting from short wave excitation can readily be calculated using the maximum position and halfwidth experimentally measured under selective excitation (see Fig. 3) and assuming the relative intensities to be proportional to the concentrations from Fig. 4. This has been done in Fig. 5. The obvious, strong deviation of the theoretical curve from the experimental (365 nm ex-

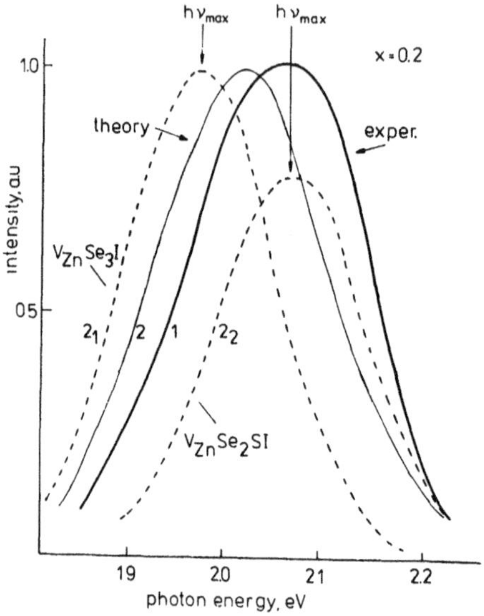

Fig. 5. Spectral distributions of photoluminescence, T=27 K; curve 1 - experimental results, excitation wavelength 365 nm, curve 2 - theory, expected from statistics (Fig. 4) by superposition of two Gaussian bands (2_1, 2_2)

citation) spectra can be interpreted as an indication of non-statistical abundances of $V_{Zn}Se_3I$ and $V_{Zn}Se_2SI$ centres in this crystal with composition x = 0.2. The only alternative explanation would be the assumption of differing hole cross-sections for both centres resulting in different excitation rates. This latter possibility has been checked by photocapacitance spectroscopy [10].

The assumption of equal recombination (emission) rates can easily be proved by measuring the luminescence decay time. In general a non-exponential decay is observed, as expected for a donor-acceptor pair recombination mechanism [2]. Fig. 6a shows results of these measurements at various spectral positions within the band shown in Fig. 3 after pulse excitation. High-lighted is the decay curve for the emission maximum. The increasing emission rate with increasing photon energy is expected again from the donor-acceptor pair mechanism [2]. Directly related to the assumption to be checked are the results given in Fig. 6b. The very close correspondance of the decay curves at the emission maxima of the two excitation narrowed lines of Fig. 4 confirms the assumption about equal transition probabilities for both centres.

Thus in conclusion one might state that very strong arguments have been given for the non-random nature of the various SA-centres in ZnS_xSe_{1-x} at least in the case of x = 0.2. In all other cases investigated [9] no evident exception has been observed from the general behaviour described, so that the highly probable case of segregation around impurities might be regarded as realized in the ZnS_xSe_{1-x}: I system. Thermodynamical considerations will hopefully show at least trends into the right direction in the next future.

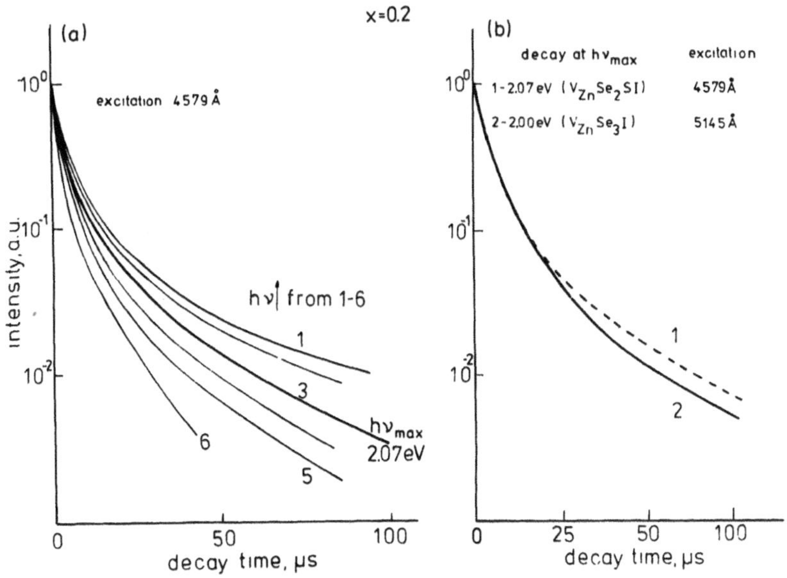

Fig. 6. (a) Photoluminescence decay curves for several photon energies in the emission band of the $V_{Zn}Se_2SI$ centre (see Fig. 3) after pulsed excitation with 4579 Å. (b) Photoluminescence decay curves for excitation at different wavelengths (resulting in the excitation of the $V_{Zn}Se_3I$ and $V_{Zn}Se_2SI$-centre respectively) and measurement of the decay at $h\nu_{max}$ of the corresponding emission bands (see Fig. 5)

References

1. H.V. Leverenz, an Introduction to Luminescence of Solids (John Wiley & Sons, New York, 1950)
2. K. Era, S. Shionoya, Y. Washizawa, J. Phys. Chem. Solids 29, 1827 (1968)
3. J. Schneider, A. Räuber, B. Dischler, T.L. Estle, W.C. Holton, J. Chem. Phys. 42, 1839 (1965)
4. J.R. James, J.E. Nicolls, B.C. Cavenett, J.J. Davies, D.J. Dunstan, Solid State Commun. 17, 969 (1975)
5. J.E. Nicholls, J.J. Davies, B.C. Cavenett, J.R. James, D.J. Dunstan, J. Phys. C 12, 361 (1979)
6. W.H. Fonger, Phys. Ev. 137, A1038 (1965)
7. J. Schneider, B. Dischler, A. Räuber, J. Phys. Chem. Sol. 31, 337 (1970)
8. M. Murcia, D. Block, D. Etienne, J.P. Fillard, J. Phys. Chem. Solids 41, 1247 (1980)
9. R. Mach, P. Flögel, L.G. Suslina, A.G. Areshkin, J. Maege and G. Voigt, phys. stat. sol. (b) 109, 607 (1982)
10. H.G. Grimmeiss, E. Meijer, R. Mach and G.O. Müller, J. Appl. Phys. (to be published)

PART V

Si/SiO_2 INTERFACE

STRUCTURE AND PROPERTIES OF THE $Si-SiO_2$ INTERREGION

K. Hübner

Sektion Physik der Wilhelm-Pieck-Universität Rostock
Universitätsplatz 3, 2500 Rostock 1, GDR

1. Introduction

The interface between silicon and its thermal oxide determines the operation and quality of various microelectronic devices. This so-called $Si-SiO_2$ interface is, however an extended interregion, which involves the structural, chemical, and physical transition from the crystalline semiconductor silicon to the amorphous insulator SiO_2 [1]. The chemical transition is characterized by an SiO_x layer (Fig. 1). The structure and properties of this interregion depend on the starting-materials, the preparation conditions, the after-treatment, and they can be influenced by measurement and application processes.

I discuss in this paper the variation of the chemical composition in the $Si-SiO_2$ interregion, the structure of different SiO_x forms, electronic properties of the material SiO_x, and typical microscopic defects of the interregion.

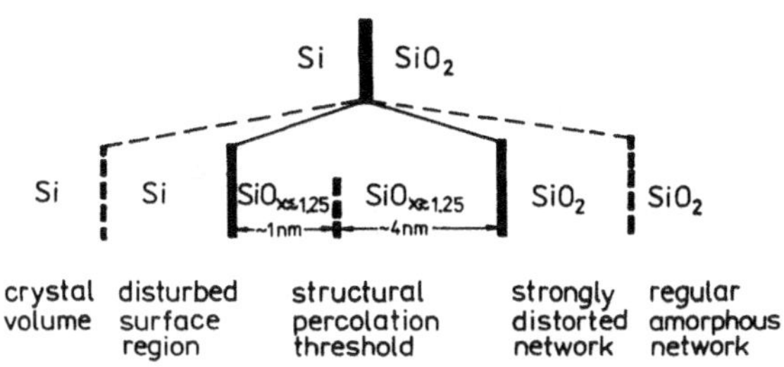

Fig. 1. The probable form of the subdivision of the $Si-SiO_2$ interregion. Between the silicon crystal with its disturbed surface region and the distorted SiO_2 network an SiO_x transition layer is located, the extension and structural percolation threshold of which is discussed in the text

2. SiO_x Transition

The results of various experimental investigations [2] show that the chemical composition x of the SiO_x layer within the $Si-SiO_2$ interregion changes more or less continuously from the value x=0 of silicon to the value x=2 of SiO_2. This chemical transition needs for its realization a finite distance.

This distance manifests itself through the microphysical properties of the material SiO_x. Therefore, distinct experimental methods may lead to different results concerning the thickness of one and the same SiO_x layer. Taking into account additionally the different resolution of distinct methods, a value of approximately 1 nm may be considered as the thickness of the SiO_x layer in which the essential transition from the electronic properties of silicon to the electronic properties of SiO_2 takes place. This does not mean, however, that the perfect stoichiometry of SiO_2 is reached at a distance of 1 nm from the silicon surface, as it will be shown below.

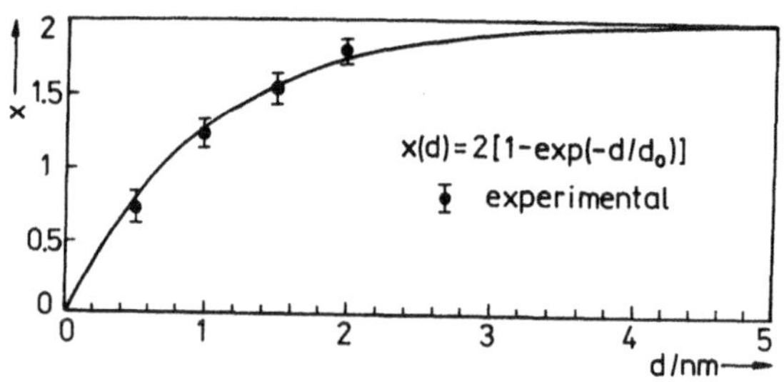

Fig. 2. Idealized profile of the chemical composition x of the SiO_x transition within the Si-SiO_2 interregion in dependence on the distance d from the silicon surface (1). Some experimental points, obtained from a corresponding analysis of angle resolved XPS data for Si (111) surfaces [3], are shown for comparison

Even more difficult than the determination of the thickness of the SiO_x transition is the determination of the profile of the oxygen concentration within this transition. From a consideration of the kinetic processes of the silicon oxidation we obtain for the relative oxygen concentration in dependence on the distance d from the silicon surface the approximative relation (Fig. 2)

$$x(d) = 2[1-\exp(-d/d_o)], \quad (1)$$

which does not include effects due to surface roughness and impurity segregation. The characteristic length d_o depends on the oxidation conditions, the orientation of the silicon surface considered and other parameters in such way that the thickness of the SiO_x layer decreases with increasing oxidation temperature and is greater for (111) surfaces than for (100) orientation. An analogous trend is known for interface states and oxide charges. This is, however, not very surprising, since these defects are immediately connected with the existence and the properties of the interregion.

3. Structure of SiO_x

For the investigation of the structure and properties of SiO_x it is useful to consider this geometrically *and* chemically disordered material as abstracted from the Si-SiO_2 interregion and to perform experiments with independent SiO_x films. With the help of the optical absorption at oblique incidence of infrared light polarized in the plane of incidence [4], we are able to detect splittings between transversal (TO) and longitudinal optical (LO) phonon modes and their chemical shifts in thin SiO_x films (Fig. 3). The measurement of TO *and* LO modes in a

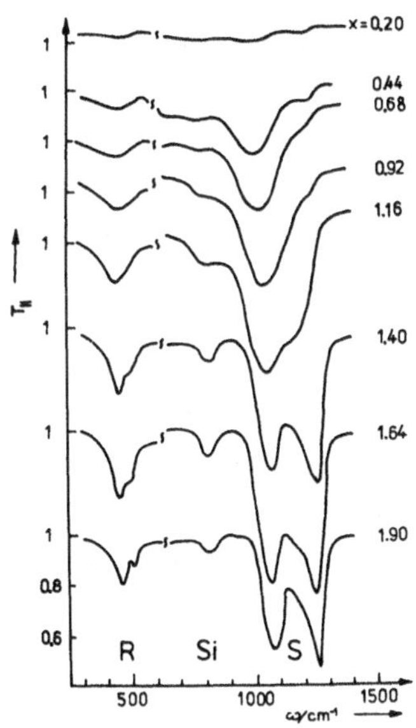

Fig. 3. Transmission ($T_{\parallel}$, in relative units) of reactively sputtered SiO_x films for oblique incidence (55°) of infrared light polarized in the plane of incidence [7]. For each spectrum the ordinate scale is the same. At 630 cm^{-1} a grid change is indicated. The three main features of the spectra are induced by the streching (S) and rocking (R) mode of oxygen and by a motion of Si atoms. The LO-TO splitting is resolved for the two characteristic oxygen modes. From the (almost linear) chemical shift of the TO peaks with x follows the random-bonding character of the structure of this SiO_x form. The relation between the stretching frequency and the bonding angle of oxygen leads to the result, that this angle decreases with decreasing value of the average chemical composition x. Further conclusions concerning the atomic structure and the scattering of charge carriers can be drawn from the LO modes

simple infrared experiment is possible due to the Berreman effect [5], which we have detected for the first time in *amorphous* thin films. Basing on cluster-Bethe-lattice calculations of the phonon density of states [6] the analysis of the experimental phonon spectra leads to very direct conclusions concerning the atomic structure of SiO_x [7].

From these conclusions as well as from the results of high resolution diffraction measurements (Fig. 4, [8]) and photoemission experiments [3] follows that all SiO_x forms investigated until now can be approximatively described with the help of a microscopic and macroscopic mixture model for the structure of SiO_x. The macroscopic mixture model assumes a simple mixture of silicon and SiO_2, whereas the microscopic mixture model is based on a statistical distribution of Si-Si

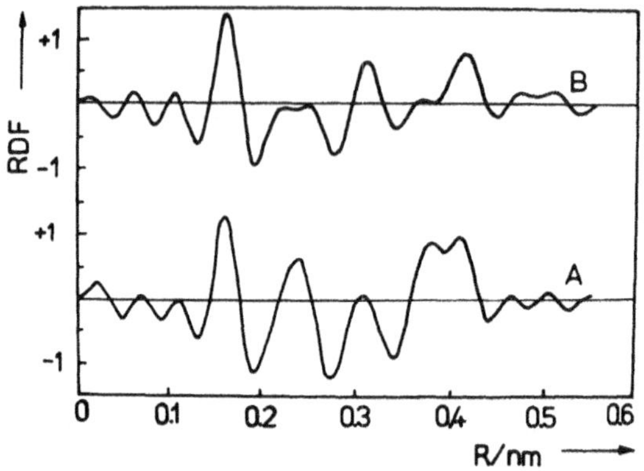

Fig. 4. Structure-dependent part of the radial distribution function (RDF, in arbitrary units) of commercial silicon monoxide (A) and of an SiO_x sample, which was prepared from the monoxide powder by means of electron-beam evaporation (B). Although the x-values of these two samples are very similar, namely $x \approx 1$, the structures are different due to the distinct preparation methods. The high resolution was obtained with the help of a high-intensity X-ray generator with rotating anode and data accumulation [8]

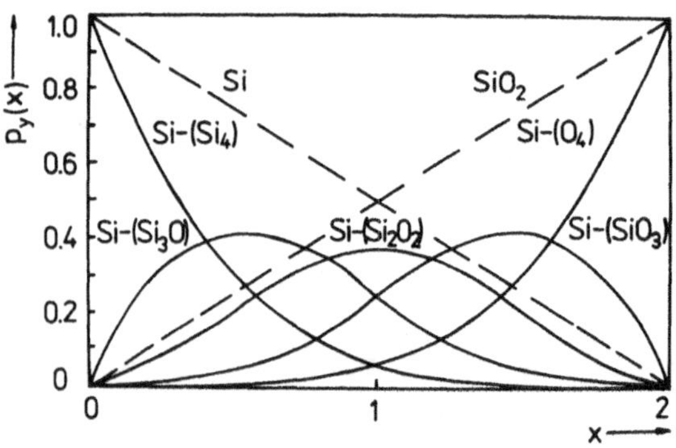

Fig. 5. Distribution probabilities $p_y(x)=(4!/(4-y)!y!)(x/2)^{4-y}(1-x/2)^y$ of the substitution tetrahedra $Si\text{-}Si_yO_{4-y}$ with y=0,...,4 of the random-bonding model in comparison with the linear interpolation of the macroscopic $Si\text{-}SiO_2$ mixture model. The macroscopic mixture model considers only $Si\text{-}Si_4$ and SiO_4 constituents, whereas in the random-bonding model also the intermediate configurations are taken into account

and Si-O bonds (random-bonding model). Both these models can be described by means of substitution tetrahedra $Si\text{-}Si_yO_{4-y}$, the distribution probabilities of which (Fig. 5) characterize the statistical model considered [9]. The comparison of the x-dependence of these theoretical distribution probabilities with the corresponding x-dependence deduced from the experimental results mentioned above yields the following relations between the structure and the preparation method of different SiO_x forms.

(1) SiO_x in the interregion of silicon and its thermal oxide as well as SiO_x forms prepared by means of electron-beam evaporation or reactive sputtering show a random-bonding behaviour.

(2) Chemically vapour-deposited films, SiO_x prepared with the help of ion implantation (oxygen into silicon) and all SiO_x forms at sufficiently high temperatures reveal a macroscopic mixture structure. The latter fact demonstrates the metastable character of SiO_x.

(3) Intermediate forms are possible. Commercial monoxide powders from distinct sources, for example, display different degrees

Table 1. Relations between methods of preparation and the atomic structure of SiO_x

SiO_x preparation method	Structure random-bonding	macroscopic mixture
thermal oxidation	x	
electron-beam evaporation	x	
reactive sputtering	x	
commercial powder preparation	x	x
chemical vapour deposition		x
ion implantation		x
high-temperature annealing		x

of dissolution into silicon and SiO_2. These results are summarized in Table 1.

4. Electronic Properties of SiO_x

Even in SiO_x forms with a statistical distribution of Si-Si and Si-O bonds, these bonds retain essentially their properties of the limiting materials, i.e., of silicon and SiO_2. This is valid, for example, for the bond length and the electronic charge density of the chemical bonds [8,9]. Therefore, SiO_x may be considered as an inhomogeneous medium on an atomic (bonding) scale, the smallest subunits of which are the Si-Si and the Si-O bonds. Total electronic properties, however, reveal an ensemble behaviour, which is characterized by the fact that they vary quantitatively and qualitatively according to the macroscopic mixture and the microscopic random-bonding model, respectively. Not only the relative numbers of chemical bonds, but also their mutual arrangement (e.g., in special sequences) determine the total electronic properties. Basing on this observation, the application of the theory of inhomogeneous media yields, for example, the real (ε_1) and imaginary (ε_2) part of the dielectric function $\varepsilon(E)$ of SiO_x in the energy range from $E \approx 0$ to approximately 9 eV in excellent agreement with corresponding experimental results (Fig. 6, [10]).

For inhomogeneous media percolation effects can be expected. A simple percolation theory leads to the prediction of a percolation threshold in the structure of random-bonding forms of SiO_x at $x_c = 1.25$, which can be indeed observed in the x-dependence of various electronic properties [9]. The physical meaning of this structural percolation threshold follows from the fact that the relative number of Si-Si bonds, which must be involved for the construction of representative bonding sequences, exceeds a critical value at x_o. The effect of the percolation threshold can be seen, for example, in the x-dependence of ε_1 and ε_2, where the

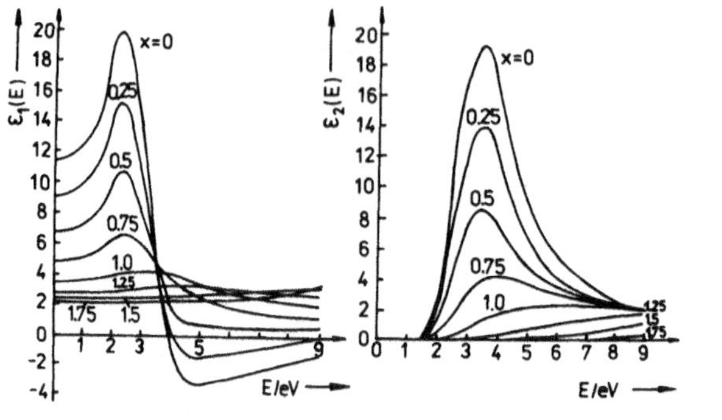

Fig. 6. Real (ε_1) and imaginary (ε_2) part of the electronic dielectric function of random-bonding SiO_x in dependence on the energy E. For the energy-dependent dielectric functions of the Si-Si and Si-O bonds we have used the corresponding macroscopic dielectric functions of Si and SiO_2, respectively, which were mixed according to the statistical distribution of the chemical bonds [10]

influence of the Si-Si bonds on the line shape vanishes only for $x \gtrsim 1.25$.

These and other electronic properties of SiO_x (the electronic valence charge density, the energy of the Si 2p level, and characteristic energies of optical constants [9]) show a strong variation with x within the range from x=0 to $x \approx 1.25$, and the electronic properties of SiO_2 are almost completely reached already at $x \approx 1.5$ [9]. Therefore, the thickness of the SiO_x transition of Si-SiO_2 interregions must be greater than commonly deduced from its electronic properties (see also Fig. 2). This conclusion is in agreement with the observation, that typical interface defects can be also more distant from the silicon surface [1]. An opposite trend in dependence on x (but with the same position of the threshold) is revealed by the absorption coefficient and the corresponding optical gap of the electronic density of states (Fig. 7), which shows a relatively weak variation within the range from x=0 to $x \approx 1.25$ due to the strong polarizability of the Si-Si bonds. This does not exclude, however, that the

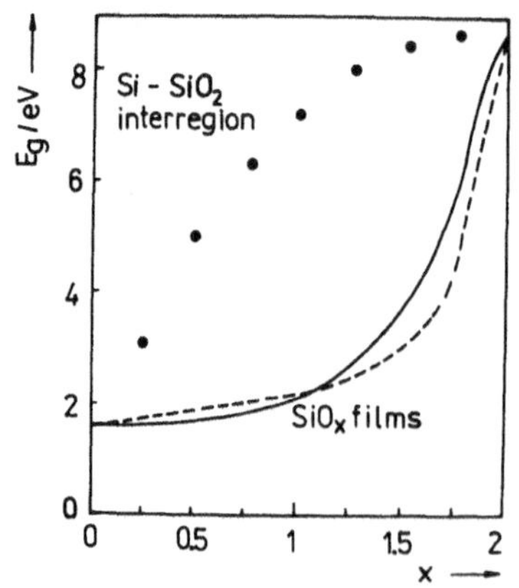

Fig. 7. Optical gap of SiO_x films with random-bonding structure (—— theoretical, determined from the electronic dielectric function shown in Fig. 6 [10]; --- experimental [11]) in comparison with the variation of the gap through the Si-SiO_2 interregion, calculated with the help of simple interface models (···, [12])

conductivity gap in the $Si-SiO_2$ region with its special gradient of the chemical coordinate x has the same barrier behaviour as described above, i.e., with a fast development of SiO_2 properties.

5. Defects of the $Si-SiO_2$ Interregion

The most important intrinsic defects of the $Si-SiO_2$ interregion are defects of the SiO_x network, which can be described in terms of a valence-alternation-pair model [1,13]. This model considers pairs of over- and under-coordinated oxygen atoms (O_3^+ and O_1^-, respectively) in connection with Si atoms and their dangling bonds. Including the most important impurities of the $Si-SiO_2$ interregion, i.e., hydrogen and sodium, this model leads for the first time to a consistent explanation of the microscopic reasons, properties and correlations of interface states, fixed oxide charges, and radiation defects of MOS sturtures [1].

The O_3^+ centers are probably the main reason of the positive insulator charges and a corresponding electron trapping. The O_1^- defects are effective trapping centers for holes and sodium atoms. For the correlation within the defect pairs, the concentration of which increases with irradiation, and from the correlation between these pairs and neighbouring Si atoms follows their relation to the interface states, which are produced by dangling Si bonds. We have detected with the help of the CC-DLTS method (constant capacitance-deep level transient spectroscopy), for example, an interface state at approximately 0.4 eV below the conduction band edge of the semiconductor, induced by electron irradiation of the $Si-SiO_2$ interregion of MOS structures [14].

The practical and technological importance of these novel results will be demonstrated elsewhere [1,15].

6. Conclusions

Our complex program of experimental and theoretical investigation of the structure and properties (phonon and electron states, defects and so on) of the $Si-SiO_2$ interregion and of the material SiO_x includes DLTS, XPS, IR, and diffraction experiments, cluster-Bethe-lattice calculations of densities of states and other methods (e.g., optical measurements, cathodoluminescence, defect theory). The main results of this paper, which reviews some of these investigations, can be summarized as follows.

(1) The $Si-SiO_2$ interregion involves an SiO_x transition layer, the thickness and profile of which are determined by the oxidation process, the surface orientation of silicon and by other parameters. The thickness of the part of the SiO_x layer in which the essential transition from the electronic properties of silicon to the electronic properties of SiO_2 takes place is approximately 1 nm.

(2) Distinct methods for the preparation of SiO_x lead to different structures, which can be approximatively described with the help of a macroscopic mixture and a microscopic random-bonding model. The SiO_x layer between silicon and its thermal oxide reveals a random-bonding behaviour.

(3) Basing on the Berreman effect transversal and longitudinal optical phonon modes of thin SiO_x films can be detected in a simple IR experiment, which leads therefore to very direct information about the atomic structure of the geometrically and chemically disordered material SiO_x.

(4) SiO_x is an inhomogeneous medium on a bonding scale with a structural percolation threshold at $x=1.25$, which can be detected in the x-dependence of various electronic properties. We must distinguish between the variation of electronic properties through the $Si-SiO_2$ interregion with its special x-gradient and the dependence of electronic properties of independent SiO_x films on their average chemical composition x.

(5) The most important intrinsic defects of the $Si-SiO_2$ interregion can be described in terms of a valence-alternation-pair model, which leads to a consistent explanation of interface states, fixed oxide charges, and radiation defects. Electron-beam irradiation of the $Si-SiO_2$ interregion of MOS structures induces a characteristic interface state at approximately 0.4 eV below the conduction band edge of silicon.

References

1. K. Hübner, Insulator Physics of SiO_2, Akademie-Verlag Berlin, in preparation
2. C.R. Helms, J. Vac. Sci. Technol. 16, 608 (1979); and in: Insulating Films on Semiconductors, ed. M. Schulz and G. Pensl, Springer Series in Electrophysics, Vol. 7 (Springer, Berlin-Heidelberg-New York, 1981) p. 19, and references therein
3. D. Schulze, F. Bechstedt, J. Finster and K. Hübner, to be published
4. K. Hübner, L. Schumann, A. Lehmann, H.H. Vajen and G. Zuther, phys. stat. sol. (b) 104, K1 (1981)

5. D.W. Berreman, Phys. Rev. 130, 2193 (1963)
6. K. Hübner, A. Lehmann and L. Schumann, in "The Physics of MOS Insulators", Proc. Int. Topical Conf., ed. G. Lucovsky, S.T. Pantelides and F.L. Galeener (Pergamon Press, New York, 1980) p. 82; and to be published
7. L. Schumann, A. Lehmann, H. Sobotta, V. Riede, U. Teschner and K. Hübner, phys. stat. sol. (b) 110, K69 (1982); 111, K103 (1982)
8. R. Engelke, Th. Roy, H.-G. Neumann and K. Hübner, phys. stat. sol. (a) 65, 271 (1981)
9. K. Hübner, phys. stat. sol. (a) 61, 665 (1980) and references therein; J. Non-Cryst. Solids 35&36, 1011 (1980)
10. K. Hübner, E. Rogmann and G. Zuther, in "Insulating Films of Semiconductors", ed. M. Schulz and G. Pensl, Springer Series in Electrophysics, Vol. 7 (Springer, Berlin-Heidelberg-New York, 1981) p. 30
11. E. Holzenkämpfer, F.-W. Richter, J. Stuke and U. Voget-Grote, J. Non-Cryst. Solids 32, 327 (1979)
12. K. Hübner, phys. stat. sol. (b) 94, K35 (1979)
13. G. Lucovsky, J. Non-Cryst. Solids 35&36, 825 (1980)
14. K. Hübner, B. Derlich and W. Ecke, to be published
15. K. Hübner et al., to be published

RADIATION DEFECTS OF THE SEMICONDUCTOR-INSULATOR INTERFACE

V.G. Litovchenko
Institute of Semiconductors
Ukr. SSR, Kiev 252028, USSR

1. Introduction

The insulator-semiconductor interface is a typical representative of the boundaries between two quite different adjacent but nonmixing phases. This fact determines the rather specific origin of the structure- and structure-impurity complex defects of the boundary.

The most important properties for such interfaces we illustrated in Fig. 1: 1) the existence of surface stress, which often causes the appearance of mechanical strain fields capable to stimulate redistribution near the boundary; 2) weakening of bonds at the surface and often a weak interface bonding which can stimulate diffusion along the interface as well as accumulation of defects at this interface leading to formation of secondary defects; 3) the accumulation of electrically charged active centers including structural defects which in turn can cause the appearance of electric fields responsible for important physical effects (e.g. two dimensional memory, two-dimensional quantum effects, etc.), and may affect the defect formation process; 4) finally the existence of adjacent but differing in properties phases with their mutual "electron-interaction", leading in some cases to mixing of properties of the neighbouring phases (not realised for each phase separately). In this review we shall briefly analyse the properties of heterogeneous layer structures *insulator-semiconductor* and *metal-insulator-semiconductor* (I-S and MIS). Main attention will be paid to the electrical structure and some optical properties will also

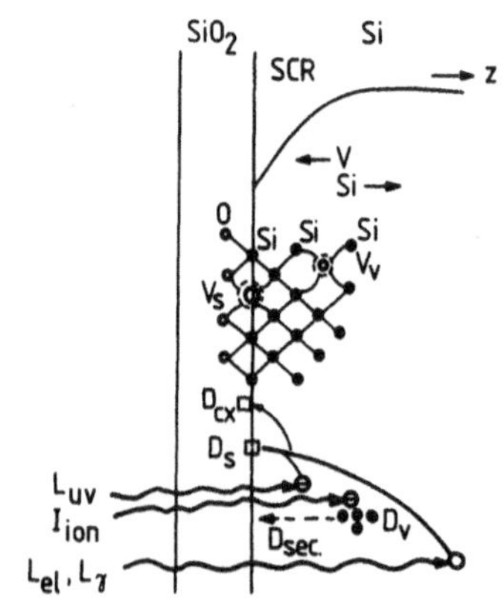

Fig. 1. Types of the surface radiation effects

be considered. Taking into account the effects at the interface itself we shall also discuss the properties in the adjacent regions of the insulator and semiconductors exposed to radiation. Most commonly γ radiation is used, but other sufficiently energetic particles and quanta could also be considered. As in the case of a uniform medium the radiation gives rise to the following effects: appearance of primary radiation defects (Frenkel-pairs) due to impact collisions of the particles or quanta with the atoms, and formation of secondary defects caused by interaction of atoms or with the impurity or structural defects induced by electro-photo-chemical reactions. The processes for heterostructures have their peculiarities [1-7], which needs a separate analysis.

2. Some Considerations on the Origin of Defects

Eariler we discussed the actual micromechanism of the formation of the intermediate thick defective layer in planar IS-systems (as the basis of the so-called three-layer model) as well as the creation of various defects in this layer. During thermal oxidation namely (which is a commonly used procedure for SiO_2-Si planar system fabrication) the Si-Si bond (1.97 eV) is broken and the substantially stronger Si-O bond (5 to 8.4 eV) is formed. As a result, at any oxidation a rather high energy E is released (of about 5 eV), which exceeds by more than two times the E_D energy of the creation of the simplest defect (in Si for a Schottky-Frenkel point defect $E_D \cong 2.2$ eV). This coincides with the value needed to start the formation of a defect ($\delta E > 2E_D$). The most important step here is the adiabatic (like impact) process during breaking up of the Si-Si bonds (especially if weakened or previously strained) under the influence of chemically active oxygen atoms and the hopping like movement of silicon atoms towards new Si-O bonds, i.e. inside of the SiO_2 (SiO_x) phase. At the same time this leaves a vacancy behind inside the silicon phase (very close to the interface, within a distance of about several lattice constants: $d_c \sim 5$-10 Å [7]). Additionally there is an intensive dissipation of the extrinsic energy δE through local creation of nonequilibrium excess phonons, but this mechanism is less important due to the small phonon energy) at least at low oxygen flux, i.e. for low gas pressure condition with the passivating oxide layer present. Following this we have a clusterisation process and the formation of the complex defects. As for the simple (primary), and the complex (secondary) defects, their built-up is much more probable near the "biographic" defects on the surface, near impurities and other distortions of the crystal lattice. For these reason the transition layer is non-homogeneous. The majority of the defects is located in the upper part of the layer (up to its amorphous) boundary.

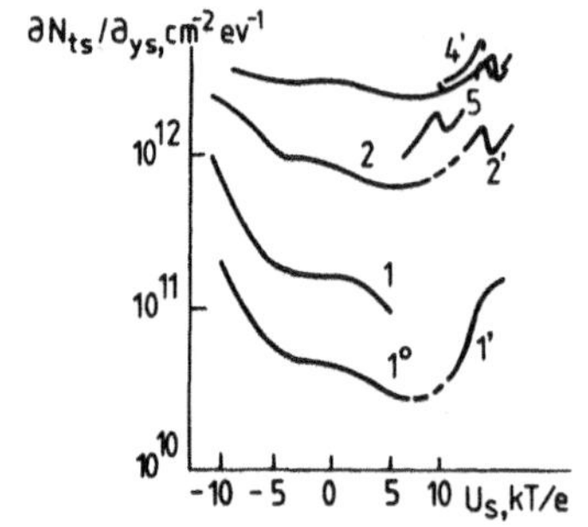

Fig. 2. Spectrum of electron states for SiO_2-Si (dry oxidation) interface, obtained by Ĉ(V), G(V) and Gray-Brown (marked by "1") methods, 1° - before irradiation, 1 - UV irradiation (3 hours, V_g = 2 V), 2 - UV (32 h - 2 V), 3 - irradiation, 10^7 rad, 4 - combined UV (32, V_g = 0) and (10^7 rad) treatment [6,7,13], 5 - plasma-mechanical carbonised oxidation $(C_2H_2O)_4$ + NO_2, 10^6 rad [13]

There are some more or less direct experimental verifications of such a model. First of all, the general look of the spectrum of the electron states connected with the surface defects for non-irradiated and irradiated MOS systems remains almost the same (Fig. 2): the U-like shape, with its minimum near the midgap, and maximum at the upper part of the gap for relatively imperfect MOS systems. From such radiation analogies we tried to classify the origin of different surface traps - as Si vacancy (V_{Si}) A - centers (association of V_{Si} with O), A' - centers (interstitial Si with O) - for shallow traps and for deep traps E x centers (association of the vacancy with P), divacancies or polivacancy associations as well as association of "deep impurities" (Au, Cu, Fe, Zn, C) with a vacancy, Fig. 3. The more complicated are the defects, the smaller is the probability of their creation. For the normal energy distribution with low $\rho(\sigma)$ - for a vacancy (with σ as the parameter of the energy dispersion) we have

$$M_n^{\nu} = N_{ij} \frac{\rho(\sigma)}{n!} B(D,R), \qquad n = 2,3,\ldots$$

For three-particle complexes the concentration is about one order of magnitude less than that for single-particle defects. Diffusion and reaction probability factor B(D,R) makes this difference even larger. Such estimations are in a good qualitative agreement with the experimental spectrum of traps (Fig. 3).

From metallographic investigations (using the Sirtl-etching to develope the defects after chemical removal of the SiO_2 layer) we have found some spots on the surface with defects very similar to those induced by radiation (Fig. 4). The picture clearly shows a number of point defects inside the Si substrate close to the interface. These

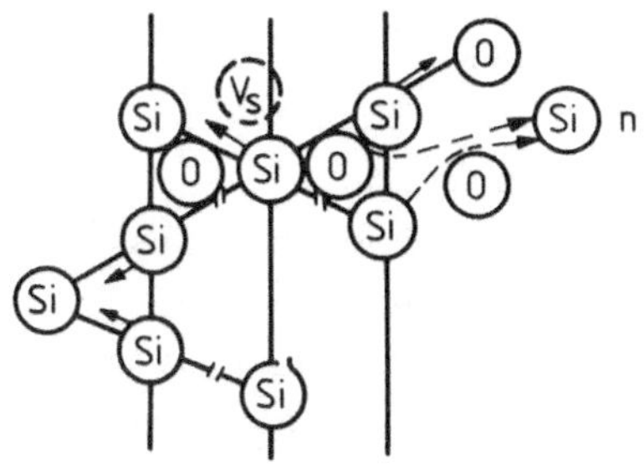

Fig. 3a. The reactions of bulk and surface defect complex creations (schematically)

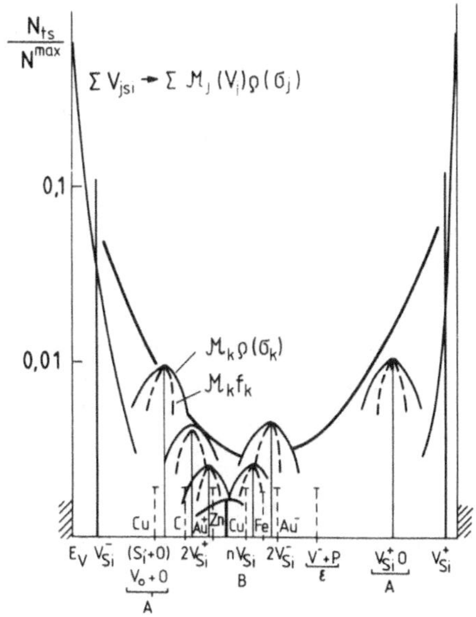

Fig. 3b. General view of the structure defect centers for SiO_2-Si boundary (dotted line-filling Fermi function for the discrete levels, solid-widened by different mechanisms, bottom-marks the different known radiative centers [6]

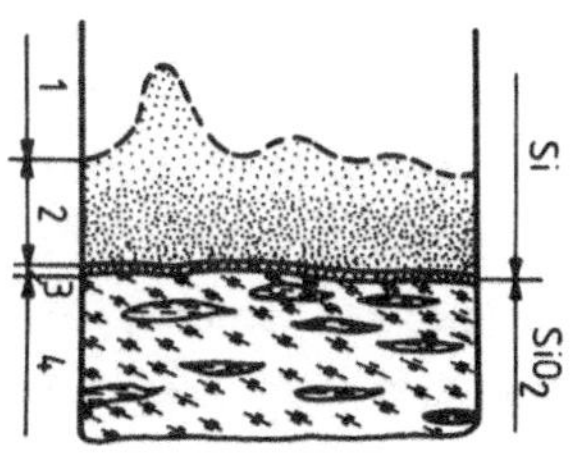

Fig. 3c. Three-layer model: Si-Si structure 1) with a detective intermediate Si layer enriched by radiation, 2) SiO_x layer with charged defects, 3) amorphous layer

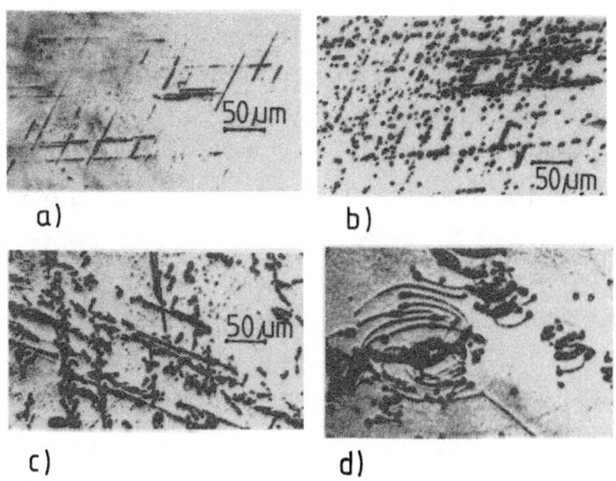

Fig. 4. The microstructure of the Si substrate after high temperature treatment (small dislocation loops near the interface), revealed by prolonged etching with Sirtl-etching (1-10 sec, 2-2 min, 3-10 min, 4 - dislocation loops, generated around a foreign inclusion probable Na, in Si near interface source of the Frank-Red type). Sample with dry thermal oxide

defects are randomly distributed over the surface and in some regions gather into groups. Their average density is about 10^5-10^6 cm^{-2} for the ordinary high temperature oxidation, and much less for the low temperature or chlorine-type oxidation. This fact is a direct proof for the chemical nature of the mechanism of defect creation.

High temperature oxidation leads to the supersaturation of the region near to the interface with points defects, which then have to relax, condense near (or create) coupled pits (near to the small randomly oriented dislocation loops, sometimes linked together). By the typical dumb-bell shape of the pits they were identified as small dislocation loops, localized in the (111) plane. These loops are apparently decorated with rather fine precipitates of SiO_2 or different impurities (Na, Fe, etc.). On the other hand relatively prolonged high temperature treatment of the sample during oxidation promotes the interaction of newly generated defects with the original grown-in defects. As a result it is possible to observe dislocation loops along the slip lines. The dislocation loops can also be formed as a result of a plastic relaxation of stress at the SiO_2-Si interface. As it was shown by electron diffraction experiments for the most destructive type of the oxidation Kikuchi lines were observed. In extremal cases this may lead to the formation even of a thin amorphous intermediate layer (at least, at some spots of the surface) [7]. A comparison of the results for various kinds of samples [7] demonstrates the important role of the oxide preparation method in the defect microstructure morphology of the Si boundary layer. High temperature oxidation is namely accompanied by the formation of a number of specific (radiation type) defects, "dots" and small dislocation loops. For dry oxidation at 1000-1150 °C their concentration amounts to 10^5 cm^{-2}. The defect generation process is intensified by water vapour in the ambient, so that the density of defects is for this case by an order of magnitude larger, than that for the case of dry oxidation. However, the conglomeration of defects into larger clusters is hampered in case of a wet ambient owing to the higher oxidation rate. Thus, the size of the clusters is much smaller, than after dry oxidation. Low-temperature methods of the oxide preparation (vacuum deposition, pyrolysis) causes small changes in the initial Si boundary layer structure. Of all the thermal grown oxides chlorine technology yields the most homogeneous oxides and boundary layers; the density of surface defects is much lower than in ordinary oxides. Besides, this preparation method gives the lowest concentration of sodium within the oxide and at the interface. Here an annihilation process of the point defects (by a reaction V + H in Si or V + Cl in

SiO_2) acts whereas gettering by impurity precipitation is absent (like Na_nSiO_2). Thus, the formation prcoess of the defective Si-SiO_2 interface layer is determined by a number of interacting factors: 1) by the chemical generation of simple defects in Si during high temperature oxidation; 2) by the formation of complex defects decorated by impurity precipitates or initial (growth) defects, and stimulated by microstress, acting around these inclusions and dislocations; 3) decorated by microcrystallites in the amorphous oxide near the boundary layer or other alien inclusions (Na_nSiO_2, etc.). In conclusion of this part we have to stress that the analysis of strong radiation treatments (like ion implantation, neutron-, gamma-irradiation, etc.) has shown actually the same kind of defects as those discussed above.

3. Influence of the Preoxidation Radiation Treatment on the Formation of the SiO_2-Si Interface

We shall now analyse the influence of the radiation-induced surface (or subsurface) defects on the interface properties of the Si-SiO_2 structures, prepared by high temperature oxidation. We used low energy ion (Ar^+) bombardment for the creation of initial surface radiation defects (ISRD) in three regimes. 1) Low doses D_R: when predominantly point defects are created, which are annealed at relatively low temperature ($\sim$ 150-200 °C); 2) Medium D_R: when large defect regions (acting as a source for the built-up of the amorphous phase) appear. During high temperature treatment many dislocations (like loops, dislocation networks, etc.) which compensate the misfits in different parts of the surface layer are created; 3) Large D_R: leading to a complete amorphisation. Critical densities of defects, the coupling of which leads to an amorphisation for Si are above 10^{13} cm^{-2} ($\sim 10^{20}$ cm^{-3}). Surface defects can be quite easily detected in each case using the selective crystallisation of organic substances (like naphtaline), Fig. 5. These defects cannot be distinguished by X-ray topography due to their rather small depth penetration ($\lesssim 10^2$ Å).

The following processes are important for thermal procedures: 1) diffusion of the defects and their recombination; 2) complexing (and annihilation of smaller defects); 3) the movement of defects and impurities from the bulk, and from undisturbed subsurface neighbouring regions. The gettering process is especially active on thermally oxidized samples with a surface, which was locally subjected to ion bombardment. For such a non-uniform surface planar gettering takes

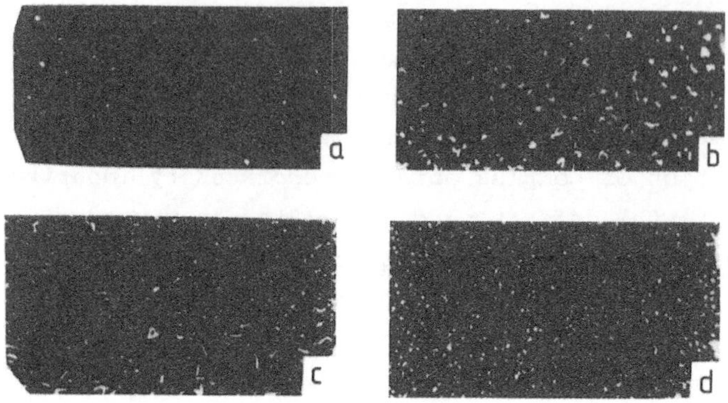

Fig. 5. Organic decoration surface defects on Si-substrate in dependence on the dose of irradiation by Ar^+ (a - initial state; b,c,d - D = $+10^{16}$, $8\cdot10^{16}$, $2\cdot10^{17}$ cm^{-})

place, and ion treated spots act like effective drains. Also there is substantial influence of the oxidation process itself, which leads to an increase in the oxidation rate in the first stage of oxidation (up to 500 Å), determining a higher homogeneity. The high efficiency of the planar gettering (diffusion length achieves 1 mm already at 800 °C) is proved by the high mobility of the surface and interface (SiO_2-Si) defects, as well as of other imperfections in general.

The process by which surface defects are driven into the disturbed (in case 3 into the amorphised) region leads to their accumulation in this region causing an increase of the oxidation rate and relatively large mechanical stresses at the edges. The latter favours the gettering of point defects and some impurities from the bulk. Therefore spikes of defect concentration are formed near the edge region, as well as a

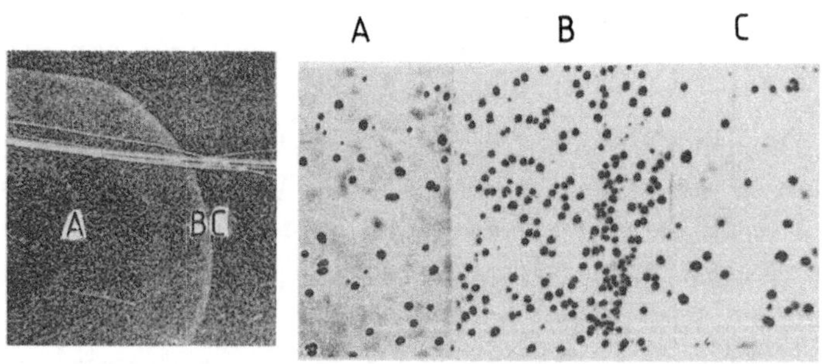

Fig. 6. X-ray topography of the Si-slab (Ar^+ treated in A region) and characteristic regions (A,B,C) after Sirtl-etching

"valley" like distribution (of several mm width) beyond the bombarded area, Fig. 6. Let us consider the defect structures of the surface

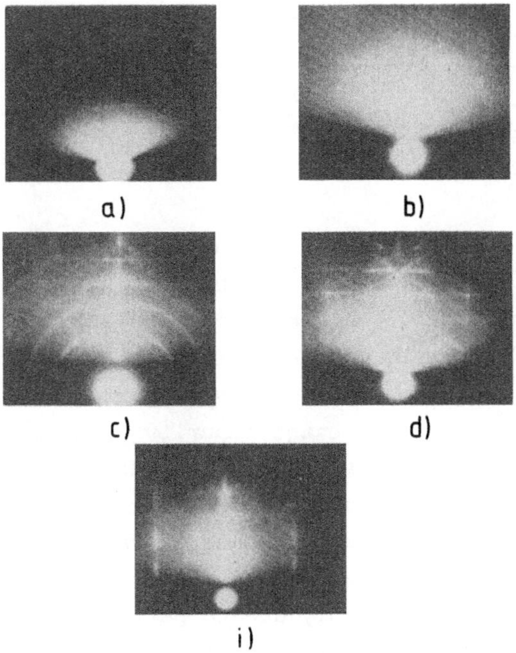

Fig. 7. Electrondiffraction patterns for initial state of Si sample. (a) after Ar^+ (10^{17} cm^{-2}, 2 keV), bombardment; (b) and preoxidation heat treatments in neutral gas ambient (N_2) at 550 °C (c), 650 °C (d), 750 °C (e)

bombarded with high doses ($\sim 10^{17}$ cm^{-2}) of Ar or O_2 ions. Here we have a completely amorphized surface before oxidation (down to a depth of 10^2 Å) with a planar microrelief of about 10^2 Å (see background halo in Fig. 7). After some time of oxidation X-ray topography already shows a reconstruction of the sub-

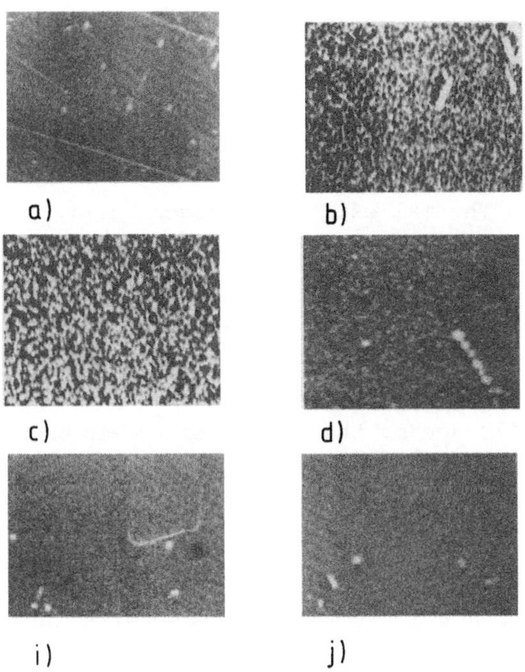

Fig. 8. X-ray topograms (Lang) Ar^+ bombarded and oxidized Si samples, in dependence on the time of oxidation in wet O_2 flow (1150 °C) for subsequent oxide layer thickness 500, 800, 1120, 2500, 4000, 6000 Å (a-j)

surface region (Fig. 8). A non-monotonous dependence of the defect density upon the thickness of the oxide layer can also be stated. The same conclusion can be drawn from the Sirtl-etching analysis in Fig. 9. It can also be seen that the defect characteristics are changed after a prolonged oxidation: more complicated shapes (as rosettes, dislocation loops, etc.) in relatively small number are formed. At a

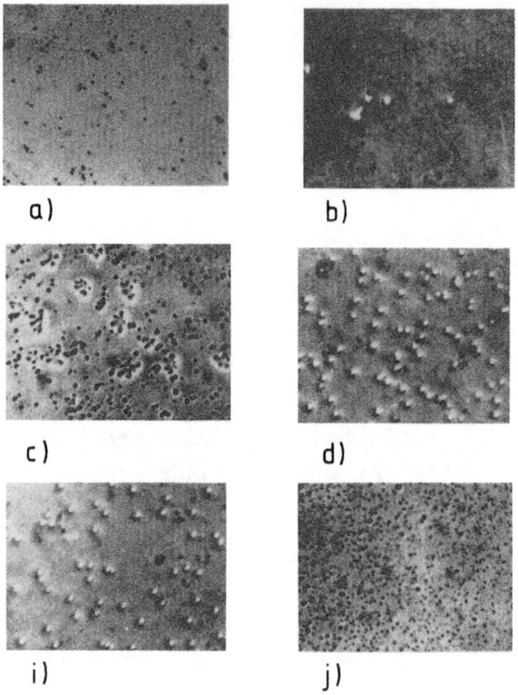

Fig. 9. Morphology of the surface after removal of the oxide layer and Secco-etching (a non-bombarded sample, b,c,d,i,j d_{ok}: 500, 1200, 2500, 4000, 6000 Å) (a-j)

(SiO_2) thickness of 1500-2000 Å the concentration of the small defects achieves its maximum. In this case misfit defects, stacking faults and impurity precipitates (usually Fe and Cu) are predominant. More prolonged oxidation leads to the disappearance of these defects, probably due to gettering and dissolution. Without the radiation treatment described, such kinds of defect properties did not appear at any regime of the oxidation. Also the ion treatment itself did not cause the very complicated defect (Fig. 7, picture taken before oxidation) and similar situation was observed after the heating procedure, which leads rather to the annealing of the defective surface layer. The thermal oxidation procedure at commonly used temperatures (T ∿ 1100 °C) will substantially change the number and type of the surface defects. The most typical defects here are non-crystalline etch pits and colonies of precipitates (which are observed as microhills). The latters are more pronounced for oxidation durations of 10^3 s, but after more prolonged oxidation they disappear.

At last we shall analyse the influence of the annealing treatments of the Si-SiO_2 structure prepared with the preoxidation irradiation (ion-beam) in neutral N_2-gas ambient. The main results can be summarized as follows (Fig. 7):

1.) The annealing in neutral ambient at 550 °C leads in the layer near the surface to an appearance of a polycrystalline phase with the lattice constant d_c = 2.36 Å (circles on the electron diffraction, pattern Fig. 7c). However, selective Sirtl-etching did not reveal any precipitates in this case.

2.) Higher temperatures (> 650 °C) changed the diffraction patterns making them typical for a more crystalline phase containing, however, some defects and mechanical stress. At the same time the number of defects in the subsurface layer increased, many pits were developed by selective etching. Thus, active defect centers, introduced by radiation ion-plasma-treatment, act on the structure of the grown oxide layer in two ways. 1) By weakening the Si-Si chemical bonds in the disordered surface layer (this is at the beginning stage, when oxidation is more intensitve) and by increasing the number of the active oxidation centers whence the structure of this part of the oxide layer becomes more homogeneous, more dense and electrically stronger. 2) With a subsequent increase of the oxide thickness and with the limitation of the O_2 diffusion flow to the SiO_2-Si interface, the oxidation rate decreases steeper for this samples, than for the non-bombarded ones. As a result the dependence $d_{ox}(t)$ becomes flatter, than $t^{1/2}$ (due to a decrease of $D(d_{ox})$). These conclusions were proved experimentally for rather clean conditions of oxidation.

4. The Influence of Backside Irradiation on SiO_2-Si Interface Defects

Another method to control the radiation defects of the Si-SiO_2 boundary layer consists of the bombardment of the back side of the Si wafer by ion beams (or high energy ion implantation) and of a subsequent thermal

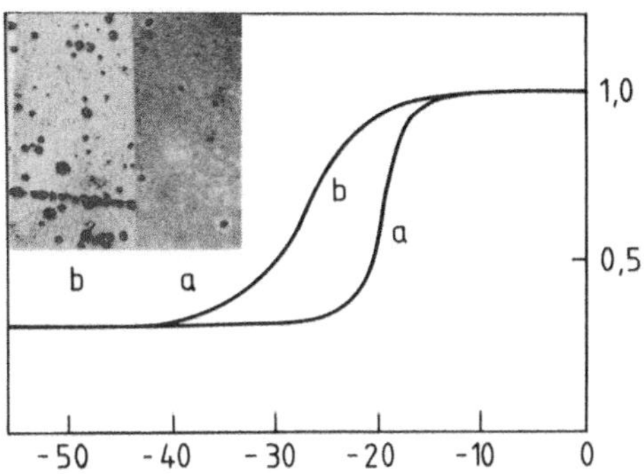

Fig. 10. Hf-C(V) for the initial (a) and backside bombarded gettered MOS systems (b) left side contains the topogramm of the defects, developed by Sirtl-etching

oxidation. The samples were analysed by X-ray topography, selective Sritl-etching technique and electron diffraction (see Fig. 10). Also high frequency C(V) method was used to examine the effectiveness of the gettering on the electrical characteristics. It was shown by X-ray that for the disturbed oxidised surfaces "white spot" inclusions appeared (a). However, a thermal treatment of the samples, bombarded by Ar_2^+ in neutral (N_2) ambient, led to the same results. Selective etching developed the impurity precipitates which appeared at the disturbed surface. Besides, we found a decrease in the number of the etch pits at the front surface (for 10^5 for the initial state to $\sim 10^3$ cm^{-2}), Fig. 10. Using electronography, second phase inclusions with a lattice constant of $\sim$2.36 Å were detected at the disturbed surface, which coincidences with the parameters of some silicon copper alloys (like CuSi). A splitting of the Kikuchi lines observed is caused by shifts of the small blocks near the surface. This data correlate with the electrical C(V) measurements in Fig. 10, the interface parameters improved remarkably.

The gettering of the point defects near the boundary of a bombarded region and a subsequent improvement of the non-disturbed area of the surface was also observed by means of the Sirtl-etching technique.

5. Electrical and Structural Parameters of the Interface in Case of γ, Electron Irradiation and UV Illumination

Changes in the properties of the interface in I-S or MIS structures manifest themselves mainly by an increase of surface state density, by variations in their spectrum,by lowering the mobility in surface regions, Fig. 1. The dose dependence of the surface state density has the form of the curve with a saturation at relatively low doses ($\leq 10^7$ rad), Fig. 11. An increase over the whole band-gap often is observed with a

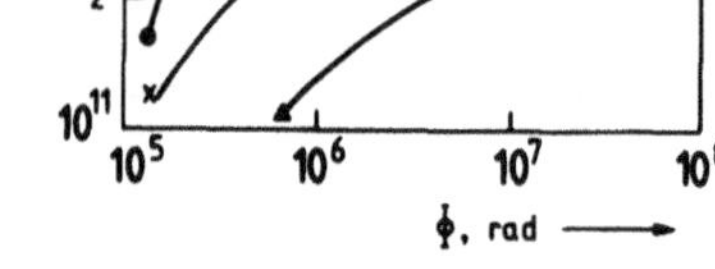

Fig. 11a. Dose dependence **of radiation induced defect densities for different IS-systems** (1 - SiO_2-Si, 2-3 - MNOS, 2-d=10 , d_2=0.5.10 Å, 3-d_1=0.5.10 , d_2=0.5.10 Å)

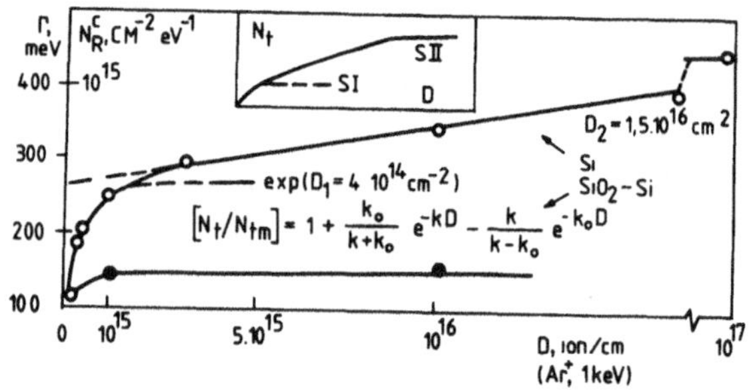

Fig. 11b. Dose dependence of the plots N_R and with - factor of ER - for Ar^+ bombardment (upper curve - free surface, bottom - SiO_2-Si)

maximum at E_c-0.17 eV [5], which can be related to surface A-centers in Fig. 1. The creation of complex radiation-induced defects is proved by the peculiarities in electrical properties of the SiO_2-Si interface. It can be seen that the density of the states rises throughout the whole region in the bandgap studied, including the region of deep levels, which are known to be associated with complex defects [1,12]. Moreover, the efficiency of their creation is considerably greater than that of the shallow levels (Fig. 1, see the maximum near the midgap). It is also possible to observe the creation of sufficiently complicated A-centers (peak at the right part of the spectrum) and other complexes, possibly $2V_{Si}$, $n.V_{Si}$-centers, which introduce levels slightly below the midgap; associations of the C- and Si-interstitials cause the peak at $E_c \cong 0.4$ eV, appearing at a carbonised SiO_2-Si interface. We obtained for chlorine oxidation the pronounced peak at $E_i = -0.08$ eV, which also can be connected to vacancy complexes or to associations like $V_{Si} + Au^-$, $V_{Si} + Cu^-$: V_{Si} + B. Such kind of complexes were experimentally identified on neutron-irradiated Si by Guldberg, Watkins et al. [14,15].

The second argument in favour of complex defects creation during irradiation is the shape of the dependences of interface deep state density versus dose, which exhibits saturation at relatively low doses. This behaviour is similar to that for bulk radiation complex defects [1,12].

Under specific (not well determined) conditions irradiation can also reduce the density of surface states as, for example, in some types of Si-MOS, GaAs Schottky-barrier systems [7,8]. Here one can speak about so-called stimulated gettering. The latter probably includes radiation enhanced diffusion and some reconstruction of the structure along the interface. A number of data obtained so far shows that radiation effects at the interface depend on the peculiarities of structure and presence of impurities.

Radiation sensitivity particularly increases in the presence of impurities like Na, O, H, etc. At the same time it can be reduced by doping with N, B, and atoms of rare earth metals. Thus the type of the impurity and the peculiarities of the crystalline structure may be either a sensitizing or a healing factor with respect to defect formation at the interface. An increase in radiation sensitivity is probably caused by an accompanying rise in mechanical strain and by an increase of the degree of disorder at the interface.

Multistage reactions for which the thermal activation is of great importance may also be involved in the processes of defect formation.

Low-temperature irradiation (77 K) can reveal several stages in the defect formation process (here under γ-irradiation) [9]. It was found that immediately after irradiation a strong rise in the built-in charge takes place depending upon the applied bias. The charge was annealed gradually by heating, and starting from some definite temperature an increase in the surface state density was observed which was partially in correlation with the field or the charge in the insulator during the irradiation. One of the conclusions which can be drawn from these results is that the surface states are not created exclusively due to the propagation of elementary defects towards the interface with their subsequent accumulation there, what might be suggested as a source of much higher effectivity (2-3 orders) of defects formation at the interface with respect to that in the bulk.

Thus the surface state formation process appears to be rather complicated. The resulting mechanism can be dependent on the bias, the irradiation conditions, and impurity- structural peculiarities of the interface. It was stated in a number of works [4-6] that the factor responsible for defect formation at the IS-interface in case of non-impact (but photochemical like) mechanism is the presence of high-energetic electrons. This was directly shown by the experiments with electron injection into the insulator under UV-illumination with bias. It was found that at least two types of additional states are created under these conditions (SI, SII), one of them being a results of a complicated interaction between the electrons and the adjacent superthin (20 Å) region of the oxide (in Au-SiO_2-Si structures). A multistage reaction was shown to be responsible for the process in which the electrons interact with the complexes like SiOH, and released atomic hydrogen interacts in turn with definite centers D (also activated by electrons D^{-*}) what leads finally to a creation of surface states. Here the following reactions take place [6]:

$$e(h\nu)_1 + SiOH \rightarrow SiO + H\uparrow$$

$$e(h\nu)_2 + D \rightarrow D^{-*}$$

$$D^{-*} + H \rightarrow D_{SII} + e$$

The parameters of such reactions are in a good agreement with the photoelectromechanical nature of their creation process.

6. Defect Types at Ion Implantation

To generalize the discussion, experiments on ion implantation will be considered together with the radiation effects. In the case of non implantation or ion-plasma surface treatments the role of surface radiation effects is even increased. This is associated with the rather high doses used and with the fact that in this case all the energy exchange takes place just at the surface. A 0.1-1 μm thick, rich in defects layer was found after such treatments [11,12]. The appearance of this layer was explained (in principle) by various mechanisms:

1) cathode sputtering of the material with a creation of surface point defects;
2) a reverse process - of rebuilding the surface with the help of the atoms moving towards the surface from the bulk;
3) a reduction of the energy of defect formation at the surface due to a reduction of the binding energy [6,9] and finally
4) secondary processes of defect formation by means of the diffusion of simple defects to the interface [10,11].

A study of the damaged surface layer was carried out with the help of a number of methods:

EPR which reflects the divacancy distribution (2V) and clustering; 3P and Σ-centers [12]; electrical measurements including C(V) in MIS structures [6-9]; optical and modulation methods [10]; electron-microscopic methods [11,12]. The experiments on the irradiation of Si through the oxide film showed that the defective layer is formed at the interface (in the surface layer $\sim 10^2 - 5 \cdot 10^3$ Å) even in the case, when ions do not penetrate to it [11,12]. Thus the mechanisms 1,2 cannot be completely responsible for the defect formation for high-energy ion beams but vice versa for low energy beams they are dominant. The third mechanism is rather probable for the surface states localized directly at the interface of within the intermediate layer. Indeed, as it was shown in a number of works on "three-layer model" [6-10] (Fig. 2b), the chemical bonds are strained in this layer. Thus, for free surfaces as for the interface the third mechanism finds serious verifications. It was shown for example in [12] that a spatial distribution of the defects in the layer depends on their type. E.g. the effective thickness of the defective layer in SiO_2-Si structure implanted with B^+ (100 keV) in Si for simple defects 3P, is about 500 Å, and for the complexes V-V approximately 150 Å. The experiments on electron irradiation did not show changes in the threshold energy of defect formation for the regions nearer to the surface [20], though a sharp rise in

A-center concentration was noted on the surface [5,15]. There is a unique evidence for various types of irradiation (electrons, γ, ion-beams) that the energies of defect formation are lower, than the threshold values. This is especially the case for UV-illumination when the bulk remains quite unchanged.

As it was mentioned above electron-stimulated reactions are here involved. This mechanism [3] appears to be more effective when a sufficient amount of energetic (excited) electrons is present near the interface (UV, γ, ion-beams, etc.). It is less probable for low energy ion beams which exhange energies for collisions. Diffusion drift mechanism with subsequent reactions is also universal. It can lead to a clusterization or annihilation of defects. Here a motion of point defects either from the bulk or back into the bulk is possible. The diffusion length for such a motion at room temperatures achieves 100-1000 Å, i.e. the thicknesses typical for implantation lengths. Such a behaviour depends on the state of the defects. In particular for Si-surface with a superthin oxide film a high dose bombardment with Ar^+ leads to a penetration of a defect layer inside the semiconductor up to $2 \cdot 10^2$ Å, which exceeds the implantation radius (50 Å). A very thin layer ($10-10^2$ Å) free of defects was observed in the experiments on electro-reflexion [10]. The penetration of the defects inside up to the opposite side of the sample was noted for high temperature technology. In most cases however the surface acts as a collector leading to an accumulation of the defects at the surface whereas the bulk becomes more perfect. Under such conditions some impurities can be bound at the surface. This is principial as the method of radiation-induced purification.

First studies of this boundary layer were carried out by means of electrical measurements [9-10], surface photoluminescence, EPR [12], and SIMS. The analysis of defect profiles in this layer obtained by means of layer by layer etching, Fig. 12, proves the important and determining role of the secondary defect formation. It was shown in particular that in the extremal case a sink of defects to the surface can lead to an amorphization of a layer at the surface while the bulk remains crystalline. The EPR, ER, SIMS, etc. data show, that in the case of usual implantation regime always a pulling-out of the defect towards surface is observed with an appearance of two distribution maxima (Fig. 12). At the same time, at high ion energies (100 keV for B: 1 meV Xe) the defective surface layer is not created at all or is very un-pronounced, what indicates a predominant role of the ionic mechanism. The increase in the surface state density is observed in

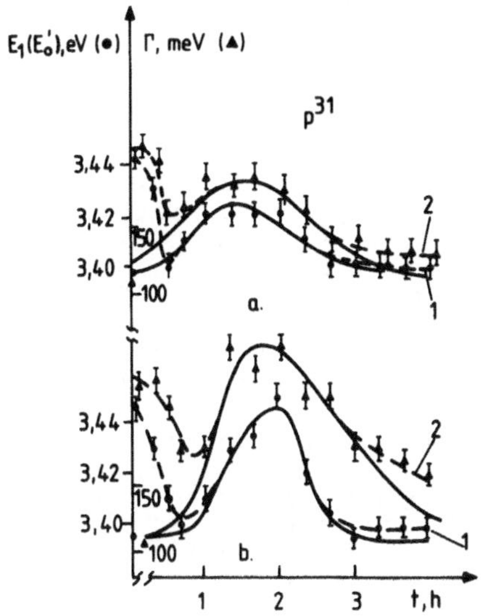

Fig. 12. Annealing-time dependence of the band critical point E_0 and with line parameters for Si, implanted P (40 keV, $0.6 \cdot 10^{13}$ cm^{-2})

these cases too, and this fact indicates the role of other (probably, electron) mechanisms. Moreover, as it was mentioned above, an increase of defect density of the surface is observed even when the energy of ions is insufficient to reach the interface [11]. Here also the electronic mechanism of defect formation is of importance. Rather direct verification of the existence of the drift-diffusion mechanism was obtained from ER experiments [12].

7. Conclusion

Analysing the radiation effects in various I-S and MIS structures three characteristic regions can be distinguished in the general case: a) the insulator layer adjacent to the interface; b) the corresponding interface; c) a thin layer of the semiconductor, adjacent to the interface. Main mechanisms which determine formation of radiation induced defects are: 1) diffusion and drift motion of point (primary) defects and high mobility contaminations from the bulk with subsequent complexing (or accumulation) near or at the interface, predominantly on the more damaged part of it,and 2) electron stimulated photo-chemical reaction causing formation of centers.

References

1. S.P. Mitchel, D.K. Wilson, Bell System Techn. J. 46, 1 (1967)
2. K.H. Zaininger, Appl. Phys. Lett. 8, 100 (1966)
3. C.V. Gvin, Appl. Phys. 40, 4886 (1969)
4. V.A. Girii, A.V. Kondraschuk, V.I. Schachovtsov, Preprint, Inst. Phys. Acad. Sci. Ukr. SSSR, 74-29 (1974)

5. R.J. Powell, G.F. Derbenwic, J. IEEE Trans. Nucl. Sci. 18, 99 (1971)
6. I.P. Kisovski, V.G. Litovchenko, R.O. Litvinov, phys. stat. sol. (a) 53, 263 (1969); Ukr. Fiz. Zh. 21, 863 (1976)
7. V.G. Litovchenko et al., Proc. Int. Conf. on Heterojunctions, (Budapest, 1971) I. p.269-274; V.G. Litovchenko, A.P. Gorban, in: Physical Basis of the MOS Systems, ed. by Naukova (Dumka, Kiev, 1978) p. 94 (in Russian)
8. A.B. Gerasimov, V.G. Litovchenko, A.G. Shillo et al., Fis. Techn. Poluprov 8, N2, 792 (1974); O.Yu. Borkovskaya, N.L. Dmitruk, R.V. Konkaova and Yu.A. Thorik, phys. stat. sol. (a) 48, K55 (1978)
9. V.Ya. Kiblik, R.O. Litvinov, V.G. Litovchenko, Ukr. Fis. Zh. 22, 1097 (1977); Microelectronica 8, N6, 534 (1979) (in Russian)
10. V.I. Gavrilenko, V.A. Zuev, V.G. Litovchenko, Fiz. Tverdogo Tela 19, 58 (1977); Fiz. Techn. Poluprov. 12, 1547 (1978) Coll.
11. D.V. McCaughan, V.T. Murphy, IEEE Trans. Nucl. Sci. 19, 249 (1972); J. Appl. Phys. 44, 2008 (1973)
12. Physical Processes in Disordered Semiconductors, Edited by Smirnov, Novosibirsk, Nauka, 1977 (in Russian)
13. V.G. Litovchenko, Y.Ya. Kiblik, S. Georgiev, Radiation Effects 66, 1, 84 (1982); Coll. Semic. Technic. Microelectronics 23, 84 (1976); 33, 39 (1981) Ed. by Naukova Dumka, Kiev (in Russian)
14. S. Guldberg, S. Phys. D.; Appl. 11, N6, 2043 (1978)
15. V.S. Vavilov, N.A. Uhin, Radiation Effects in Semiconductors, Ed. by Nauka, Moscow, 1969 (in Russian)

ANALYSIS OF Si/SiO_2 INTERFACE DEFECTS BY THE METHOD OF TERM SPECTROSCOPY

H. Flietner

Zentralinstitut für Elektronenphysik der AdW der DDR
1199 Berlin, Rudower Chaussee 5, GDR

The Si/SiO_2 interface is found to be a very thin transition region of the order of 1 nm between Si and SiO_2 [1,2]. Such a thin matching layer between two different lattices is enabled by defects lowering the stress of bond complexes by generating broken bonds. So it is a fundamental question to ask for the defects and their nature at interfaces. The number of defects is very low ($< 10^{13}$ cm^{-2}), so only the sensitive electrical methods can be used for their analysis. Electrical methods sui generis do not have a selectivity to the atomic-chemical character so we need a method which enables us to discern defects of different character. This method we call term spectroscopy.

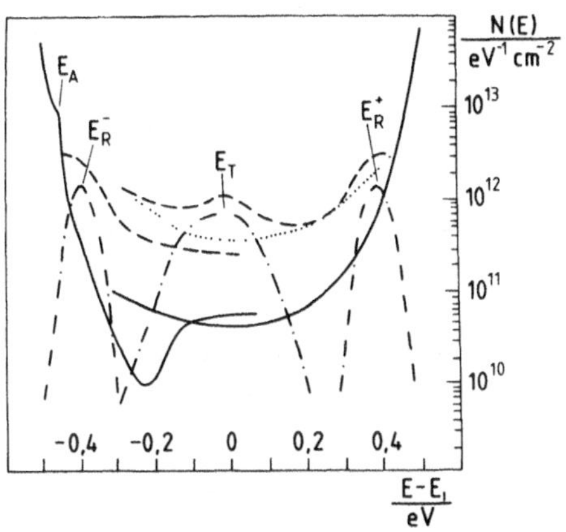

Fig. 1. Termspectroscopy = separation of states into groups demonstrated for B-implanted, not annealed Si/SiO_2-IF. — total distribution for not implanted p- and n-samples; --- total distribution for implanted samples; ··· intrinsic group of states for implanted samples; -·- Gaussian groups of states for implanted samples

Figure 1 shows an example of such kind of analysis. The continuum of states inside the gap of Si is separated into groups of states of different physical and chemical character [3-6]. Investigations on a wide range of surface preparations - etched, thermally and anodically oxidized, plasma treated, implanted- showed that the following groups may be separated out of the continuum of the interface states [7]:

- $N_{intr} = N_{BD}$ (Bond Distortions) + + N_{DB} (Dangling Bonds)

  These intrinsic states bring about the U-shaped distribution, being the continuation of the bonding/antibonding symmetry of the band states into the gap for localized states [8]. BD and DB are separated by their different technological behaviour shown in

Table 1.

- N_{extr} = states whose distribution is not showing the symmetry of the Si-band structure are called therefore extrinsic.
- N_{RD} = Radiation Defects.
- N_{ox} = Oxide fixed charge.

Characteristics of the rechargable IF-states are shown in Table 1. From the coincidence of the characteristic data for differently prepared samples we may conclude that it is the same kind of defect which exists at these so much differently prepared surfaces. Therefore only a few kinds of defects cited above are to be expected at Si/SiO_2-interfaces. This is remarkable for a suchwise disturbed system.

Table 1. Characteristics of IF-states

$N_{BD} \approx 10^{13}$ cm^{-2} independent of technology
$N_{DB} \approx 10^{12}...10^{9}$ cm^{-2}, $E_o = -(0.06...0.15)$ eV

	N_{extr}/cm^{-2}	E_t-E_i/eV	σ/eV
HF-treated, ordinary surf.	$10^{12}...10^{13}$	$\sim$0.06	0.12
therm. oxid.	$10^{10}...10^{12}$	0.05...0.18	0.12
implanted (not annealed)	$\sim 10^{11}$	$\sim$0.04	0.08
	N_{RD}/cm^{-2}	E_t-E_i/eV	σ/eV
implanted (not annealed)	10^{11}	$\pm$0.38	0.03

Another conclusion may also be drawn from these investigations. During all of the preparations the groups of states are changing only by changing their total number. The character of the distribution remains constant. This means, all the states are equivalent with respect to the influence (field, temperature, chemicals etc.) effecting the change. There is no other way of realizing such behaviour than to have the same total energy for each state of the group. And this is a crucial statement connecting physical and chemical properties of the defects.

The total energy of a defect is the sum of the one electron energy and a modified (by many electron interaction) lattice energy: $E_{total} = E_e + E_l$. The distribution of states shows a spread in E_e, for the Gauss-distributed states given by σ, and for the DB ground state = = bonding state distribution $\approx \frac{3}{8} \Delta E = 0.44$ eV [3,7]. The larger the width of the distribution of states the larger the spread in lattice energy or the scatter of atomic configuration of the defect.

Let us begin with the intrinsic defects. A defect of such kind may be modellized by a molecule picture like Fig. 2a. Bond streching gives rise to bonding/antibonding states in the gap either in the picture of complex band structure [8] (Fig. 2b) or by Bethe-lattice calculations [9] (Fig. 2c, d). Without taking into account the total energy no distinction can be made between the two groups BD and DB. The variation of d from d_o to ∞ describes both types of states. But stretching one bond does not leave the surrounding unchanged. If we think of a Si-tetrahedron stretched uniformly, the situation is stable only as long as the total energy of this complex does not exceed the equilibrium energy by one Si-Si binding energy. Otherwise the breaking of one bond is feasible. Therefore each bond of this tetrahedron complex is stretched no more than loosing one quarter of its binding energy. This is the threshold value dividing the two groups of BD- and DB-states and we get insight too into their different technological behaviour according to Table 1.

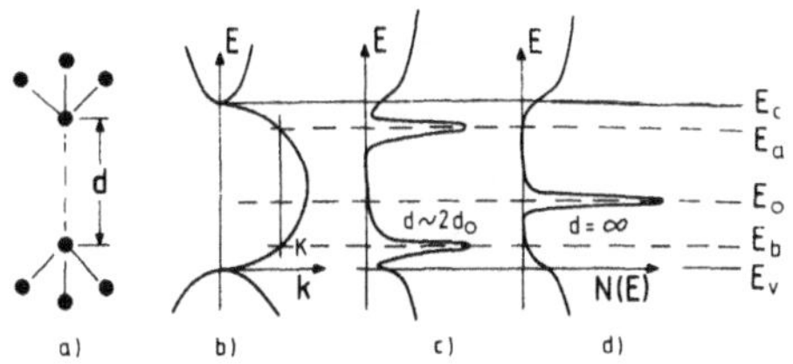

Fig. 2. (a) Model for intrinsic defect; (b) Complex E(k)-curve in the two band model; (c,d) Local density of states by Cluster-Bethe-Lattice-Calculations for intrinsic defects like a)

If the gap ΔE is considered to be a measure of the Si-Si binding energy E_{Si-Si}, a surface gap $\Delta E_s = \frac{3}{4}\,\Delta E$ should exist reflecting the loss of binding energy by stretched bonds. These stretched bonds give rise to BD-states appearing as exponential tails of the bands. Inside the surface gap broken bonds bring about DB-states. The experimentally determined width of the DB states (given above) corresponds well to this estimate. Furthermore this width is equal to $\frac{1}{4}\,E_{Si-Si}$ - 0.45 eV. This fits also well into this picture.

This line of considerations allows also to understand how may be achieved to lower the IF-energy by the generation of defects and to increase the degrees of freedom for the matching of different lattices. Suchlike also the short distance for matching Si to SiO_2 becomes understandable. Defects are of essential meaning for the oxidation process and in general for the thermodynamic behaviour.

Throughout the whole region of the DB-distribution there is no preference for any atomic configuration. This situation we call complete disorder contrary to complete order (the same atomic configuration for each defect of the group). Complete disorder is described by a constant distribution in the imaginary (because of localized states) k-space. This is the complementary case to a δ-function (equivalent to $\sigma = 0$), which describes the case of order. The experimental confirmation that our description makes sense may be drawn from Fig. 1: A structure disturbing influence does not change the character of the distribution.

Some short remarks now to RD and extrinsic defects. They have considerable smaller width of their distribution given by σ. In the light of the just performed considerations the conclusion may be drawn that the atomic configuration of these defects is much more stable. With respect to RD we think therefore of vacancies at subsurface layers, bearing in mind the intrinsic character of these states. The more stable configuration of the extrinsic defects is thought to be brought about by a Si-O contribution to the back bonds.

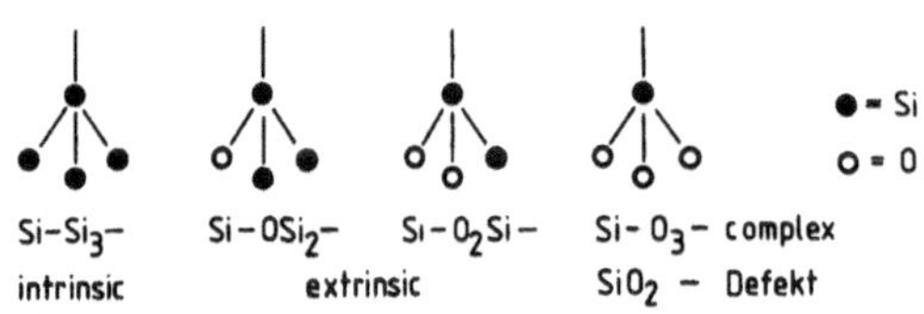

Fig. 3. Models for different dangling bond defects

Breaking bonds in the process of oxidation gives rise to a series of general DB-defects like those shown in Fig. 3. These complexes are intermediary products originated by the oxidation process. During oxidation a certain number of each of these defects is maintained. Cooling effects a freezing of the defects. The defects with at least one Si-Si back bond have a good contact to the Si-substrate (rechargable centers). The SiO_3-complex is thought to be the fixed charge center [10].

But the defects are also characterized by their behaviour during chemical and technological processes. Table 2 gives a small section of results. Some features of DB and extrinsic defects are:

- DB and extrinsic defects exhibit similar behaviour during some generation and annealing processes (thermal oxidation, radiation influence, hydrogen annealing). From this behaviour we concluded that the centers are in both cases no foreign atoms. Especially for the extrinsic centers this was the first hypothesis. This and the above cited characteristics support centers like Fig. 3. A certain number of these defects is generated by the oxidation process, either by high temperature process or treatment with oxidation agents.

Table 2.

Ordinary surface: Influence of etching components

p-Si, 111, $1000^{(1)}$ resp. $10^{(2)}$ Ωcm

	$i^{(1)}$	$HV(v)^{(1)}$	$HF(l)^{(1)}$	$+ NH_4OH^{(2)}$	$H_2O_2^{(2)}$	$HNO_3^{(2)}$
N_{DB}	3	3	3	10	30	$100 \cdot 10^{11}$ cm^{-2}
N_{extr}	-	1.2	10	-	-	$100 \cdot 10^{11}$ cm^{-2}

Therm. oxid. surface: Influence of pretreatment

n-Si, 111, 10 Ωcm, dry oxid (1200 °C, 150 nm)

1: 48 h - rinse of deionized H_2O before oxidation

2: 70 min HF(v) - treatment

	as oxid.(1,2)	+ AlT (1)	(2)
N_{DB}	4	1	$1...2 \cdot 10^{11}$ cm^{-2}
N_{extr}	5	-	$1.5 \cdot 10^{11}$ cm^{-2}

Therm. oxid. surface: Influence of PMA (Post-Met. Anneal)

n-Si, 111, 10 Ωcm, wet oxid. (1180 °C, 500 nm), H_2 - Anneal

		WC + PMA	AlT + WC + PMA	
N_{DB}(met=free)		4	1	$4 \cdot 10^{10}$ cm^{-2}
N_{extr}	met	2.1±0.9	0.2	$1.2\pm0.6 \cdot 10^{10}$ cm^{-2}
	free	6.8±4.0	0.2	$5.8\pm5.0 \cdot 10^{10}$ cm^{-2}

- Low temperature treatments bring about reactions with the defects existing. Changing the number of defects means in this case a reaction like D + A ⇄ DA or with more agents involved, suchwise eliminateing or generating the electronic level of the defect. Atomic hydrogen is especially effective in lowering the number of DB and extrinsic centers. This behaviour and likewise the time dependence and the influence of different atmospheres on the number of states suggest that the number of states is determined as the equilibrium value of certain reactions. One main difficulty is the determination of the active agents involved. But a first step for the thermodynamic description is done in recieving the total number of defects involved in the reaction by term spectroscopy.
- HF acts to generate extrinsic defects. But also N_{DB} is slightly increased. According to the former statements HF is changing the reac-

tion equilibrium being an additional active agent. The different behaviour of DB and extrinsic defects comes from the different reactivity and electro-negativity of the respective centers. Especially the extrinsic defects are much more sensitive to chemical reactions. This sensitivity is especially annoying at free surfaces.

As a concluding remark to this short survey we may state that term spectroscopy has given some interesting insight into the phyical-chemical properties of the Si/SiO_2 interface. Especially the existence of only a few kinds of distinct defects and the possibility to describe the disorder effects connected with the defects bring about some progress in understanding the properties of interfaces. There are remarkable similarities to amorphous substances like a-Si [11,12]. Though we now understand much better the electronic and chemical behaviour, many questions still remain open. One of the most prominent of these may be to get a better understanding of the electronic and energetic properties of the complexes presented in Fig. 3. The reader surely noticed that all defects treated here are dangling bond defects but only for one group we used the name DB. The reason is historical. But with a progress in understanding we could use the notation $D_x = Si{-}O_xSi_{3-x}$, $x = 0,\dots,3$. But this will not be before understanding for example the difference between D_1 and D_2 and the reason why we found only one group of extrinsic states.

References

1. S.I. Raider, R. Flitsch, IBM J. Res. Dev. 22, 294 (1978)
2. C. Hollinger, Appl. Surface Sci. 8, 318 (1981)
3. H. Flietner, N.D. Sinh, phys. stat. sol. (a) 47, 533 (1976)
4. H. Flietner, W. Füssel, N.D. Sinh, phys. stat. sol. (a) 43, K99 (1977)
5. W. Füssel, H. Flietner, K. Heilig, Physik der Halbleiteroberfläche 9, 165 (1978)
6. F. Fenske, Dissertation A, Humboldt-Univ. zu Berlin (1980)
7. H. Flietner, Physik der Halbleiteroberfläche 13, 157 (1982)
8. H. Flietner, Rostocker Phys. Manuskripte 5/II, 59 (1979)
9. K.L. Ngai, C.T. White, J. Appl. Phys. 52, 320 (1981)
10. S.I. Raider, A. Berman, J. Electrochem. Soc. 125, 629 (1978)
11. D. Weaire, Energy Bands. In N.B. Hannay (ed.): Treatise on Solid State Chemistry 1, 43 (1974)
12. J. Mort, J. Knights, Nature 290, 659 (1981)

PART VI

IMPLANTATION, ANNEALING AND RADIATION EFFECTS

THEORETICAL ASPECTS OF LASER ANNEALING

J. A. Van Vechten

IBM Thomas J. Watson Research Center
Yorktown Heights, N.Y. 10598, USA

The athermal enhancement of the point defects migration by electronic excitation (e.g. by optical excitation) has been established for several years [1]. The athermal, CW-laser induced gliding of dislocations subject to external stress, e.g. due to misfit with adjacent layers, has also been previously established [2,3]. Because the athermal dislocation gliding was observed to have an extremely abrupt threshold, it was concluded that it might result from some "new cooperative phenomenon", perhaps analogous to that occurring at the superconducting transition [2]. More recently, CW-laser irradiation of AlGaAs structures, very similar ones to those studied by Monemar, et al., except that they are free of dislocations, has been shown [4] to produce an enhancement of laser luminescent efficiency of as much as 80 %. Measurement and calculation show the material is not heated above 300 C. Evidently this occurs by inducing a rearrangement of the point defect complexes that were introduced into the device by prior crystal growth and processing; the laser output contains a new line about 90 meV below the band-to-band recombination of the devices prior to irradiation. This new line is most likely due to donor acceptor pair recombination at complexes in the alloy similar to the antisite-defect-single-donor (or single-acceptor) complexes predicted by Van Vechten [5] and observed by Reynolds et al.[6]. (Other affirmation of antisite complexes in GaAs has been provided from several sources [7-9].) This athermal defect rearrangement effect also exhibits a sharp threshold that seems to be essentially coincident with that observed by Monemar et al.

Although not all authors have agreed that pulsed laser annealing is an athermal process, experimental evidence that it is, has been pervasive since the pioneering work of Khaibullin and coworkers [10]. This author and his coworkers have noted [11-13] additional evidence that the effect, by which amorphous Si and other semiconductors can be made single crystals of remarkable quality, or defective crystals may be releaved of strain and defects, by excitation with a submicrosecond pulse of laser or other ionizing radiation (within a restricted energy density range), cannot be any simple thermal process, such as melting and re-crystallization. The direct measurements of lattice temperature by Raman scattering [14-16] by Compaan and coworkers demonstrates that the annealing effect is not only athermal, but occurs without heating the atomic lattice substantially more than 400 C, and certainly not to the normal melting temperature, 1412 C. One should note that these Raman measurements, which require correction to the raw Stokes-anti-Stokes ratio for differences in the optical absorption and Raman cross-section at the two frequencies [17], were calibrated not only with furnace heated samples and samples actually melted with a CW laser, but also by direct measurement of the correction factors using a variable frequency dye laser and the principle of Time Reversal Invariance [16].

This author has previously pointed out [13] that certain X-ray diffraction [18] and charged particle emission [19] data, which had been offered in support of the notion that pulsed laser annealing involves heating the lattice to the normal melting point, thermal melting, rapid crystallization and quenching, actually contradict that notion.

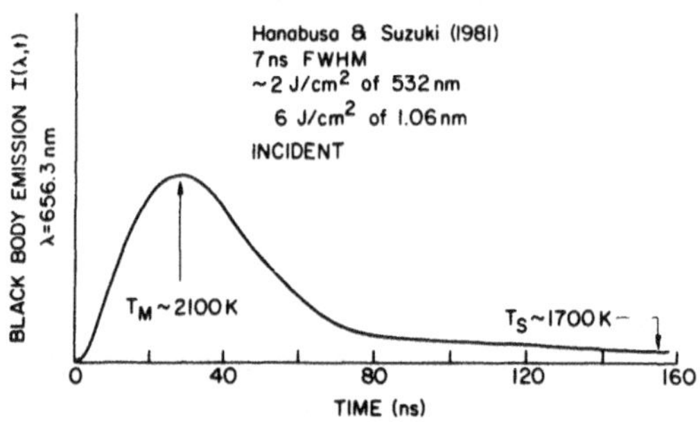

Fig. 1. Thermal radiation observed by Hanabusa and Suzuki [21] in their laser sputtering work. The fluence used here is 10 times that used for laser annealing. Note the 20 ns delay between the peak of the laser pulse and the peak of the surface temperature. This is consistent with an estimate of the effective heating depth, due to motion of photoexcited carriers at they loose energy to the lattice, in molten Si of 2 μm. The maximal temperature is also consistent with that effective heating depth

(More will be published about this elsewhere.) Other studies of ion and atom emission during pulsed laser irradiation affirm the nonthermal nature of that effect both for fluences used for annealing [20] and at much higher fluences, where the sample is severaly damaged and actually heated to about 1850 C as determined by black-body emission (see Fig. 1). At the fluences used for annealing no black-body emission is observed [21], although band-to-band luminescence is observed and shows a width indicating a temperature [22] in rough agreement with the Raman measurements. This band-to-band transmission persists [23,24] for more than 1 μs which must imply large densities of electrons and holes remain unrecombined for that period. Earlier black-body measurements had show that the pulsed laser damage threshold for Si occurs when the maximal temperature reaches 1500 K, about 200 °K below the melting point [25]. Recent time resolved measurements of the work function [26] also show that it reaches 4.3 eV, a value that may be deduced for molten Si from X-ray emission data [27], only above the single shot damage threshold. Below that threshold the work function shifts from 4.6 eV to 5.2 eV on the Si <111> as surface state pinning is eliminated and the material goes to the flat-band condition. Additional evidence that the normal molten phase does not form, comes from 4-point probe measurements of electrical conductivity [28], optical transmission [29], and dopant redistribution [30].

Having established what does not happen during laser annealing, one would wish to understand why the energy of the ionizing radiation is not simply converted to heat in the lattice before and annealing can occur, so that the sample would be thermally shocked and quenched, and how the annealing does take place. This author originally proposed [31] that the annealing occurred due to the athermal enhancement of point defect migration and the athermal gliding of dislocations in the presence of a hot dense plasma of excited electrons and holes. The role of optically induced dislocation glide was affirmed by the observation that the crystallization is not epitaxial [32], dislocations in the substrate bend away from the annealed region. The ubiquitous SiO_2 precipitates do not have their structure continued in the annealed region either. It was proposed that a combination of effects including dielectric screening, plasmon decoupling from phonon modes, and velocity dependence of electron-phonon scattering matrix elements accounted for the slowing of the rate of transfer from the excited carriers to the lattice. There is no doubt that such effects occur [33,34] although their magnitude may not be sufficient [35] to account for the small degree of lattice heating that has been observed. It has also been clearly demonstrated that the electron temperature is driven past 2000 K with laser pulses only 10 % of the annealing threshold [36]. Due to the thermionic emission at higher electron temperatures, the sample charges up to the point that positive ions are stripped off in numbers equal to the electron yield [37]; for this reason it is difficult to infer the electron temperature when laser pulses approach the annealing threshold. It was further proposed that, at sufficient density of excited carriers, the transverse acoustic phonon modes go to zero frequency and the material undergoes a weakly first order phase transition to a state that may be called liquid, because it cannot resist a shearing stress, but is very different from the thermal molten phase obtained above 1412 C. Finally, it was proposed that the reason why the surface of the material does not become very hot is that the plasma of carriers diffuses rapidly into the bulk of the sample carrying the energy that was absorbed with it.

That initial proposal has been modified in view of new theoretical insights and more extensive experimental data. Significant changes in our understanding include: 1) realization that the plasma of excited carriers will contract [38,39], rather than expand, when induced with a sufficient gradient and density due to a counteracting gradient of total vibronic states available; 2) realization that energy is re-radiated away from the region in which it is absorbed (note earlier observations of this radiation and of the carrier confinement effect in [23]); 3) realization that the excited carriers within the self-confined layer are not likely a thermal distribution of free electrons and holes, but are better described as Frenkel excitons [13], which accounts for the observation that neutral atoms emitted from the surface of laser irradiated Si emit 5 eV photons [21] and that the surface layer expands its lattice constant [40] as if it were heated to about 1000 C,but this electronic expansion is tightly confined to the surface [18], which could not be produced by a thermal gradient consistent with conservation of energy and the known thermal conductivity of Si (Fig. 2). (Note that thermal model calculations [41] require $dT/dz = 0$ at the surface, $z = 0$, where generally T is greatest, when the laser is off because they then have no source of heat and thermal radiation from the surface is generally negligible. As no heat flows in or out from the surface, the temperature gradient must be zero there. Thus the gradient of atomic spacing from which [18] inferred the T profile in Fig. 2 must have a nonthermal origin). 4) realization that this dense layer of Frenkel excitons could undergo a boson condensation to a superfluid state [12,14,42]. Note that this last point brings us back to the conclusion that the very abrupt threshold for CW-laser induced dislocation glide [2] and for the CW-laser annealing of Salathé et al. [4] requires a cooperative, many-body effect similar to that which produces superconductivity. Additional evidence for such a cooperative phenomena, for both pulsed and CW laser irradiation,

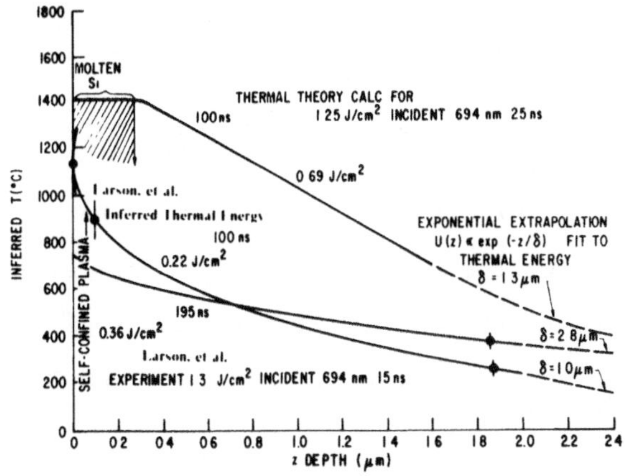

Fig. 2. Comparison of the thermal gradient inferred by Larson et al. [18], from their time resolved X-ray scattering measurements of atomic spacings, which they interpretted as if thermal expansion were the only cause of a change in atomic spacing, with the thermal model of pulsed laser annealing of [41]. The profile is in much better agreement with that predicted by the author in [40]

comes from observations of linear surface ripple patterns which appear to be persistent standing waves [43]. For a discussion of Frenkel excitons degenerate with band states see [44] and [45].

References

1. J.C. Bourgoin and J.W. Corbett, Rad. Eff. 36, 157 (1978) and therein
2. B. Monemar, R.W. Potemski, M.B. Small, J.A. Van Vechten and G.R. Woolhouse, Phys. Rev. Lett. 41, 260 (1978)
3. J.H. Basson and J.A. Van Vechten, Phys. Rev. B 23, 2032 (1981)
4. R.P. Salathé, H.H. Gilgen and Y. Rytz-Froidevaux, IEEE J. Quant. Electron. QE-17, 1989 (1981)
5. J.A. Van Vechten, J. Electrochem. Soc. 122, 419 and 423 (1975); in "Handbook on Semiconductors" ed. S.P. Keller (North-Holland, Amsterdam, 1980) p. 1
6. D.C. Reynolds, C.W. Litton, R.J. Almassy and G.L. McCoy, J. Appl. Phys. 51, 4842 (1980)
7. R.Y. Wagner, J.J. Krebs, G.H. Strauss and A.M. White, Solid State Comm. 36, 15 (1980)
8. N.K. Goswami, R.C. Newman and J.E. Whitehouse, Solid State Comm. 40, 473 (1981)
9. J. Lagowski, H.C. Gatos, J.M. Parsey, M. Kaminska and W. Walikiewcz, Appl. Phys. Lett. 40, 342 (1982)
10. I.B. Khaibullin, E.I. Shtyrkov, M.M. Zaripov, R.M. Bayazitov and M.F. Galjautdinov, Rad. Eff. 36, 225 (1978) and therein
11. J.A. Van Vechten, R. Tsu, F.W. Saris and D. Hoonhoutn, Phys. Lett. 74a, 417 (1979)
12. J.A. Van Vechten and A.D. Compaan, Solid State Comm. 39, 867 (1981)

13. J.A. Van Vechten, in "Laser and Electron-Beam Interactions with Solids",ed. B.R. Appleton and G.K. Celler (Elsevier Science, New York, 1982) p. 49
14. A. Compaan, A. Aydinili, M.C. Lee and H.W. Lo, ibid. p. 43
15. H.W. Lo and A. Compaan, Phys. Rev. Lett. 44, 1604 (1980)
16. A. Compaan, H.W. Lo, M.C. Lee and A. Aydinli, Phys. Rev. B 26, 1079 (1982)
17. J.B. Renucci, R.N. Tyte and M. Cardona, Phys. Rev. B 11, 3885 (1975)
18. B.C. Larson, C.W. White, T.S. Noggle and D. Jills, Phys. Rev. Lett. 48, 337 (1982)
19. J.M. Liu, R. Yen, H. Kurz and N. Bloembergen, Appl. Phys. Lett. 39, 755 (1981)
20. T. Nakayama, N. Itoh, T. Kawai, K. Hashimoto and T. Sakata, Rad. Eff.Lett. 67, 129 (1982)
21. M. Hanabusa and M. Suzuki, in Op. Cit. Ref. 13, p. 559
22. M.C. Lee, A. Aydini, H.W. Lo and A. Compaan, J. Appl. Phys. 53, 1262 (1982)
23. N.G. Nilsson, Phys. Cripta 8, 166 (1973) and therein
24. D.H. Lowndes, G.E. Jellison and R.F. Wood, in Op. Cit., Ref. 13, p. 76, Fig. 3
25. A.A. Grinberg, R.F. Mekhtiev, S.M. Ryvkin, V.M. Salmanov and I.D. Yaroshetskii, Sov. Phys. Solid State 9, 1085 (1967)
26. R.T. Williams, M.N. Kabler, J.P. Long, J.C. Rife and T.R. Royt, in Opt. Cit. Ref. 13, p. 97
27. C.F. Hague, C. Senemaud and H. Ostrowiecki, J. Phys. F: Metal Phys. 10, L267 (1980)
28. M. Yamada, H. Kotani, K. Yamazaki, K. Yamamoto and K. Abe, J. Phys. Soc. Japan 49, Suppl. A 1299 (1980)
29. H. Kotani, M. Yamada, K. Yamamoto and K. Abe, Solid State Comm. 41, 461 (1982)
30. A.V. Dvurechenskii, T.N. Mustafin, L.S. Smirnov, H.D. Geiler, G. Gütz and U. Jahn, Phys. Stat. Sol. 63a, K203 (1981)
31. J.A. Van Vechten, J. de Phys. 41, C4-15 (1980)
32. W. Porter, D.L. Parker, T.W. Richardson and E.J. Swenson, Appl. Phys. Lett. 33, 886 (1978)
33. A.V. Nurmikko and B.D. Schwartz, J. Vac. Sci. and Techn. 21, 229 (1982)
34. S.S. Yao, J. Buchert and R.R. Alfano, Phys. Rev. B 25, 6534 (1982)
35. E.J. Yoffa, Phys. Rev. B 23, 1909 (1981)
36. M. Bensoussan and J.M. Moison, J. de Phys. 42, C7-149 (1981)
37. J.M. Moison and M. Bensoussan, J. Vac. Sci. Techn. 21, 315 (1982)
38. M. Wautelet and J.A. Van Vechten, Phys. Rev. B 23, 5551 (1981)
39. A. Aydinli, H.W. Loo, M.C. Lee and A. Compaan, Phys. Rev. Lett. 46, 1640 (1981)
40. J.A. Van Vechten, Jpn. J. Appl. Phys. 21, L125 (1982)
41. R.F. Wood and G. Giles, Phys. Rev. B 23, 2923 (1981)
42. M. Nagy and M. Noga, Czech. J. Phys. B 31, 1358 (1981)
43. J.A. Van Vechten, Solid State Comm. 39, 1285 (1981)
44. W. Hanke and L.J. Sham, Phys. Rev. B 21, 4656 (1980)
45. J. Rogan and J.E. Inglesfield, J. Phys. C. Solid State Phys. 14, 3585 (1981)

RADIATION METHODS FOR CREATION OF HETEROSTRUCTURES ON SILICON

N.N. Gerasimenko

Novosibirsk, USSR

The physical aspects for the creation of heterostructures on silicon and the control of its properties by radiation methods are considered in the present lecture.

First of all let us determine the expression "heterostructure on silicon". The following structures can be attributed to this class:

1) Structures of semiconductor-dielectrics, in particular structures as SiO_2 and SiO_x $(x < 2)$.
2) Semiconductor films on silicon substrate, for example structures as Si-SiC and Si-GaP.
3) Silicide films on silicon.

Radiation treatments in our particular case are irradiation by particles and quanta for changing structure of material and/or stimulation reaction of formation of new phases. At the same time, it means special methods of annealing of radiation damage using electromagnetic radiation in wide range of wavelength.

We will consider the peculiarities of radiation effects in $Si\text{-}SiO_2$ system from the point of view of structural rearrangements in SiO_2 films.

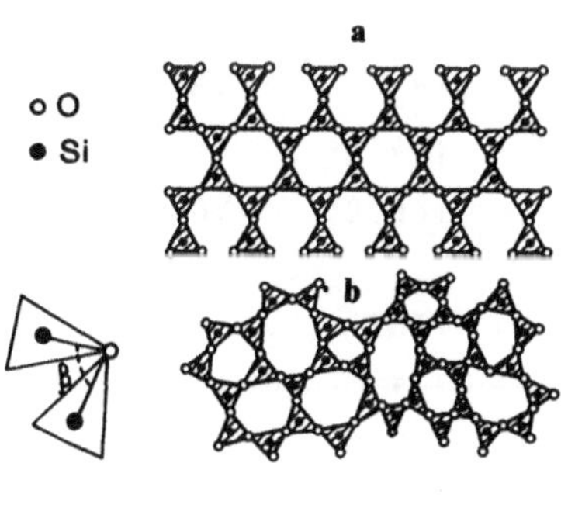

Fig. 1. Structure of silicon dioxide in crystal (a) and amorphous (b) states

The structure of crystalline and amorphous silicon dioxide is shown in Fig. 1. It is seen, that the basis of this system is silicon-oxygen tetrahedron which is connected with the neighbouring one through the so called bridge oxygen atom. The tetrahedra form rings which build a three-dimensional structure. The size of the rings determines the density of silicon dioxide because the empty space inside the rings depends on size of the rings. The size of the rings is connected with the angle β

(Fig. 1) between the chemical bonds of silicon and oxygen atoms.

The change of angle β (and consequently the change of density) will result in changing of the vibrational frequency of Si-O bonds (the band ∿1100 cm^{-1} in IR absorption spectrum). This effect can be demonstrated comparing properties of amorphous SiO_2 and its various crystalline modifications (see Table [1]).

Both elastic and nonelastic interaction between the bombarding particles and the silicon dioxide are responsible for its structure changes, which can be observed using IR spectroscopy.

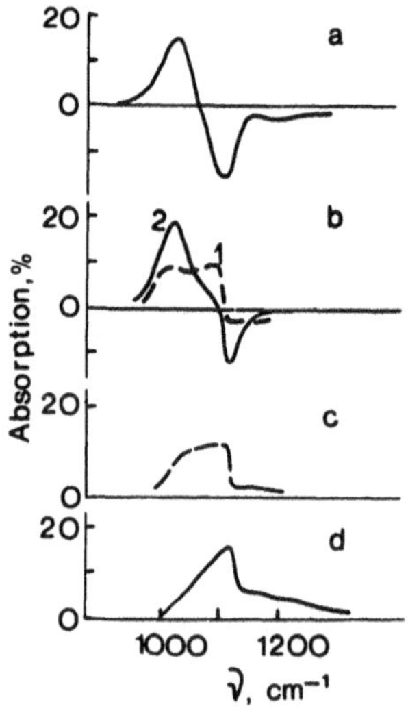

Fig. 2. IR absorption spectra (in comparison) SiO_2 films, irradiated by: a) ions B^+ (33 keV) and P^+ (100 keV); 10^{15} cm^{-2}, thickness of film d = 0.2 μm; b) protons (40 keV): 1 - 10^{15} cm^{-2}, 2 - 10^{16} cm^{-2}, d = 0.9 μm; c)-d) electrons (500 keV) and γ-rays, 10^{18} cm^{-2}, d = 0.9 μm

In Figure 2 [2] IR absorption spectra of irradiated and non-irradiated Si-SiO_2 systems are presented.

Heavy particle bombardment (in this case the radiation effects are caused mainly by elastic interactions leading to displacement of atoms of the matrix) as well as light particle - electrons and protons or

Table 1.

	Density g[cm^{-3}]	n	β	ν [cm^{-1}]
Koesite	2.87	1.596	139°	1077
α-quartz	2.65	1.549	143.9°	1078
α-crystoballite	2.33	1.485	146.8°	1095
α-trydymite	2.27	1.473		1106
Glass	2.20	1.458	150°	1098

γ-rays (when ionization leads to bond-breaking and, consequently, to structural changes) were used in these experiments. It is seen that ion bombardment leads to decreasing of the number of Si-O bonds and at the same time to decreasing of the size of rings in which silicon-oxygen tetrahedra are united. The last fact shows up in the shift of the absorption band toward the longer wavelength. The irradiation by electrons and γ-rays leads to opposite effect. The results of proton bombardment are similar to the results of electron bombardment at small doses or ion bombardment at high doses. In the last case the defects produced by elastic processes accumulate and they become more important in changing of the structure. From the analysis of the results of treatment by light particles and γ-rays we can conclude that irradiation can increase the structural quality of silicon dioxide films. Probably this is because of the nonideality of the structure of as-grown films.

The shift of the discussed band to the longer wavelenghts shows us that the size of the rings in which silicon-oxygen tetrahedra are combined decreases and consequently the empty space inside the film also decreases, thus silicon dioxide becomes more dense. This densification by irradiation can be used for the production of the optical waveguides in silicon dioxide [3].

The structure of radiation defects in SiO_2, its number and space distribution can be determined by different methods. One of the most informative methods is the electron spin resonance. The spectra corresponding to oxygen-vacancy (the centre E_1') and to free oxygen atom (the centre DI) are presented in Fig. 3. These centres disappear during annealing at temperature less than 400 °C (Fig. 4 [4]). In the same temperature interval the densification produced by heavy ion bombardment disappears. When the densification was produced by light particles irradiation it disappears at rather high temperature ∿900 °C.

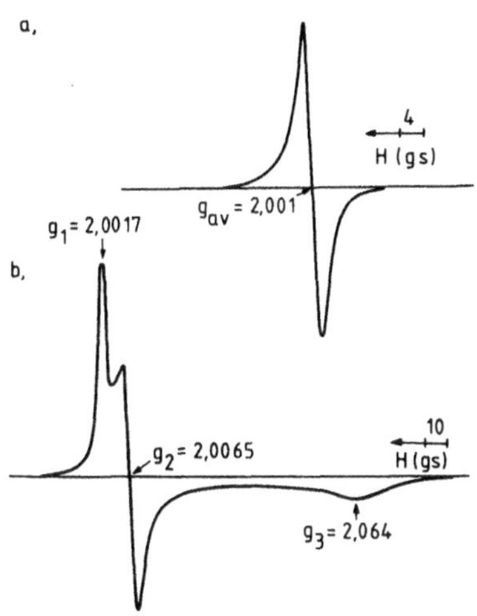

Fig. 3. ESR spectra of E_1' (a) and DI (b) centres in SiO_2

The heavy ion bombardment produces not only point defects. Heavy ions penetrating in the material create disordered region at the end of ion track. During a very short time (∿10^{-11} sec) inside this region very high temperature (∿10^4 °C) and pressure exist. For this reason we can expect principal rearrangement of the structure inside such a region. In fact, ion bombardment leads to creation of inhomogeneity of amorphous SiO_2

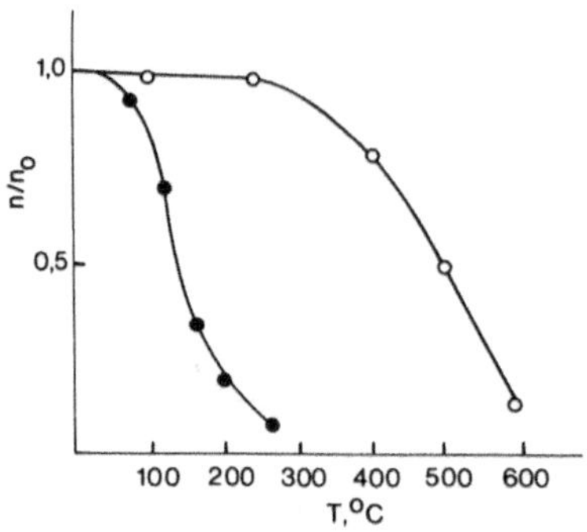

Fig. 4. Isochronal annealing of paramagnetic centres in the helium atmosphere: o- E_1'-centres; ●- DI-centres

as shown in Fig. 5 [5]. By electron microscopy using gold-carbon replica and by removal of the irradiated material it was shown that inclusions with different properties (different density) appear inside the irradiated layer. But the irradiated layer remains still amorphous. During consequent annealing these inclusions can play a role of nuclei for crystallization. The crystallization leads not only to crystoballite formation (in the region of maximum of elastic energy losses for bombarding ion) but also to creation of coesite (high pressure phase) close to the interface.

The effect of lowering the temperature of crystallization in SiO_2 films after irradiation was found earlier by Pavlov and coworkers [6].

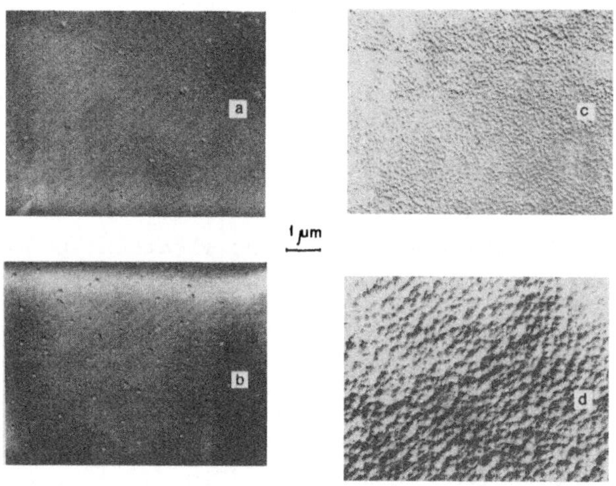

Fig. 5. Microphotos received by replica method (x 10000) SiO_2 films: non-irradiated (a) and irradiated by Ar^+ ions with energy 40 keV and doses (cm^{-2}): b) - 10^{13}; c) - 10^{15}; d) - $5 \cdot 10^{15}$

An important peculiarity of the structural changes of SiO_2 caused by irradiation is the formation of damaged region close to the surface.

In Figure 6 [7] this effect is shown. It turned out, that during ion bombardment independently of the energy of the ions a damaged layer is formed which absorps water vapor, hydrogen from the air creating bonds Si-H and Si-OH.

Formation of these layers, in our opinion, is a general process during irradiation of compounds, this is connected with the fact that in the compound different atoms have different diffusivity and volatility. Gradient of the electrical fields at the surface and the fields of elastic strains also help this process.

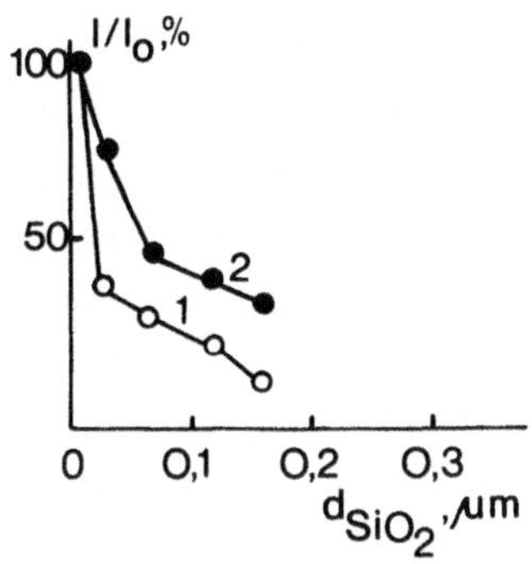

Fig. 6. Changing of the intensity of Si-OH (o) and Si-H (•) absorption band during stripping of SiO_2, irradiated by P^+ ions, 10^{15} cm^{-2} and annealed 500 °C

Using irradiation one can control the properties of heterostructure Si-SiO_2.

We will describe only certain possibilities of the use of radiation for this purpose. It is well known, that as a consequence of irradiation in the Si-SiO_2 system positive charges are induced. This charge build-up causes a shift in the C-V characteristics of such systems. On the other hand, if the Si-SiO_2 system is irradiated with electrons at elevated temperatures the build-up of the charges in the Si-SiO_2 system will not occure and a following irradiation at room temperature will not cause shift in the C-V characteristics (Fig. 7 [8]).

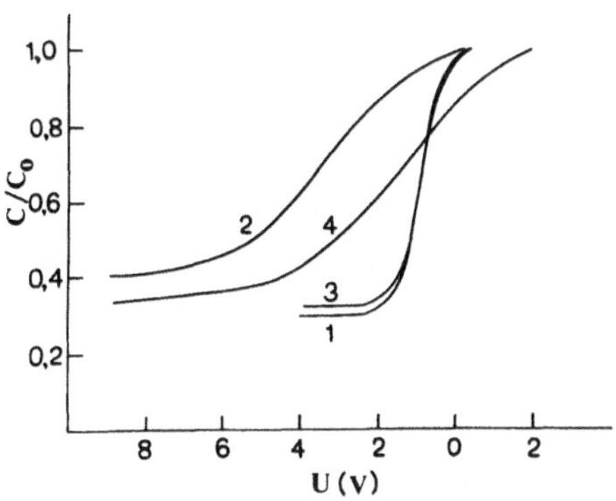

Fig. 7. C-V curves of Si-SiO_2 structure - nonirradiated (1) and irradiated by electrons with energy E = 3.5 MeV under the conditions: (2) 10^{15} cm^{-} , 20 °C; (3) 10^{15} cm^{-2}, 300 °C; (4) 10^{15} cm^{-2}, 300 °C, 10^{15} cm^{-2}, 20 °C

An important factor of failure of submicron devices is the lateral inhomogeneity of the build-up charge. Measurements carried out with mercury-probe show that the build-up charge changes from point to point along the surface. In Fig. 8 the fluctuations of charge, conductivity and capacity of the Si-SiO_2 system is shown, the measurements were carried out using mercury-probe the size of which is represented by a dot in this figure. Figure 8a shows data for the non-irradiated structure, Figure 8b is given for the case of irradiation with electrons with dose of 10^{15} cm^{-2}, energy 3.5 MeV at room temperature and Fig. 8c presents the results of the same irradiation carried out at 300 °C. It is well seen, that in the last case the fluctuations are practically eliminated - at least in the range of the resolution of the

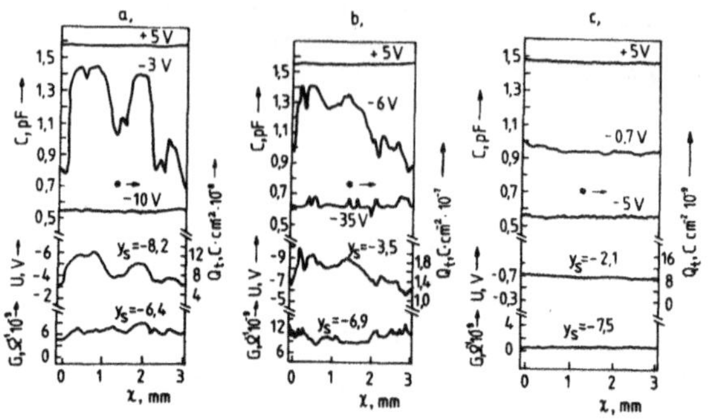

Fig. 8. Fluctuations of capacitance (C), charge (Q) and conductivity (G) in $Si\text{-}SiO_2$ structures-nonirradiated (a) and irradiated by electrons, E = 3.5 MeV and dose 10^{15} cm^{-2} at 20 °C (b) and 300 °C (c)

mercury-probe.

The analysis of data shows that in contrast with the irradiation at room temperature (when there are mainly introduced electron traps) in the case of "hot" irradiation traps are created both for electrons and holes with comparable characteristics and concentrations.

Using radiation methods it is possible to control not only the structure (and consequently the electrophysical characteristics) of the $Si\text{-}SiO_2$ system but its composition too.

The best demonstration of this possibility gives us the results of experiments of the irradiation of the $Si\text{-}SiO_2$ system with boron ions.

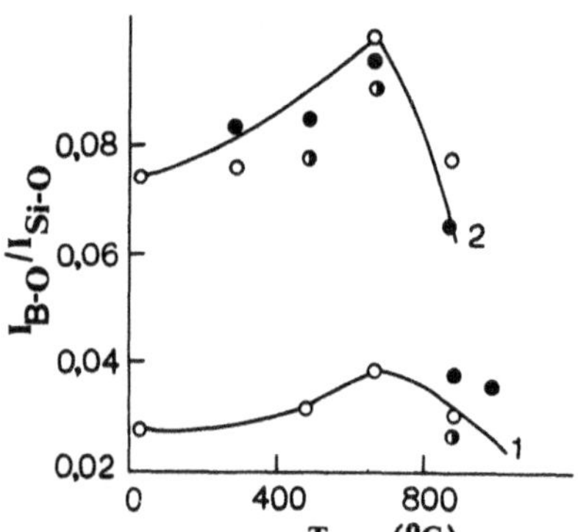

Fig. 9. Dose dependence of the intensity of the absorption band 245 nm (B_2-band) upon ion bombardment of quartz glass: (1) B^+, 100 keV; (2) He^+, 550 keV

Using such a treatment with annealing it is possible not only to introduce B-O bonds into SiO_2 (amount of which can be determined by IR spectroscopy) (see Fig. 9 [9]) but also to change the stoichiometry of the film by creating excess of silicon. This means, using bombardment of boron it is possible to obtain layers of SiO_x $(x < 2)$ in a controlled way. The enrichment in Si of the film SiO_2 after bombardment with boron is demonstrated with data shown in Fig. 10 [10].

It is seen that beginning with the dose of boron when its concentration reaches the concentration of the atoms of the matrix the concentration of the B-2 centers increases rapidly. The B-2 centers con-

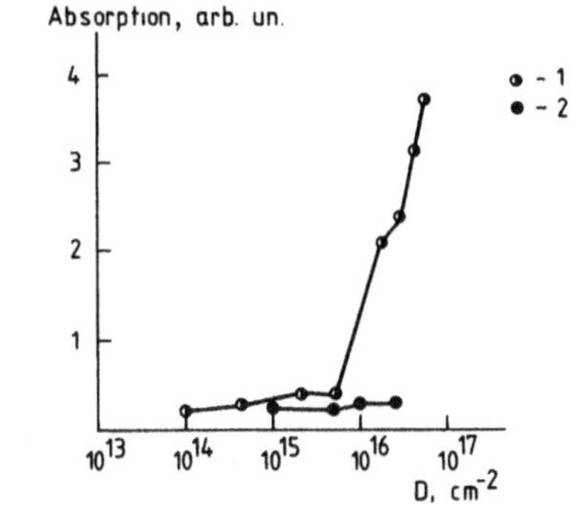

Fig. 10. Changing of the number of B-O bonds during annealing of SiO_2 films irradiated by B^+ ions with doses: (1) - $1 \cdot 10^{17}$ cm^{-2}; (2) - $2 \cdot 10^{17}$ cm^{-2}

nected with oxygen vacancies can be detected by UV spectroscopy at 245 nm. In Fig. 11 scheme of formation of oxygen vacancies in irradiated SiO_x ($x < 2$) is presented.

Using boron bombardment of SiO_2 when enrinchment of Si occures, it is possible to create optical waveguides in SiO_2 films on Si substrates.

The refractive index of such films increases greatly because of the silicon enrichment (Fig. 12 [11]).

Applying ion bombardment one can not only control the composition of heterostructures but create new heterostructures as well.

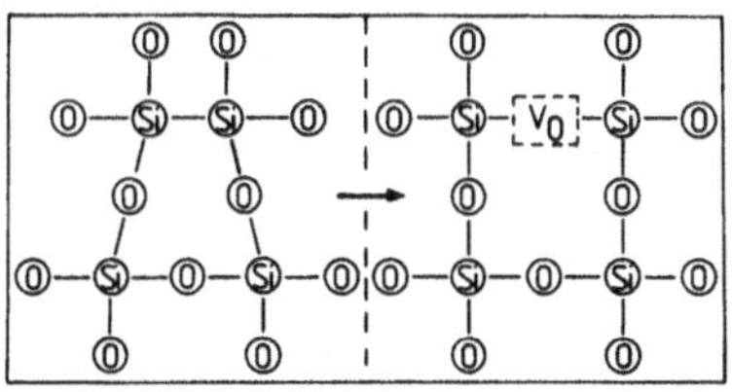

Fig. 11. Scheme of the creation of the oxygen vacancy under irradiation of SiO_2

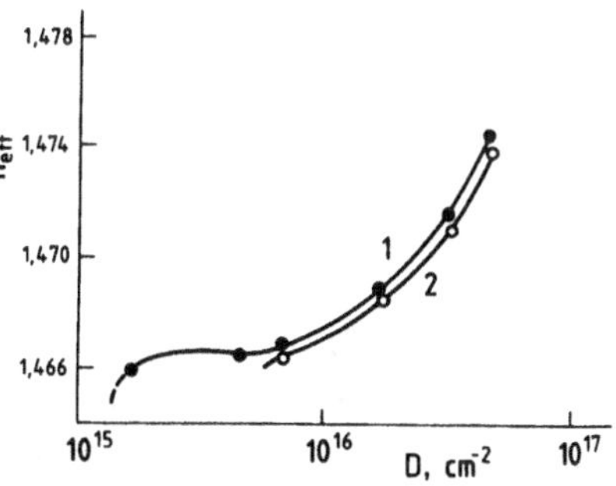

Fig. 12. Dose dependence of the effective index of refraction under B^+ irradiation (100 keV): 1 - silica glass; 2 - Si-SiO_2 structure

The method of creation such structures by ion bombardment was called ion synthesis and expecially for silicides production - ion beam mixing. By ion synthesis it is possible to make such heterostructures which are too hard or simply impossible to produce by other methods.

The first most successful example of using the ion synthesis was the production of single crystal of silicon carbide films on silicon substrate (Fig. 13 [12]). It was shown that the density of interface states is relatively small in such a system ($< 10^{11}$ cm^{-2}). At the last time very good quality heterojunctions GaP-Si were produced by the new version of ion synthesis - ion - liquid epitaxy [13].

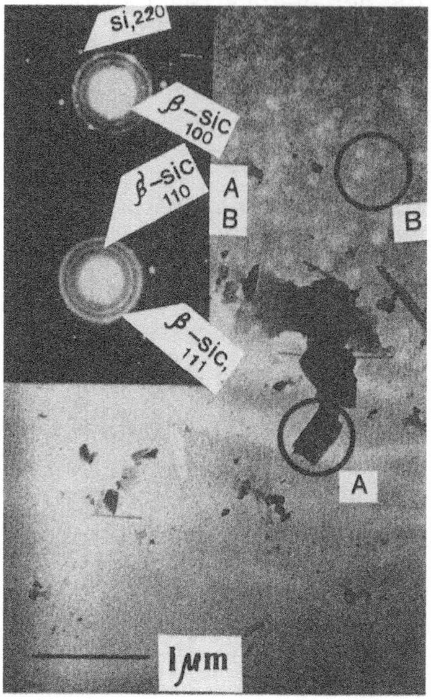

Fig. 13. TEM photo of the SiC film produced by C^+ ion, E = 30 keV, 10^{17} cm^{-2} on Si substrate

In most cases when radiation methods are used for creation of heterostructures it is necessary to use subsequent heating to perfect the structure. The advantages of pulse heating methods (laser, electron beam annealing) are well known, but the practical use of such methods are still limited because of their serious technical troubles. In some cases in our opinion microwave annealing is more respective because of its good lateral homogeneity and its easy control [14]. The most interesting field for using of such a method is the annealing of systems containing conductive layers. In this case mostly these layers will be heated because of the skin-effect.

In conclusion we have to underline that the problem of control of the structure and the composition of heterosystems cannot be solved successfully without paying attention to the existence of electric and mechanical strain fields close to interface. Some effects connected with such fields were described in Prof. Litovchenkó's lecture [15]. Unfortunately, there is very little information about the influence of these fields on radiation effects.

We will show anyway two exciting examples which can demonstrate the important role of this factors in radiation effects in heterostructures.

In Figure 14 [16] the results of investigation of mechanical stresses in Si-SiO_2 systems under irradiation by fast electrons are shown. It is seen, that these stresses are dose dependent and correlate with the changes in the IR absorption spectrum of SiO_2.

Figure 15 [17] demonstrate the effect of electric field on defect redistribution. In this experiments the information about defects was obtained by photo e.m.f. method.

This figure shows that defects redistribute in the p-n junction electric field.

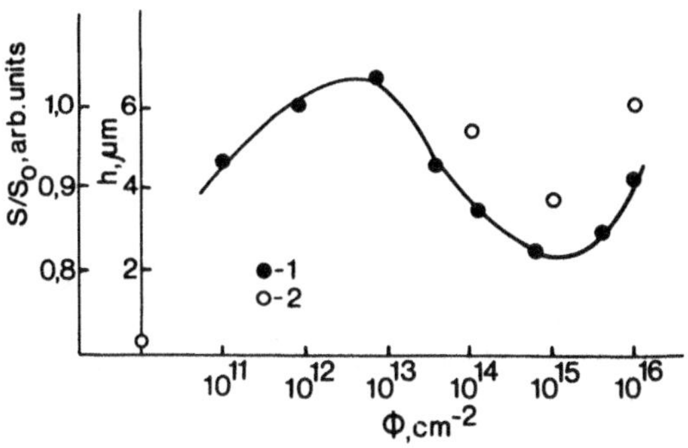

Fig. 14. Dose dependence of the bending h (1) and intensity of IR band (1100 cm^{-1}) of $Si-SiO_2$ structures irradiated by fast electrons

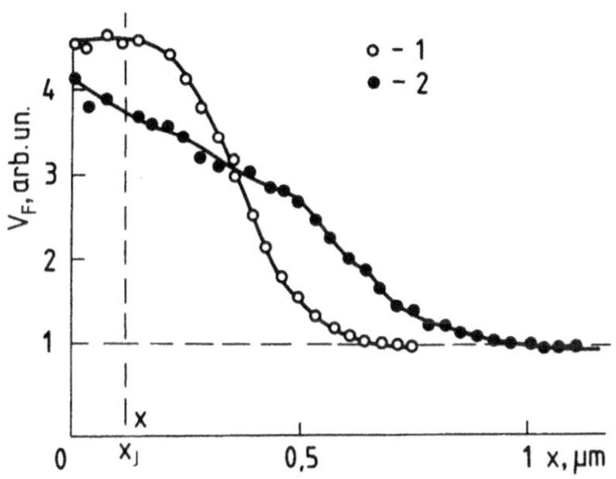

Fig. 15. Depth distribution of photo e.m.f. for samples with n^+p junction (1) and without it (2)

Results presented in this lecture show that at the very moment we have enough information about radiation processes in heterosystems to use radiation methods for creation of such systems and for control of their properties.

References

1. A.G. Revesz, Phys. Rev. Lett. 27, 1578 (1971)
2. N.N. Gerasimenko, Proc. Int. Conference on Ion Implantation (Boulder, 1976) p. 671
3. K. Aritomo, T.I. Kegami, T. Nishimura, K. Masuda, S. Namba, Anneal Report of Namba Lab. (Osaka Univ. Press, 1972) p. 18
4. B.I. Vikhrev, G.P. Lebedev, N.N. Gerasimenko, Microelectronica 6, 71 (1977)
5. N.N. Gerasimenko, V.I. Ivanchenko, G.P. Lebedev, Neorganicheskie Materialy 4, 691 (1978)
6. P.V. Pavlov, E.V. Shitova, E.I. Zorin, N.A. Genkina, Crystallographia 18, 609 (1973)
7. N.N. Gerasimenko, T.I, Kovalevskaya, G.M. Tseitlin, Zhurnal Prikladnoi Spectroscopy 28, 881 (1978)

8. N.V. Vershinina, N.N. Gerasimenko, phys. stat. sol. (a) 72, 1 (1982)
9. N.N. Gerasimenko, P.E. Kandyba, Z.V. Panova, V.Z. Petrova, Electronnay Technika Ser 3, 2, 55 (1977)
10. N.N. Gerasimenko, G.M. Tseitlin, S.V. Vasiliev, phys. stat. sol. (a) 62, K169 (1980)
11. N.N. Gerasimenko, V.G. Pankin, K.K. Svitashev, G.M. Tseitlin, Pisma v Zhurnal Technicheskoi Physiki 4, 582 (1978)
12. N.N. Gerasimenko, O.N. Kiznetsov, L.V. Lezheiko, E.V. Lubopytova, L.S. Smirnov, F.L. Edelman, Microelectronica 3, 467 (1974)
13. N.N. Gerasimenko, L.V. Lezheiko, E.V. Lubopytova, L.V. Sharanova, A.G. Sharshunov, A.Ya. Shik, U.V. Shmartsev, Physika i Technika Poluprovodnikov 15, 1088 (1981)
14. A.V. Rzhanov, N.N. Gerasimenko, V.I. Obodnikov, S.V. Vasiliev, Pisma v Zhurnal Techichiskoi Phiziki 7, 1221 (1981)
15. V.G. Litovchenko, this issue
16. G.A. Zhukova, V.N. Mordkovich, Proc. Int. Conference on Radiation Effects in Silid State (Tbilisi, 1979) p. 829
17. N.M. Morosov, D.I. Tetelbaum, P.V. Pavlov, E.I. Zorin, Proc. X Conference on the Interaction Charge Particles with Solids, Moscow, 1979

ION BEAM GETTERING IN GaP

H. Klose and M. Griepentrog
Sektion Physik, Humboldt-Universität zu Berlin
1040 Berlin, Invalidenstr. 42, GDR

1. Introduction

Gettering of fast diffusing elements by defects and defect complexes in the bulk, interfaces and surfaces of the semiconductors in vicinity of space charge regions of electronical devices leads to an increase of the minority carrier lifetime and to a decrease of electrically active deep center concentrations, respectively.

Different gettering procedures are applied in silicon technology today. Such gettering methods can be characterized by the mechanism of defect production or by the nature of the migration and binding type. Damage regions can be produced by mechanical treatment, ultrasonic, misfit growth in epitaxial processes, bending of wafers, laser irradiation, highly doped regions, planar layer growth and last but not least by ion implantation.

In the case of ion beam gettering electrically inactive atoms in the lattice like Ar or Kr and/or active atoms like B in Si or Al in GaP are introduced by ion implantation following a special annealing procedure. The advantages of the ion beam gettering compared to other gettering methods are

(i) the high reproducibility of the damage production and concentrations of implanted ions;

(ii) the introduction of electrically inactive and/or active atoms inside the crystal lattice and in surface regions;

(iii) the high vacuum conditions;

(4i) the damage production through and in planar layers and the lateral resolution of damage regions;

(5i) the lower annealing temperatures in comparison with the planar and diffusion doping gettering;

(6i) the combination with other implantation or annealing steps.

The ion beam gettering is already a part of silicon technology [1] and especially the gold gettering was intensively studied in the last years [2], whereas in the case of $A_{III}B_V$-compounds this gettering procedure was well investigated only for Cr in GaAs [3]. Recently short notes were published on the glass gettering in GaP LED structures [4] and on the copper gettering in epitaxial GaP material [5].

Already low copper concentrations change the recombination parameters of the GaP [6] and the electrooptical characteristics of LED's. From the viewpoint of the ion beam gettering copper in GaP performs all conditions for an effective gettering, because it has a high solubility and a big diffusivity, and is a tracer impurity element in nearly all technological treatments.

2. Experimental Details

2.1. Preparation of Samples

Samples for all investigations were prepared from (100) gallium-phosphide crystals grown by means of the LEC technique, excluding a possible effective gettering in the misfit range of the epitaxial material. Ion implantations were performed from the polished front side of the wafers with the special 350 keV accelerator of the Humboldt University. After ion implantation copper vacuum deposition took place on the backside with a thickness between 1000 and 3000 Å. During a following temperature treatment under flowing nitrogen copper atoms diffused into the bulk and the gettering occurred.

2.2. Measurements

The technique of secondary ion mass spectrometry has been used in order to measure the concentration depth profiles of gettered copper in the ion damaged regions and the profiles of implanted atoms, respectively. In most experiments the secondary-ion analyzer SMI 300 (Cameca) was applied. The primary oxygen ions had an energy of 5.5 keV. For the detection of copper the $(^{69}Ga^{63}Cu)^+$ cluster ion was analysed instead of $^{63}Cu^+$ ion because of the better signal to noise ratio. Control measurements were made on mass number 63 and in special cases with the high resolution microanalyzer ims 3f, too. The depth calibrations were carried out by measuring the sputtering time and the sputtering crater depths by means of an electromechanical depth profiler (Talystep, Taylor and Hubson). Implanted profiles of copper served as standard for calibration. In Figure 1 implanted profiles are shown.

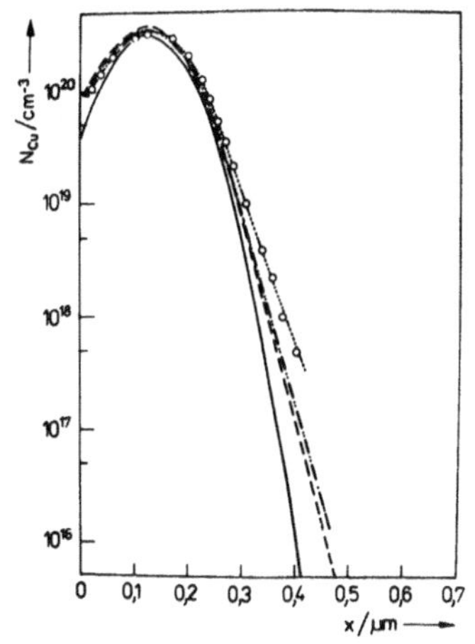

Fig. 1. SIMS depth profiles of ^{63}Cu implanted and calculated profile in GaP
ooo by means of SMI 300 with oxygen primary ions;
—··— by means of ims 3f with oxygen primary ions;
---- by means of ims 3f with cesium primary ions;
—— calculated profile with R_p and ΔR_p after [7]

In principle SIMS analysis can also give information about the depth profiles of the implanted elements for producing the ion damaged regions. Depth profiles of implanted matrix elements can not be measured by SIMS. In the case of arsenic implantation the $^{75}As^+$ ion signal was used for analysis. For matrix elements the depth profiles were calculated by LSS theory using the data from [7]. Measurements of depth profiles of noble gas elements are very difficult, because of the low secondary ion yield. In the case of ^{40}Ar implantation also mass interferences can take place (e.g. the ^{40}Ca have a high secondary yield). In special investigations it has been shown, that for the determination of Ar implanted profiles the cluster ion $(^{69}Ga^{40}Ar)^+$ can be used with good results [8]. For measuring of krypton depth profiles the ^{83}Kr signal was used.

3. Results

The copper gettering in ion damaged regions in GaP samples was shown in [5] by means of SIMS and RBS data for the first time. Copper atoms decorate the damage regions produced by ^{40}Ar ion implantation in case of front and back side gettering, respectively. In Figure 2 the redistribution of implanted copper atoms in ^{40}Ar and in ^{63}Cu ion damaged regions is shown. Depth profiles have a maximum in distance of more as ΔR_p from R_p values and a typical surface peak. In comparison with the backside gettering the same profile of gettered copper is got in argon implanted regions.

For the study of gettering mechanism the annealing time and temperature dependence of the ion beam gettering in damaged layers containing argon were investigated. The results are represented in Fig. 3.

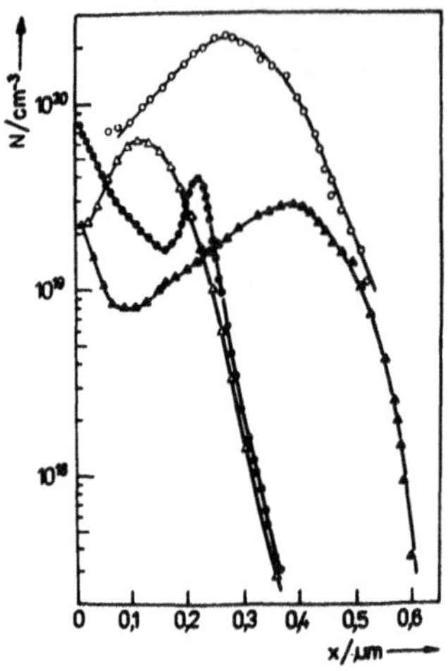

Fig. 2. SIMS depth profiles of the front side gettering of Cu in Cu or Ar implanted regions.
ΔΔΔ ^{63}Cu, D = 1.10^{15} cm^{-2}, E = 200 keV
ooo ^{40}Ar, D = 5.10^{15} cm^{-2}, E = 200 keV
●●● ^{63}Cu, distribution after annealing for 1 h at 600 °C without Ar ion implantation
▲▲▲ ^{63}Cu, distribution after Ar and Cu implantations and the same annealing procedure

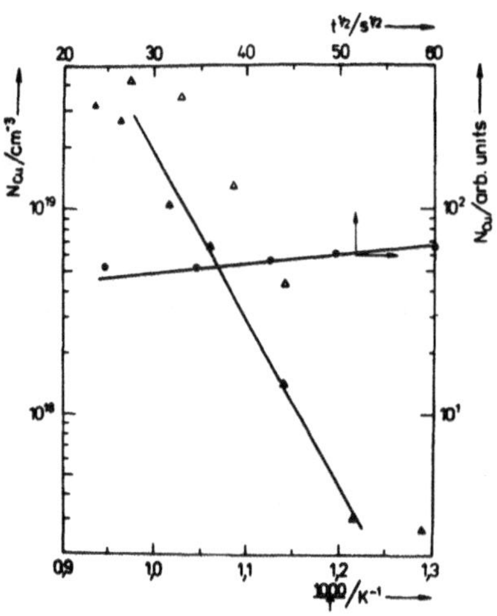

Fig. 3. Time and temperature dependence of Cu back side gettering in ^{40}Ar ion implanted regions (D = = 2.10^{15} cm^{-2}, E = 300 keV, d_{Cu} = = 3000 Å on backside):
▲▲▲ annealing time 1 h, average copper concentration
ΔΔΔ values from [5]
●●● annealing temperature 600 °C, whole gettered copper concentration

These curves can be described by a time and temperature dependence of gettered copper concentrations in the following expression:

$$N_{Cu}(t_a,T_a) = \text{const.}\ \exp\{-E_A/kT_a\} \cdot \exp\{At_a^{1/2}\}$$

The data of Figure 3 yielded for the characteristical activation energy E_A = 1.62 eV and for the "migration diffusion" coefficient D_m(600 °C) = = $1.04 \cdot 10^{-7}$ $cm^2 s^{-1}$ in correspondence with earlier measurements [5] assuming that the thickness of the sample L and D_m can be expressed by $A = D_m^{1/2} \cdot L^{-1}$. Deviations from this expression are evident for lower annealing temperature before efficient gettering occurs and for upper temperature region produced by different competing procedures of gettering, annealing of radiation damage and decomposition of the lattice in the surface regions. Different gettering effects in presence of various complex structures in GaP lattice are supposed. The possible experiment proving this assumption is the damage production by P or Ga ion implantation. The gettered copper depth profiles of ^{31}P and ^{69}Ga implanted samples show an effective gettering effect, a dependence of implanted ion species and in both cases a maximum of gettered copper concentrations lying some ΔR_p behind the R_p value.

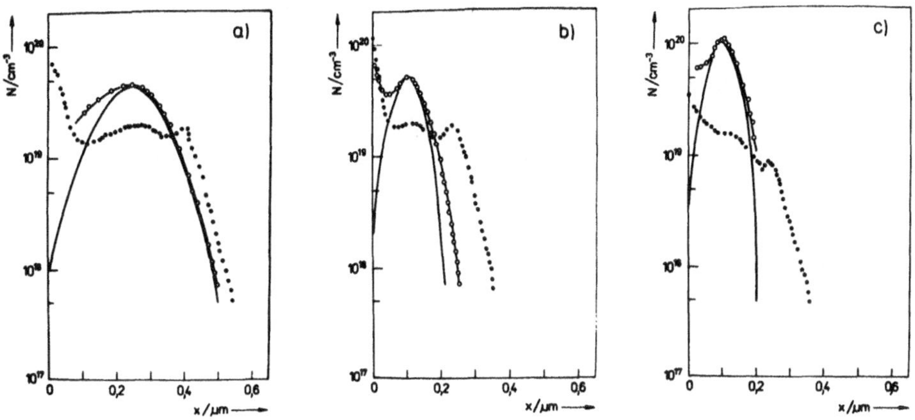

Fig. 4. SIMS depth profiles of gettered Cu in dependence on implanted ion species annealed for 1 h at 600 °C (—— calculated profiles of implanted ion species, ooo measured depth profiles of implanted ion species, ●●● gettered copper depth profiles). a) ^{40}Ar implanted sample (E = 300 keV, D = 1.10^{15} cm^{-2}); b) ^{75}As implanted sample (E = 250 keV), D = 1.10^{15} cm^{-2}); c) ^{83}Kr implanted sample (E = 228 keV, D = 1.10^{15} cm^{-2})

Figure 4 demonstrates the dependence of copper gettering on implanted ion species. In all cases two maxima appear in the copper depth profiles. From this experiments it can be concluded that

(i) the middle copper concentration per unit area decreases in the range of ($R_p \pm \Delta R_p$) with the mass number of implanted atoms;

(ii) the second maximum of gettered copper has a distance of R_p value of about 0.04 ΔR_p per mass number of the ΔR_p value implanted ion species.

4. Conclusions

The experimental investigations of the ion beam gettering of copper in GaP outlined in this paper indicate that

(i) copper atoms decorate the ion implanted damage not only of noble gas atoms but also other atoms and especially damage region caused by copper ions themselves;

(ii) in comparison with the gettering of process-induced midgap recombination centers in Si [9] and Cr in GaAs [3] the determined activation energy could characterize the binding of copper atoms in damage regions, whereas the migration coefficient D_m agrees with the diffusion coefficient for this temperature;

(iii) the differences of the copper gettering between $^{31}P^+$ and $^{69}Ga^+$ implanted regions could be caused by the dependence of depth of the Ga vacancies and/or by stoichiometric disturbances like in ion implanted GaAs [3], because the mass ratio of P and Ga amounts to about 0.45;

(4i) the dependence of gettered copper atoms on the mass number implanted ions can be produced by special radiation damage and the different depths of dislocation network.

Acknowledgements

The authors would like to thank the Werk für Fernsehelektronik Berlin for support of investigations and expecially their SIMS group for special measurements of copper depth profiles by means of ims 3f.

References

1. W.R. Hunter, L. Eprath, W.D. Grobman, G.M. Osburn, B.L. Crowder, A. Cramer, H.E. Luhn, IEEE Journ. Solid State Cir. SC-14, 275 (1979)
2. D. Lecrosnier, J. Paugam, G. Pelous, F. Richou, M. Salvi, Journ. Appl. Phys. 52, 5090 (1981)
3. T.J. Magee, J. Hung, V.R. Deline, C.A. Evans, Appl. Phys. Letters 37, 53 (1980);
T.J. Magee, H. Kawayoshi, R.D. Ormond, L.A. Christel, J.F. Gibbons, C.G. Hopkins, C.A. Evans, D.S. Day, Appl. Phys. Letters 39, 906 (1980)
4. B.W. Wssels, El. Letters 15, 748 (1979)
5. M. Griepentrog, H. Kerkow, H. Klose, U. Müller-Jahreis, Proc. Int. SIMS Conf. 3, 322 (1981), Budapest

6. H.G. Grimmeiss, B. Monemar, L. Samuelson, Solid State El. 21, 1505 (1978);
R.N. Bhargavy, P.M. Harnack, S.P. Herko, P.C. Mürau, R.J. Seymour, Journ. Lum. 12/13, 515 (1976)
7. A.F. Burenkov, F.F. Komarov, M.A. Kumakhov, M.M. Tomkin, Tablizy parametrow prostranstrennosvo raspredelenija ionnoimplantirovannych primessej. Minsk 1980 (in russian)
8. W. Frentrup, M. Griepentrog, Gemeins, Jahresbericht 1981, ZfK Rossendorf, Dresden (to be published 1982)
9. C.T. Sah and C.T. Wang, Journ. Appl. Phys. 46, 1767 (1975)

PART VII

PANEL DISCUSSION

PANEL DISCUSSION

Led by J.A. van Vechten and by J.I. Nishizawa
Written by J.A. van Vechten
IBM Research, Yorktown Heights, New York, USA

Introductory Remarks

I am again reminded of Dr. Szép's introductory remarks to this School and to the first question that was posed to me as I started my graduate studies at the University of Chicago. That first question was, "What knowledge is worth knowing?" I think that was a very good introduction to an education in semiconductor science, as were Dr. Szép's remarks a good introduction to this School. This afternoon, if things get dull, there are several controversies which we can discuss. These include the question whether or not the Si self-interstitial, Si_i, exists in any significant quantity and constitutes the dominant native defect during the high temperature processing of Si. As many of you may know, I am one of those who believes that this is not true and that vacancies, V_{Si}, are the only important native defects in Si. I furthermore believe that the "interstitial type" dislocation loops and stacking faults, which are commonly observed, are formed by reactions of vacancies. There are those among you who, for reasons I am sure you are willing to defend, believe the Si_i's become the dominant native defect at high temperatures. However, in the same spirit as Dr. Szép's introductory remarks, I would like you to consider the question, "Is it really worth deciding that question?"

Another controversial question which has been raised at this School whether or not Green's Function Theory is so much superior to all other methods of electronic structure calculations that only such calculations should be admitted to scientific discussion. In considering such questions, I suggest one consider whether the object is to make a quick estimate, to make a reliable estimate, or to achieve "maximal rigor". I believe "degree of rigor" is a hard quantity to measure; I also recall the basic theorem of mathematical logic which states, "If any inconsistency is allowed in any logical system, then any conclusion can be derived from any premise". (While presenting this theorem in class, Bertrund Russel was once challenged to prove he was the Pope if given that 2 = 1. He replied, "The Pope and I are two, 'this true, but if two be one, then the Pope and I be one.") Therefore, I suggest that one does not necessarily achieve a more reliable estimate with a technique that elects not to make simplifying assumptions or not to fit certain empirical constants, but chooses instead to consume larger quantities of computer logic and memory space for greater periods of time.

I will talk tomorrow on the mechanism of pulse laser annealing and you may know that I am convinced that it is not true that that involves only normal thermal heating, melting and rapid crystallization followed by a very rapid quenching. If any of you wish, that can be debated today also.

There is also the question whether or not the EL2 defect is the isolated antisite defect, or a complex involving an antisite defect, or something else. That may be viewed as part of the larger question whether or not defect complexes are really what we look at almost all the time or whether we look mainly at isolated point defects. It has been my opinion since 1975 that essentially all deep level defects are complexes and not isolated point defects. This is now becoming a majority view, but if it is not your view, that is something that can be debated also. It may debated even if there is a minority of one who wants to hold that position.

To add a little structure to this discussion, I thought we could make up a tableau which shows in one direction various material and technologies. On the other axis there are processes, for example, crystal growth dissolution, and etching (see Fig. 1). I suggest it would be useful to consider what would be the impact on a technology of an improvement in our understanding of one of the processes. Then one should also consider whether or not the particular tools that each of us knows how to use is apt to make an impact in that area. For example, there is at present no established theory of how it is possible to grow Si crystals dislocation free. The accepted theory for describing how most solids grow at low degrees of supercooling from a melt is the Frank-Burton-Cabrera Theory of the screw dislocation. But, as you probably know, it is standard to grow Si totally dislocation free--and that means zero dislocations. Obviously the Frank-Burton-Cabrera Theory is not at all applicable in this case.

Now, consider the following. For bi-polar devices, dislocations are a big problem, which has been pointed out by several speakers here. I might go to the people who have money and say, "Give me $\$10^7$ and I will tell you how it is possible to grow Si dislocation free." They would likely say, "Well, we already know how to grow Si dislocation free so go back and exist on the money you are getting now." On the other hand, we can consider that GaAs generally is not grown dislocation

MATERIAL TECHNOLOGY / PROCESS	Xtal Growth Dissolution Etching	Diffusion	Insulation Oxidation	Ohmic Contact	Schottky Barrier Formation	Gettering	Annealing Passivation	Degradation Lifetime	Theory, Prediction Enhancement
Si logic 4 µm		+	+		+				+
logic 1 µm		+	+		+				+
memory			+						+
photovolt	+		+				+	+	+
III-V Laser		+	+	+			+	+	+
LED			+						+
logic	+	+	+	+	+		+	+	+
memory	+	+	+						
photovolt	+	+	+				+	+	+
II-VI Laser	+	+		+			+		+
photovolt							+	+	+
Other									+
IV-IV									
Other									

Fig. 1.

free and is not available dislocation free. I can say, "Well, give me $\$10^7$ and will tell you how to grow GaAs dislocation free." Then we must consider whether it is important to accomplish that. Many people believe that the GaAs logic and other devices will be made in an FET MIS type structure for which dislocations are not a serious problem, or so they say when they are telling me why they are not giving me the $\$10^7$. They might argue that, while it is certainly possible to make bipolar devices in GaAs, that is not a particularly desirable enterprise. So they may give me $\$10^5$, not $\$10^7$, to tell them how to grow it dislocation free. I am sure that we can think of a similar discussion with the people who have money in each of our organizations that could occur over each of these processes. Hopefully we can have a lively discussion today about what it is we might seek to learn more about and what possible impact it could have on these technologies. Conversely, we may discuss what it is needed to improve the development of the technology in each of our organizations. Please note that the answer to that question may be different in one organization from another.

To this end, I will ask you to fill in the tableau for yourselves. At the end we will take a group vote to obtain relative ratings of each of these processes and materials for the group as a whole. (This is what is presented in Fig. 1.)

Finally, we should consider suggestions for our future work. There are three topics which I wish to suggest. One is the question of the coefficients of thermal expansion, which has been noted to be important for lattice matching to relieve misfit stress. The problem is that one can match at the growth temperature, room temperature, or at some temperature in between, but not over the complete range of temperatures. Of course, we are interested in chemical trends in these coefficients of thermal expansion. I mentioned earlier doping effects and the view that they are important to those coefficients. Second is the mechanism of pulse beam annealing, which I think is in a very early stage of development. Third is the mechnism of sputtering and etching.

Professor Nishizawa has suggested two more topics. One is the question of the concentrations of various point defects in III-V materials and their effect on devices. The other is the segregation of point defects to dislocations, stacking faults, and grain boundaries.

Dr. Maria Kaminska (Warsaw, Poland):
"Passivation of EL2 with Hydrogen"
I wish to report some data about the hydrogenation of crystalline GaAs and the effect this has on the EL2 level of GaAs. We (Professor Gatos, his group at MIT and myself) did hydrogenation of GaAs and measured the concentration of different electron traps before and afterwards. A hydrogen plasma was used and the sample was kept at 300 C for two hours. We found that the hydrogenation process decreases the EL2 concentration by more than a factor of five. It looks like hydrogen goes into one of the components of EL2 defect, saturates bonds and makes the EL2 level disappear. The other electron traps which are normally present in GaAs, that is the 0.54, 0.40, and 0.32 eV levels, were not influenced by the hydrogenation process.

Dr. I.C. Szép (Budapest, Hungary):
I have an explanation for this process. You may simply have driven off As from the surface, therefore, your equilibrium has been reshuffled such that if the EL2 defect is an As antisite defect, its concentration is reduced when the sample is depleted of As.

Kaminska: But the effect was measured to a depth of 1 µm and I do not think that you can deplete that much depth in two hours at 300 C. We had reference samples kept at the same temperature for the same time in vacuum and observed no change in EL2 concentration. It really looks like there is an influence of the hydrogen on the EL2 defect, but not the other electron traps.

Van Vechten:
In Si, vacancies are known to interact with hydrogen interstitials (which diffuse as donors), H_i^+. The H_i^+ precipitates into the V_{Si}^- saturating at least some of the dangling bonds there. In GaAs, the Ga vacancy, V_{Ga}, is thought to be negatively charged and the As vacancy, V_{As}, is thought to be positively charged. One would expect the H_i^+ to interact with the V_{Ga}^- to a much greater extent than with the V_{As}^+. If the EL2 defect involves, for example, an As antisite plus a Ga vacancy, probably something else too, then one would expect the H_i to interact with that complex and at least change its electronic structure. The fact that the H had such an effect on the EL2 level and not on other levels which were observed, I think would imply that the EL2 is the only one which involves the Ga vacancy. In general, GaAs is Ga rich so there are not that many Ga vacancies around to form complexes. Maybe that is why the EL2 is the dominant one which involves Ga vacancies.

Speaker Not Identified:
I have a concern about the practicallity of the H passivation of EL2 in microelectronic devices. H is very mobile, and may be more than normally mobile in the presence of very strong electric fields and currents that are present in microelectronic devices, where dimensions are also small. Therefore, I am concerned that the H might diffuse out. It might even be dangerous as a source of explosive gas during processing.

Kaminska:
I do not maintain that the passivation of EL2 will solve all problems in GaAs. That electron trap trends to make GaAs semi-insulating. We know that it is important to be able to make GaAs semi-insulating. Simply removing all the EL2 traps from the wafer might not be good. However, I do think this result may be important for identifying the nature of EL2.

Prof. Dr. K. Tempelhoff (Berlin, GDR):
"Self Interstitials in Si"
I wish to cite two experiments which support the hypothesis that it is the interstitial, Si_i, which is the dominant native defect. In the as-grown crystal during the crystallization, the equilibrium concentra-

tions, $[V_{Si}]$ and $[Si_i]$, were frozen in. We have these equilibrium concentrations also after the growth in the room temperature sample, except for perhaps a small rim around the crystal about 2 µm from the surface. Experimentally we do not find vacancy clusters unless we quench rapidly from the growth temperature. This implies that during the crystallization the main defect disorder is due to Si_i.

At temperatures lower than the melting point, it may be that the ratio of $[Si_i]$ to $[V_{Si}]$ is less than at the melting point. However, as I mentioned in my talk, we have done quenching and annealing experiments. After quenching from 1000 C, vacancy clusters were found but after annealing such material for ten minutes at 800 C no more vacancy clusters were found. These are two observations which I think support the interstitial hypothesis.

Van Vechten:
We observe interstitial type dislocation loops and stacking faults but not isolated interstitials. Electron radiation experiments even at 2K which have led me to believe that the Si_i is quite unstable because no Si_i could be found although V_{Si} and remnants of Si_i, such as Al_i, were found. There is also the fact that the measured value of the transfer-energy required to displace a Si from its lattice site with electron irradiation experiment is about 12 eV. The estimated value of the enthalpy of formation of the species which diffuses at high temperature is much less than 4 eV; I believe it is 2.4 eV. I find it difficult to believe that both V_{Si} and Si_i can coexist. At least at low temperatures the Si_i's seem to be extremely unstable. They manage to get out of the crystal even at 2 degrees. I find it difficult to understand how the Si_i could become stable as one raises the temperature.

Tempelhoff:
I do not deny that many processes in Si do involve V_{Si}. I agree there may be some effects that can only be interpreted with vacancy diffusion. But the majority of the disorder defects after the melting point are in our experimental point of view the Si_i's. I have said no more. (It may be that is some experiments there is no equilibrium in the $[V_{Si}]$ or $[Si_i]$.) This may be the case for the laser quenching experiment reported by A. Chantre in Amsterdam. We find that the equilibrium $[V_{Si}]$ is an order of magnitude less than that of $[Si_i]$ at the melting point.

Van Vechten:
Is it not true, though, that you do not see any isolated Si_i's? Instead, you see dislocation loops or stacking faults, which you believe were

created from the precipitation of the Si_i's.

Tempelhoff:

We do not observe isolated Si_i. Instead, we take a value $[Si_i] = 10^{16} cm^{-3}$ at the melting point from the diffusion literature. From our analysis of the vacancy clusters, which we do see in our TEM, we estimate $[V_{Si}]$ to be an order of magnitude less than this, about 10^{15} cm^{-3}.

Van Vechten:

People who do EPR or ODMR or other techniques, which I think unambigously indentify vacancies, have failed to find isolated Si_i. However, they do find Al_i, B_i, Ga_i and other interstitials that result from electron radiation. They have also seen certain complexes, which are sometimes called an interstitialcy, e.g., the C-Si interstitialcy and other defects that Prof. Ammerlaan has described, but they have not seen the isolated Si_i, which is hypothesized as the dominant defect for diffusion at high temperature.

Tempelhoff:

That is right. We have not seen the isolated Si_i. But we have made our investigations in very pure, vacuum-floating-zone Si, which has very low impurity levels. The levels of C and of O were lower than 10^{14} cm^{-3}. It was not similar to the material which I have shown in my talk where I described the precipitation of C and O on the stacking faults, dislocations and so on.

Van Vechten:

Would you agree with me that this is the question which is worth solving?

Tempelhoff: Yes.

Van Vechten:

Why would you agree with me on that?

(No reply.)

It might be more useful at this point to discuss the question whether or not it is worthwhile to settle that question. Experiments can be done which I believe would unambigously settle that question. I think in particular about the grazing incidents electron defraction experiments such as has been done by Spence's Group in Arizona and in Japan. By using grazing incidents electron defraction in a high resolution high vacuum TEM system,one can observe that motion of steps at the surface of a sample which must recede or advance as the crystal expands or contracts as the temperature changes. In this way we can distinguish an increase in $[V_{Si}]$, which causes the steps to advance, from an increase in $[Si_i]$, which causes the steps to recede. This experiment is more sensitive than a so-called Simmons-Balluffi experiment (measuring expansion or contraction of the crystal and simultaneously the lattice constant by X-ray diffraction), which is the standard way of determining whether vacancies or insterstitials are introduced into a metal sample. This experiment clearly can be done but it is not easy and will cost time and money. Therefore, we must question whether we really need to know if vacancies or interstitials are the dominant native defect at high temperature in Si. Some people say we do not need to know the origin of Si better than we already do to make 4 μm logic or memory devices. Is it adequately parameterized by models that involve either only vacancies or only interstitials. Now the question arises do we need to understand the diffusion any better to make 1 μm devices? In my opinion, the answer to that is yes. Others may have a different opinion.

Szép:

"Stress in the Crystal Growth of Thin Films, as Distinct from Bulk Crystals"

I would like to go to the question of crystal growth. I think that the problems may be very different in the case of bulk crystal growth and in the growth of thin films. Technologists are now much interested in techniques such as molecular beam epitaxy, MBE, and metal-organic chemical vapor deposition, MOCVD. I think there must be a difference in the local fields acting on the atoms that the building up thin films with these techniques and in those acting at the growth interface of one of the techniques used to grow bulk crystals. When the behaviour and structure of defects are investigated, these differences should be kept in mind.

Van Vechten:

Before we try to answer that would you indicate in which technologies would you be willing to spend money to get a better answer than the one you have now.

Szép:

I am sure not for Si technology. That is such an elaborate and highly developed field that we would be reluctant to switch over to anything new. But in the case of the III-V's, I am sure that a lot more money should be spent. As we have seen, we have a lot of problems with defects.

Van Vechten:

Would you think that most important for lasers, or LED's, or memory, or logic, or photovoltaics?

Szép:

I would think that in lasers the precipitates are a big problem. And in the photovoltaics also.

Van Vechten:
I think your question was directed primarily to Professor Nishizawa.

Prof. J.I. Nishizawa (Sendai, Japan):
In my opinion, the cases of Si and of the III-V compounds are largely different. In the case of Si a great part of the field of point defect science may be of academic interest only. This is because usually the Si_i and V_{Si} have only a very slight effect on the electrical properties and only complexes are important for diffusion and propagation of dislocation, and so on. They have no direct effect. Therefore, engineers of Si devices have little interest in the question whether the high temperature native defect is Si_i of V_{Si}. However, I personally believe in the importance of these point defects. But in the case of the III-V compounds, it is obvious that the importance of point defects is extremely large. Therefore, we are simply forced to study that problem directly.

In the case of Si, we have had a very serious discussion over ten years about which is better - whether to have perfect crystals or to have crystals with some density of dislocations. On one side, they say that if we have some density of dislocations, the material becomes very good. The other people say the dislocation are very bad. About five years ago, a conclusion was reached. Now virtually every one in Japan believes that it is important to have perfect (dislocation free) crystals, particularly for large-scale integration, LSI and VLSI. If we increase the concentratin of oxygen, [O], in the crystal, then we increase the total number of segregated particles in the single crystals. This is connected with an increase in the leakage current in diodes and we also have a lowering of the breakdown voltage. Therefore, as we approach 1 μm integrated circuits, if we have dislocation loops larger than 1 μm, these will introduce inhomogenities in the VLSI circuits. In the VLSI circuit every element should be completely the same. If we have some kind of segregation, we will have deviations in the material which will damage the performance of the integrated circuit. Therefore, by all means, in the case of the Si VLSI, it is necessary to reduce the concentration of dislocations and also of segregated precipitates. This is very clear.

"Gettering of Point Defects in Si"
On the other hand, there is the question of gettering action. That is an interesting problem that is now being much discussed in Japan. Some Japanese companies use the technology of sand blasting the backside of the wafers. (Others use ion implantation or mechanical scratching.)

They found that they can obtain a gettering in this way. The result is that the quality of devices made on the front side is much improved. They say that the gettering action is produced by dislocations induced in the backside of the wafer beneath where the devices are formed. Therefore, it may be necessary to have some density of dislocations in the substrate beneath the active layer in order to maintain a high level of quality on the front surface necessary to give good quality to the front surface. On the other hand, these dislocations may easily propagate toward the front surface where they would damage the quality. Some people suggest to introduce oxygen into the crystals. This will precipitate into small particles of SiO_2. These particles can be used for precipitate hardening. That is, they will prevent the propagation of dislocations. Nowadays, the Japanese believe some amount of oxygen is necessary to keep dislocations from moving. Therefore, Japanese companies have a tendency to introduce a controlled and limited amount of oxygen into the Si crystals. I think this should be limited to the substrate. We should have dislocations in the back of the substrate to provide gettering action. Deep level impurities may be gettered at these dislocations in the back of the substrate and we must avoid stress which would cause these dislocation to propagate to the top. Finally, we can say if we dope with some oxygen we can form precipitates which will prevent the dislocations from coming to the top. Therefore on top we may grow very pure perfect crystals. And the quality of devices may be very good.

I think this is one of the ideal distributions of the concentrations of impurities. Now in Japan, many companies like to use Si crystals prepared by the Cz method.

Sony uses a technology which is just the Cz method with a particular method to control melt convection during growth. The crucible used in Cz is made of fused Silica, there is always some C. The crucible dissolves in the molten Si introducing oxygen which is distributed in a convection pattern. It is not uniformly distributed.

Mr. Hoshi invented a method in which the application of a very strong magnetic field is used to control the convective flow in the melt. As I understand it, some of the Si atoms in the melt are ionized so that there is an electrical current produced by the convective flow which will couple with the imposed magnetic field. Therefore, it is possible to control the flow with the field. Of course, one can also apply an electric field between the crystal seed and the melt to produce a DC current. This can also be used to help control the convective flow. Using these methods, very uniform incorporation of the O and the C into the Cz crystal has been achieved.

However, I believe that in the future the floating-zone, FZ, method of growing Si will become widely used to produce material for VLSI. If we want to reduce the concentration of O then this methods seems much better. If we control the vapor pressure of the ambient in an FZ system, then we can control the concentration of O very easily. In the future I think there is a large probability that FZ material will be used for VLSI.

Now I would like to talk about the present situation in Japan regarding crystal perfection. We have worked hard and solved bad problems to produce some nearly perfect single crystals. But to make many devices, we need to have heterostructures. Thus, we have the problem of growing one perfect crystal on top of another perfect crystal of a different type and we must minimize the defects at the interface where they join. The problem of defect complexes at the interfaces is another difficult one. How we can have contacts between two materials. That is not yet solved.

Fortunately, we can learn something about this problem from the study of crystal growth of a single material, the study of homoepitaxy, which has been studied for a long time and is better understood. The heteroepitaxy of layers of, for example, different III-V materials that have the same crystal structure but different lattice constants is not that much different from the homoepitaxy of a new layer growing on top of an old layer of the same material. Therefore, we are learning about this type of heteroepitaxy from the traditional science of crystal growth. Then finally there is the problem of the heteroepitaxy of two materials with completely different structure. This problem is very far from being solved. However, such heterostructures are widely used in present day devices. Examples are those in Schottky diodes and passivation films. It is clear we have very large problems to solve in the future. Anyhow, we cannot now get completely perfect crystals with insulating layers at the top surface. This is a very difficult problem to be solved.

Van Vechten:
You have made the remark that you are using in Japan technologies that you do not perfectly understand. The question is how much more is it worth investigating these technologies, for example, the gettering technology that you described.

Nishizawa:
I have no general answer. Now in Japan there is not much interest in the inhomogeneous distribution of O, but I think in the future the control of the distribution of impurities will be very important.

Van Vechten:
Do you believe that the magnetic field in the Cz growth provided a very homogeneous distribution of O?
Nishizawa:
Yes. This is one solution.
Van Vechten:
That was in Si, is it used also in GaAs?
Nishizawa:
No. I think that in the case of the GaAs, no way has yet been found to obtain a nice distribution of impurities.
Van Vechten:
I think that many people find that the distribution of dopants in GaAs and the electrical mobility and other properties of GaAs commercial wafers vary considerably across the wafer in a pattern that shows the striation typical of those Cz wafers of Si for which such techniques have not been used.
Nishizawa:
I think the case of the III-V's is different from the case of Si so I do not want to discuss the III-V's in this context. I think that the future technology of III-V's should be photochemistry.

Tempelhoff:
"Remarks on Gettering"
I want to make a remark about the intrinsic gettering. In the first years of Si technology, all crystal growers would make Cz with a low concentration of O. But in the latter years it has been shown that the intrinsic gettering (the precipitation of O as SiO_2 in the wafers and the gettering of the various impurities at these precipitates) is an important advance in device technology.

You have heard from Professor Nishizawa's summarization of four techniques for gettering which are used all over the world, not just in Japan, but also in Germany and in the USA. The first and simplest is intrinsic gettering. (Gettering by sand blasting, ion implantation, scratching, etc. on the back side of the wafer is called "extrinsic gettering" as distinct from gettering by action of impurities originally grown into the wafer material.) Impurities from the first row of the Periodic Table (C, N, and O) may be precipitated into small particles, particularly SiO_2. I have shown TEM micrographs of two kinds of crystals, one with grown-in defects (precipitates) and one without grown-in defects especially in Cz material, but also in FZ Si. After high temperature annealing, these show that only in the middle of the wafer are precipitates present. At the surface of the wafer is a denuded zone without defects where the O diffused out. I have also shown micrographs showing the effect of double annealing. The first anneal nucleated very small precipitation centers. The O and other impurities can precipitate in the next higher temperature annealing step on these centers. You can invert this procedure and have the high temperature anneal first and after this the low temperature annealing step. This is a procedure you can use with wafers which have a large O concentration in the Si device technology.

Van Vechten:

Would you agree that what you have described also serves the purpose of precipitate hardening. That by causing the O to precipitate you provides sites in which dislocations propagating from the perimeter of the chip once it has been diced are stopped and do not continue into the interior of the chip where you wish to make devices.

Tempelhoff:

Yes, but I wish to repeat that the initial wafers are free of dislocation and they have only small roots after the dislocation.

Dr. Regina Mach (Berlin, GDR):

"Utility of II-VI Materials"

I think I am the only one in this room working on II-VI compounds. From my point of view the II-VI's are the most suitable materials for a whole lot of purposes. They have the largest range of band gaps. You can tune from the infrared to the ultraviolet if you go from CdTe to ZnS. You can have great luminescence efficiency and you can make photovoltaic and solar cells, and you can make large area electroluminescence displays. They have one great advantage in comparison to the III-V, that the surface is very stable. You can make very good and stable Schottky diodes, you can tune the barrier heights with the variation of the metal. Because the surface is very stable, native defects do not pin the Fermi level at the surface or play any other role to control the Schottky barrier height. For this reason, we can obtain a very wide range of Schottky barrier heights and also make good ohmic contacts. This stability of the surfaces of II-VI compounds also makes the problems of passivation and insulation much less serious than with other semiconductors.

The only remaining rather fundamental problem with II-VI materials is that of forming p-n junctions. That is one of the problems to be addressed in our School. I would like to ask Dr. Van Vechten and the other participants why it is difficult to make p-n junctions in wide band gap II-VI materials? It is known that all these compounds have very shallow acceptor levels. Yet it has not been possible to make many of them low resistivity p-type. I think this is really the only serious problem preventing these materials from a wide-spread use in semiconductor laser applications. A lot of other applications do not necessarily demand p-n junctions. But it is a fundamental question too whether there are possibilities to make, e.g., ZnS highly p-type.

Nishizawa:

I am another person working with II-VI compounds. Toward the end of this year we will publish a paper describing the production of blue LED's based on II-VI compounds, ZnS and ZnSe. I agree with Dr. Mach that II-VI compounds have not received the attention that they deserve. However, I do not believe they can compete with Si for VLSI because they have a lower carrier mobility than Si. That is a classic problem in semiconductor technology.

About 5 years ago my group worked on HgTe and published some papers. We found that the mobility of that material changed by a factor of 10^4 as we varied the vapor pressure in the crystal growth system. It is obvious that the mobility and other properties of II-VI's is greatly dependent upon the initial concentrations of point defects that are grown into these materials.

We have found similar variations in the quality of other II-VI compounds with the vapor pressure during crystal growth. We have also succeeded in producing p-n junction by such methods.

I would also like to point out that II-VI compounds are very cheap. Therefore, I think they will be widely used for purposes other than VLSI.

However, I do think that there are unsolved problems regarding the formation of Schottky barriers and passivation films on II-VI's. As in other materials, these involve the formation of complexes of point defects near the surface or interface. As devices become smaller, the distance between metal contacts and the active region of the device become smaller. Thus the diffusion of metal from the contact into the semiconductor becomes an increasingly important problem. This is now a great problem in Si technology; I my opinion, when you get to the same dimensions with II-VI's, you will encounter the same problems with passivation.

Mach:

I must say that I have a lot of experience with Schottky barriers on II-VI's and I find them really very much better than III-V compounds. I find no pinning of the Fermi level with native defects at all. I can tune the barrier heights by variation of the metal only. I can explain all the characteristics that we measure with the very simple contact theory.

Van Vechten:

To answer Dr. Mach's question regarding the difficulty in forming p-n junctions, I note that I have published several papers and a handbook chapter on the subject of selfcompensation by native point defect, particularly vacancies and antisite defects. The basic problem is that to produce a low resistivity p-n junction, we must move the Fermi level from near the valence band edge on one side (p) to near the conduction band edge on the other side (n). As the band gap gets larger, this energy across the Fermi level must be swung gets correspondingly larger. The problem with doing this simply by doping under near equilibrium conditions is that the material can itself generate native defects, such as vacancies. These have ionization levels, can be ionized, and will tend to compensate the effect of the extrinsic ionization levels of the dopants. For example, for the problem of making ZnS p-type, the material may generate S vacancies, V_S. These are donors. Once created at the cost of some energy, which my papers tell you how to estimate and which can also be inferred from experiment, the V_S can be ionized at the cost of some more energy, the ionization energy which I have estimated from experiments and which can be calculated in principle with methods such as the Green's function methods of electronic structure calculation. If the sum of these two energies is less than the energy regined by allowing the electron, that was promoted to the conduction band when the V_S was ionized, to fall to the present Fermi level, which you may have driven down towards the valence band edge by introducing acceptors, then this selfcompensation reaction will be energetically favorable and will tend to occur. By near equilibrium means, the Fermi level cannot be driven any deeper than the sum of the energy of formation plus ionization of that native donor defect for which this sum is least.

Of course, many industrial processes are not near equilibrium. Laser annealing is a prime example of one that is not. But there are other methods of crystal growth which are also far from equilibrium. One problem with such methods, however, is that of degradation. If the material is far from equilibrium, then, during subsequent processing and operation of the device, it may be expected to revert towards the equilibrium condition.

Mach:

Low temperature crystal growth methods such as metal-organic chemical vapor deposition, MOCVD, can be useful when working against these equilibrium restrictions.

Van Vachten:
Indeed, it is generally easier to stay away from equilibrium when working at low temperatures.
Nishizawa:
I note that often during crystal growth and generally during subsequent processing, compounds are heated without protection against the chance that the material may generate defects by evaporation. In both III-V and II-VI compounds, the group V or group VI element generally has a higher vapor pressure. This tends to make the material non-stoichiometric evaporation. The group of Prof. Gatos, as described by Dr. Kaminska, has obtained results similar to ours. Companies that provide 30 % of the world's supply of GaAs use our technology to maintain stoichiometry. It has an important effect on the quality.
Van Vechten:
An exception to that rule is CdS, which dues evaporate congruently. That is why it can be used so cheaply to make solar cells. It can simply be spray deposited onto sheets of Cu that have been reacted to give them a surface Cu_2S film to make the heterostructure p-n junction solar cell.

Kaminska:
"Role of O in GaAs"
I would like to note the difference between the behaviour of O in Si and in GaAs. Whereas, in Si O produces SiO_2 precipitates and passivating layers and so forth, O does not react strongly in GaAs. The main role of O in GaAs is to react with any Si that may be present. During GaAs crystal growth, this reaction tends to keep the Si out of the GaAs crystal. This means that the initial crystal is less doped (with Si from the SiO_2 crucible that is generally used), but the intrinsic gettering with SiO_2 precipitates, that was described for Si, cannot work for GaAs.

We think it is good to keep Si out of GaAs. We found its presence increased the concentration of dislocations and of electron traps, such as EL2. The quality of our material was improved when we added O to the melt in the form of Ga_2O_3. In addition to reducing the Si concentration, this greatly reduced the dislocation density.

Speaker Not Identified:
"Green's Function Method"
I do not like the way in which the question about the Green's function method was posed. The "Green's function theory" is really not a theory;

it is rather a technique for solving the problem of the electronic structure of a deep level once the potential has been given. The real problem is to obtain the potential. The Green's function method is a very powerful tool, but the real question is how to determine the potential. Especially for the deep levels, I think this problem is unsolved.

Most of the theories which are called "self-consistent" use some sort of potential which is based on the local approximation, which is certainly not valid at these levels, where the energy differences in question are extremely small compared with the differences in atomic levels. I think that these methods that use the local approximation cannot describe the potential accurately enough that either the Green's function or any other method of solving for the electronic structure can give a result sufficiently accurate (0.1 eV) to be useful. Would Dr. Van Vechten comment on that.

Van Vechten:

I must say that I have not been overwhelmed by the local density approximation results either - particularly when I note that it predicts the band gap of Si to be 0.6 eV. But it does seem to get the lattice constant and the bulk modulus fairly accurately.

Would any other theorist or other person like to remark on that point?

(No.)

Van Vechten:

"Effect of Stress Fields"

We might return to another of the problems posed by Dr. Szép, that of the effect of the large stress fields, that generally are present in thin films, on point defect complexes and on line defects.

Of course, there is literature on this subject already. In particular, there is the work of John Matthews showing how misfit stress can be used to expel dislocations from a growing film by causing them to bend over. When they bend over they grow out of the film and are not continued in the material grown subsequently. This is one method by which dislocation free layers can be grown.

Those of us that are interested in the problems of the energy of formation of point defect complexes ought to consider how those energies are affected by stress fields. We should also consider how the orientation of the complex with respcet to the junction could affect the operation of a device. I might refer you to my 1980 Kyoto International Physics of Semiconductors Conference paper on the effect of cluster orientation on the DLTS signal.

Note that many elegant stress alignment experiments have been reported. Most defect complexes have a dipole moment so it will make a difference how they are oriented with respect to the junction. The performance of a finished device will depend on that orientation, in addition to the concentration of defects. Moreover, there are things that we can do to modify the generally large stress fields at the various junctions. We might use graded junctions of buffer layers for example.

Mach:
"Laser Annealing Dislocations out of Surface Films"
Regarding this question of the dislocation in surface films and the subject of laser annealing. Can laser annealing be used to reduce the density of dislocations in thin film layers? Has the following experiment been performed? One can take a sample and stress it uniformly from top to bottom so that dislocation run all the way through. Then the top surface can be laser annealed. Only a thin layer at the top would be affected. If there is normal melting and recrystallization, then I would expect the dislocation would continue from the substrate to the top surface. But if there is a weakening of the bonding strength due to the high electronic excitation, then I would expect the dislocations to be annealed out of the surface layer.

Yes, indeed you are right. There was a nice experiment done by Porter, Parker, Richardson and Swenson (to which I refer in my paper at this School), which did exactly what you suggest with a Si wafer. By bending a initially perfect wafer, they produced a large density of dislocations running entirely through the wafer. With pulsed laser annealing, dislocation were expelled from a zone on the top surface of the sample of order 10 µm deep. This is much deeper than the depth to which anybody suggests the material could have been converted to any kind of liquid. As you very astutely noted, if the material had been simply melted, it would have, at best, regrown epitaxially so that the dislocations in the substrate would have been continued in the regrown layer. Clearly the volume that is affected by the laser is much larger than that which could have been converted to any sort of liquid phase.

Therefore, one must conclude that, as you have suggested, the effect of softening the lattice with electronic excitation is to allow the dislocations, which are non-equilibrium defects, to be expelled from the crystal. The threshold for that expulsion of dislocations is much lower than the threshold to convert the surface layer to a cool, electronic liquid. The optically induced dislocation glide threshold observed in GaAlAs by Monemar et al., was about $2x10^5$ W/cm^2, whereas the powers used for pulsed laser annealing 10^8 or more W/cm^2. That is why the dislocations are expelled from a depth much greater than that in which there is evidence of a liquid phase of any sort, such as redistribution of dopants.

Prof. Dr. H. Flietner (Berlin, GDR):
"Dislocations in Surface Films"
I think that the properties of dislocations at interfaces are very much influenced by surface films. Strain is generally inhomogeneous throughout the surface. This makes the problem of predicting behaviour or analysing data more difficult than Dr. Van Vechten implied in his earlier remarks. For the example of epitaxial layers which we have been discussing in this School, point defect complexes associated with the interface may strongly affect the local strain field and, thereby, affect both threading and misfit dislocations. I think that at the present it is not possible to make any quantitative prediction for dislocations at interfaces.
Van Vechten:
Do you think this problem is worth working on or is it too complicated to give one any hope that progress could be made?
Flietner:
What means could you suggest to obtain information about the inhomogeneity of the strain in the plain of the interface?
Van Vechten:
Optical birefringence microscopy. That is an optical microscope fitted with crossed polarizers so that light is passed only where a strain field has caused the plain of polarization to rotate. The resolution of this technique is that of the optical microscope but it is very easy to see individual dislocations provided they are not too closely packed. Note that the strain field of a dislocation, or precipitate, is generally much larger that the core of the dislocation or precipitate, - the image of a dislocation will appear to be 100 µm across.

Nishizawa:
"Heterostructures using Coincidence Plains"
I would like to return to the problem of heterostructures between materials with different crystal structures. One of many possible approaches is to find a coincidence between some subset of lattice sites in one lattice might match up with every fourth lattice site in the other. In such a case the interface between the two materials might

contain few defects and dislocation.

Van Vechten:

A leading example is the growth of Si on sapphire, Al_2O_3. If it were not for the coincidence plain, there would be a 15 % mismatch.

Nishizawa:

"Reduction of Dislocation Density by Addition of O"

We found the concentration of dislocations grown into III-V films was reduced when we introduced O into the vapor over the film. The introduction of O seemed to prevent the dislocations in the substrate from threading up into the growing layer. It may be that the O produces precipitates to which the dislocations from the substrate are attracted so that they join together there and are not continued in the over layer. I believe Prof. Quiesser and Prof. Gatos have similar results.

"Fine Precipitates of SiO_2 in Si"

Recently we in Japan have studied the precipitation of very fine grains of SiO_2 in Si. We find that if there is no N, then it is very difficult to precipitate small grains of SiO_2. We conclude that Si_3N_4 acts as a nucleation site for SiO_2. I believe the critical concentration of N required for good nucleation is 10^{11} cm^{-3}. When we introduce large concentrations of O without also introducing that much N, then we find no nucleation of SiO_2 precipitates.

I think this illustrates the complex interaction among point defect complexes and lattice defects. In our studies of the effect of variation of vapor pressures on the quality of crystals, we find that a change which produces only a small signal in optical absorption (indicating a small concentration of point defects) may have important effect of the crystalline defects. I think that many manhours of effort will be required to sort out these problems. In particular, the sudden reduction in the concentration of dislocations in GaAs when we introduce enough O vapor into the growth system is a difficult and important problem.

MECHANICAL STRESS INDUCED DEFECT CREATION IN GaP

G. Ferenczi, L. Dózsa and M. Somogyi

Research Institute for Technical Physics
1325 Budapest, Újpest 1, P.O.Box 76, Hungary

ABSTRACT

The concentration of a donor like deep level - T_7 - about 0.8 eV from the conduction band has been correlated to the amount of accumulated mechanical stress in GaP p-n structures. Under forward bias stressing the initial concentration of T_7 decreases. Depending on the local strain field pattern new levels appear. If compressive strain field is dominant within the p-n junction, the level created is 0.49 eV below the conduction band. If the strain field is absent two levels - 0.71 eV and 1.18 eV below the conduction band - are created. We tentatively identified the latter two levels as the two charge states of P_{Ga} - phosphorus antisite defect. Both defects act as non-radiative recombination centres, hence responsible for the performance degredation of GaP LED's.

1. Introduction

Continuing efforts are devoted to further improve the lifetime of semiconductor light sources. Phenomenologically the creation and motion of dislocations have been identified as a main degradation mechanism. (It has been verified that the immediate area around dislocations is non-radiative.)

TEM studies [1-3] revealed that beside the formation of <110> oriented dislocations, the strongly degraded areas always contain <100> oriented dislocation loops.

The <110> oriented dislocations propagate via glide mechanism, the <100> oriented dislocation loops grow by a non-conservative climb mechanism.

The use of external uniaxial stress especially with the simultaneous application of forward current proved to be the most straightforward mean to create <110> dislocations in GaP [4] and in GaAlAs [5] structures. The formation of <110> dislocations is always accompanied by the creation of <100> dislocations as well. More recently it has been established that internal stresses in device structures are also sufficient to create these dislocations [6,7].

These observations provide sufficient circumstantial evidences to link internal or external strain fields with the motion of dislocations, hence degradation. TEM studies, however, are not able to reveal the microscopic process behind these particular dislocation reactions. It was found that DLTS [8] is an excellent method to follow the dislocation reactions and identify the defects which are responsible for the increased non-radiative recombination rate.

3. Experimental

The present investigations were carried out on commercial GaP LED's produced by VPE and LPE techniques. The diodes were mounted on TO-18 headers and mechanical stress was applied perpendicular to the p layer via a copper needle of 100 μ radius. The applied stress never exceeded 10 kbar. Forward current was simultaneously flowing through the samples. DLTS experiments were carried out using a high sensitivity lock-in type spectrometer [9] in its Differential DLTS [10] operation mode.

3. Results

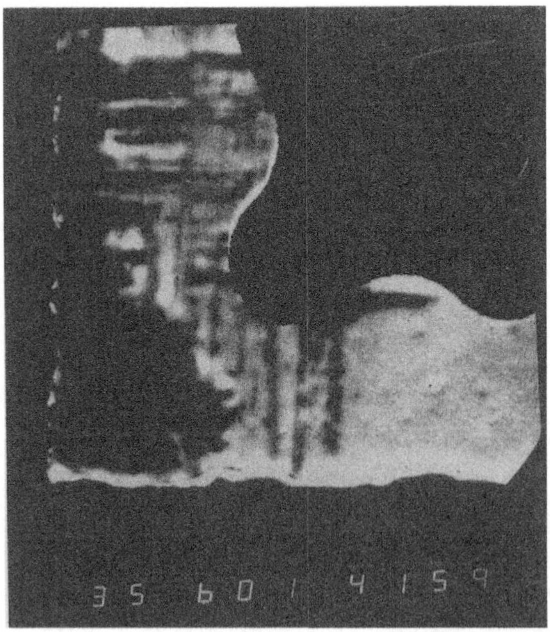

Fig. 1. EBIC image of a VPE grown GaP: N LED after mechanical stress of 10 kbar was applied perpendicular to the p layer and simultaneously 5 A/cm^2 forward current was flown through it for 3 hours. The <110> dislocation network was absent prior to stressing

The simulataneous application of mechanical stress and 5 A/cm^2 forward current produced a <110> dislocation network in the samples as it is illustrated on Fig. 1.

DLTS experiments proved that the creation of the dislocation network is linked to the formation of a donor like deep level, 0.8 eV below the conduction band.

The same level - T_7 - was observed previously [11] in samples where no intentional external mechanical stress was applied. Those samples, however, were differently manufactured. An octogonal opening on the Si_3N_4 cover layer was used as diffusion mask. The thermal expansion coefficient difference between the Si_3N_4 cover layer and GaP is sufficient to create a strain field responsible for T_7 formation.

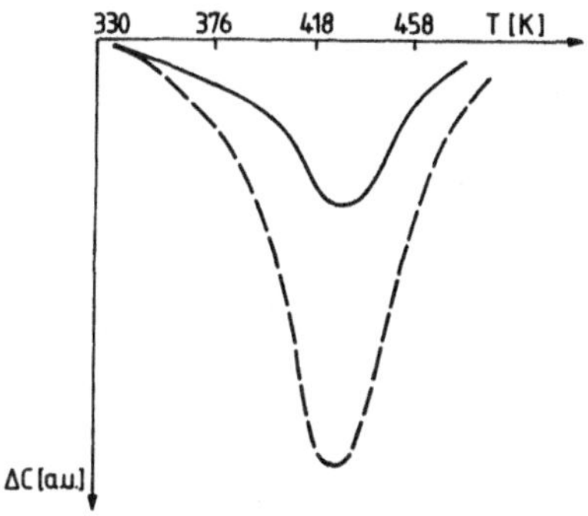

Fig. 2. DLTS signature of T_7 in the function of stressing time (LPE sample). Full line: stressed for 1h (10 kbar, 5 A/cm²); broken line: stressed for 3h (10 kbar, 5 A/cm²)

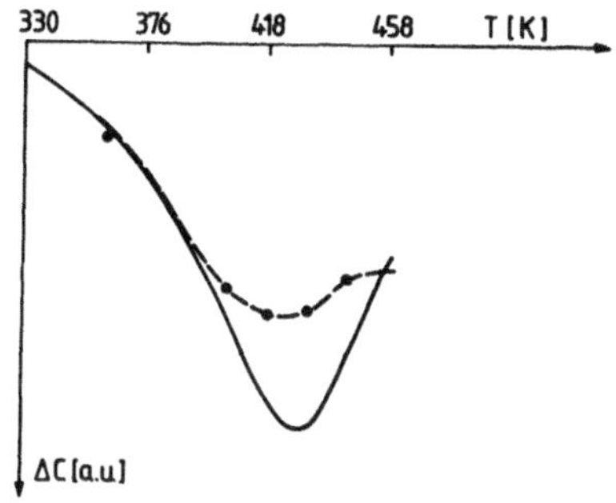

Fig. 3. The change of the DLTS signature of T_7 with ageing (50 A/cm², 50 hours). Full line: before ageing; broken line: after ageing

The concentration of T_7 is a function of stressing time: Fig. 2. After simultaneous mechanical and current stressing the samples were wire bonded and aged at room temperature at 50 A/cm² d.c. As a result of ageing the initial concentration of T_7 decreased but the DLTS curve form changed as well: Fig. 3.

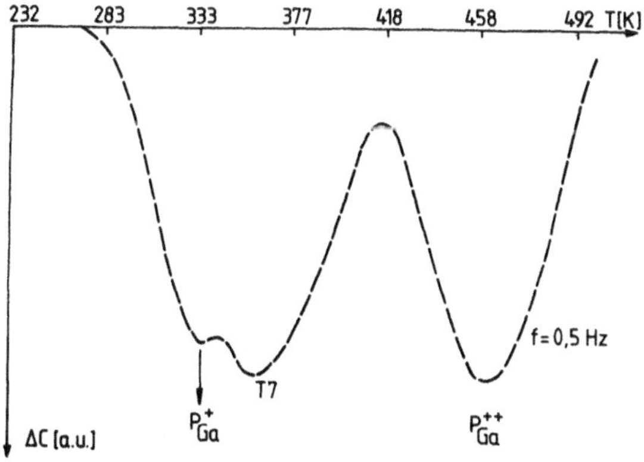

Fig. 4. The complete DLTS recording of the T_7 complex on an aged diode (previously mechanically stressed). Three peaks are clearly resolved

More careful analysis showed that as a result of forward current stressing

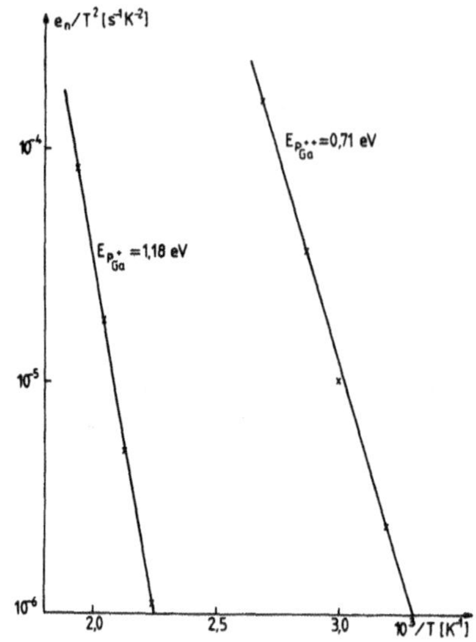

Fig. 5. The Arrhenius plots of the new peaks

two new peaks, appeared as illustrated on Fig. 4.

The activation energies of the new peaks were also determined: Fig. 5.

Due to difficulties in resolving the first peak the accuracy of the determination of the activation energy is less then usual

$$E_{P_{Ga}^{+}} = 0.71 \text{ eV} \pm 80 \text{ meV}$$

The temperature range for the Arrhenius plot of the second level is also limited which is due to the depths of the level. To pursue the measurement at even higher repetition frequencies would destroy the sample:

$$E_{P_{Ga}^{++}} = 1.18 \text{ eV} \pm 50 \text{ meV}$$

The determination of the capture cross sections was also attempted. The shallower level is a strong recombination centre with very high capture cross sections, hence only lower limits were determinable.

$$\sigma_n > 10^{-15} \text{ cm}^2$$

$$P_{Ga}^{+}$$

$$\sigma_p > 10^{-13} \text{ cm}^2$$

For the deeper level

$$\sigma_n = 10^{-17} \text{ cm}^2 \qquad P_{Ga}^{++}$$

The two donor type levels appear simultaneously, hence it is tempting to identify it with different charge states of the same defect. Unfortunately both levels are near midgap, and the shallower one is a strong recombination centre, further T_7 partially masks the shallower level. Our measurements suggest that both levels have the same introduction rate, equal concentrations were not unanimously establsihed. The introduction of the twin peaks were not observed on samples with

Si_3N_4 mask. In those samples a level, 0.49 eV below the conduction band, was introduced with ageing.

4. Discussion

The activation energies of the twin-peaks correspond within experimental error to the theoretically predicted activation energies of the singly and doubly ionised state of the phosphorus antisite defect P_{Ga} [12]. From photoquenching of the ESR signal of the paramagnetic state of P_{Ga} the second ionisation energy is (1.1 ± 0.1) eV [13]. ODMR experiments [14] establish the same threshold energy for the doubly ionised state. It has to be noted that the magnetic resonance experiments were carried out on semiinsulating or p-type materials, our experiments were done on n-type samples. There is no previous experimental data on the first ionisation energy of P_{Ga}.

The correspondence to the theoretically predicted activation energies of P_{Ga} may be circumstantial. The identification is more positively supported by the dislocation climb mechanism in a binary semiconductor.

Our experiments suggest that T_7 is a measure of the accumulated mechanical strain in the crystal. Whether T_7 is due to unsaturated dangling bonds or point defects around the dislocation core is hard to tell.

In any case the energy liberated at non-radiative recombination events will advance the non-conservative climb motion of the <100> dislocations. Along the climbing dislocation point defects - vacancies and interstitials - from both sub-lattice might be generated. As it was pointed out in [15] the following asymmetric steps are possible:

$$V_{Ga} + dc \rightarrow P_I \tag{1}$$

$$V_P + dc \rightarrow Ga_I \tag{2}$$

$$Ga_I - dc \rightarrow V_P \tag{3}$$

$$P_I - dc \rightarrow V_{Ga} \tag{4}$$

where dc means dislocation climb.

The actual concentration of point defects in the crystal will determine which of these processes are taking place.

The point defects created during dislocation climb are in thermodynamic non-equilibrium, hence the free energy of the crystal is increased. One possibility to lower the crystal energy is the reaction:

$$V_{Ga} + P_I \rightarrow P_{Ga} \quad (5)$$

Theoretically Ga_P might be created in the same way, this was, however, not seen experimentally.

Reaction (5) is favourable only if there is no chemical potential to balance deficiency in one of the point defects.

In case of a compressive strain field there is a large demand for vacancies. The Si_3N_4 mask creates just such a compressive strain field in the open part of the device during operation. Hence the creation of the V_P type T_6 [11] instead of the twin-peak.

Similar observation was made in [7]. In external stress free material the P_{Ga} creation is energetically favourable, this is why we believe that the twin-peak observed in the present experiments corresponds to P_{Ga}.

5. Conclusions

GaP LED's where mechanically stressed under forward bias. A trap, 0,8 eV below the conduction band, was created. Its concentration is proportional to the amount and time the stress was applied.

Removing the external mechanical stress but forward biasing the diodes a twin-peak appears. The activation energies correspond to the theoretically forcasted values for the first and second ionisation energies of P_{Ga}. The non-conservative dislocation climb model during ageing makes this identification plausible. As a consequence the performance degradation is not due directly to the formation of the dislocations but to the creation of P_{Ga} during dislocation climb. The first ionisation state P_{Ga}^{+} is an effective non-radiative recombination centre. The increase of P_{Ga}^{+} concentration fully accounts for the observed reduction of the quantum efficiency.

We believe this is the first case that both charge states of an antisite defect were experimentally determined.

Acknowledgement

The authors are grateful to Mr. A. Toth for supplying the EBIC results. This work was partially supported by the Werk für Fernsehenelektronik, Berlin (GDR).

References

1. P.M. Petroff, O.G. Lorimor and J.M. Ralston: J. Appl. Phys. 47, 1583 (1976)
2. P.W. Hutchinson, P.S. Dobson, B. Wakefield and S.O'Hara, Solid-State Electron. 21, 1413 (1978)
3. G.R. Woolhouse, B. Monemar and C.M. Serrano, Appl. Phys. Lett. 33, 94 (1978)
4. M. Iwamoto and A. Kasami, Appl. Phys. Lett. 28, 591 (1976)
5. T. Kamejima, K. Ishida and J. Matsui, Jap. J. Appl. Phys. 16, 233 (1977)
6. N. Shimano, Y. Kawai and M. Sakuta, J. Appl. Phys. 51, 1227 (1980)
7. M.J. Robertson and B. Wakefield, J. Appl. Phys. 52, 4462 (1981)
8. D.V. Lang, J. Appl. Phys. 45, 3014 (1974)
9. G. Ferenczi and J. Kiss, Acta Phys. Hung. 50, 289 (1981)
10. G. Ferenczi, P. Krispin and M. Somogyi, J. Appl. Phys. to be published (1983)
11. G. Ferenczi, IEEE Trans. on Electron. Dev. ED-28, 421 (1981)
12. M. Scheffler, S.T. Pantelides, N.O. Lipari and J. Bernholc, Phys. Rev. Lett. 47, 413 (1981)
13. U. Kaufmann, J. Schneider, R. Wörner, T.A. Kennedy and N.D. Wilsey, J. Phys. C. 14, L951 (1981)
14. K.P. O'Donell, K.M. Lee and G.D. Watkins, Solid State Commun. to be published
15. E.R. Weber, H. Ennen, U. Kaufmann, J. Windscheif, J. Schneider and T. Wosinski, J. Appl. Phys., to be published

INDEX OF CONTRIBUTORS

Lecture Notes in Physics

Vol. 136: The Role of Coherent Structures in Modelling Turbulence and Mixing. Proceedings 1980. Edited by J. Jimenez. XIII, 393 pages. 1981.

Vol. 137: From Collective States to Quarks in Nuclei. Edited by H. Arenhövel and A. M. Saruis. VII, 414 pages. 1981.

Vol. 138: The Many-Body Problem. Proceedings 1980. Edited by R. Guardiola and J. Ros. V, 374 pages. 1981.

Vol. 139: H. D. Doebner, Differential Geometric Methods in Mathematical Physics. Proceedings 1981. VII, 329 pages. 1981.

Vol. 140: P. Kramer, M. Saraceno, Geometry of the Time-Dependent Variational Principle in Quantum Mechanics. IV, 98 pages. 1981.

Vol. 141: Seventh International Conference on Numerical Methods in Fluid Dynamics. Proceedings. Edited by W. C. Reynolds and R. W. MacCormack. VIII, 485 pages. 1981.

Vol. 142: Recent Progress in Many-Body Theories. Proceedings. Edited by J. G. Zabolitzky, M. de Llano, M. Fortes and J. W. Clark. VIII, 479 pages. 1981.

Vol. 143: Present Status and Aims of Quantum Electrodynamics. Proceedings, 1980. Edited by G. Gräff, E. Klempt and G. Werth. VI, 302 pages. 1981.

Vol. 144: Topics in Nuclear Physics I. A Comprehensive Review of Recent Developments. Edited by T.T.S. Kuo and S.S.M. Wong. XX, 567 pages. 1981.

Vol. 145: Topics in Nuclear Physics II. A Comprehensive Review of Recent Developments. Proceedings 1980/81. Edited by T. T. S. Kuo and S. S. M. Wong. VIII, 571-1.082 pages. 1981.

Vol. 146: B. J. West, On the Simpler Aspects of Nonlinear Fluctuating. Deep Gravity Waves. VI, 341 pages. 1981.

Vol. 147: J. Messer, Temperature Dependent Thomas-Fermi Theory. IX, 131 pages. 1981.

Vol. 148: Advances in Fluid Mechanics. Proceedings, 1980. Edited by E. Krause. VII, 361 pages. 1981.

Vol. 149: Disordered Systems and Localization. Proceedings, 1981. Edited by C. Castellani, C. Castro, and L. Peliti. XII, 308 pages. 1981.

Vol. 150: N. Straumann, Allgemeine Relativitätstheorie und relativistische Astrophysik. VII, 418 Seiten. 1981.

Vol. 151: Integrable Quantum Field Theory. Proceedings, 1981. Edited by J. Hietarinta and C. Montonen. V, 251 pages. 1982.

Vol. 152: Physics of Narrow Gap Semiconductors. Proceedings, 1981. Edited by E. Gornik, H. Heinrich and L. Palmetshofer. XIII, 485 pages. 1982.

Vol. 153: Mathematical Problems in Theoretical Physics. Proceedings, 1981. Edited by R. Schrader, R. Seiler, and D.A. Uhlenbrock. XII, 429 pages. 1982.

Vol. 154: Macroscopic Properties of Disordered Media. Proceedings, 1981. Edited by R. Burridge, S. Childress, and G. Papanicolaou. VII, 307 pages. 1982.

Vol. 155: Quantum Optics. Proceedings, 1981. Edited by C.A. Engelbrecht. VIII, 329 pages. 1982.

Vol. 156: Resonances in Heavy Ion Reactions. Proceedings, 1981. Edited by K.A. Eberhard. XII, 448 pages. 1982.

Vol. 157: P. Niyogi, Integral Equation Method in Transonic Flow. XI, 189 pages. 1982.

Vol. 158: Dynamics of Nuclear Fission and Related Collective Phenomena. Proceedings, 1981. Edited by P. David, T. Mayer-Kuckuk, and A. van der Woude. X, 462 pages. 1982.

Vol. 159: E. Seiler, Gauge Theories as a Problem of Constructive Quantum Field Theory and Statistical Mechanics. V, 192 pages. 1982.

Vol. 160: Unified Theories of Elementary Particles. Critical Assessment and Prospects. Proceedings, 1981. Edited by P. Breitenlohner and H.P. Dürr. VI, 217 pages. 1982.

Vol. 161: Interacting Bosons in Nuclei. Proceedings, 1981. Edited by J.S. Dehesa, J.M.G. Gomez, and J. Ros. V, 209 pages. 1982.

Vol. 162: Relativistic Action at a Distance: Classical and Quantum Aspects. Proceedings, 1981. Edited by J. Llosa. X, 263 pages. 1982.

Vol. 163: J. S. Darrozes, C. Francois, Mécanique des Fluides Incompressibles. XIX, 459 pages. 1982.

Vol. 164: Stability of Thermodynamic Systems. Proceedings, 1981. Edited by J. Casas-Vázquez and G. Lebon. VII, 321 pages. 1982.

Vol. 165: N. Mukunda, H. van Dam, L.C. Biedenharn, Relativistic Models of Extended Hadrons Obeying a Mass-Spin Trajectory Constraint. Edited by A. Böhm and J.D. Dollard. VI, 163 pages. 1982.

Vol. 166: Computer Simulation of Solids. Edited by C.R.A. Catlow and W.C. Mackrodt. XII, 320 pages. 1982.

Vol. 167: G. Fieck, Symmetry of Polycentric Systems. VI, 137 pages, 1982.

Vol. 168: Heavy-Ion Collisions. Proceedings, 1982. Edited by G. Madurga and M. Lozano. VI, 429 pages. 1982.

Vol. 169: K. Sundermeyer, Constrained Dynamics. IV, 318 pages. 1982.

Vol. 170: Eighth International Conference on Numerical Methods in Fluid Dynamics. Proceedings, 1982. Edited by E. Krause. X, 569 pages. 1982.

Vol. 171: Time-Dependent Hartree-Fock and Beyond. Proceedings, 1982. Edited by K. Goeke and P.-G. Reinhard. VIII, 426 pages. 1982.

Vol. 172: Ionic Liquids, Molten Salts and Polyelectrolytes. Proceedings, 1982. Edited by K.-H. Bennemann, F. Brouers, and D. Quitmann. VII, 253 pages. 1982.

Vol. 173: Stochastic Processes in Quantum Theory and Statistical Physics. Proceedings, 1981. Edited by S. Albeverio, Ph. Combe, and M. Sirugue-Collin. VIII, 337 pages. 1982.

Vol. 174: A. Kadić, D.G.B. Edelen, A Gauge Theory of Dislocations and Disclinations. VII, 290 pages. 1983.

Vol. 175: Defect Complexes in Semiconductor Structures. Proceedings, 1982. Edited by J. Giber, F. Beleznay, J. C. Szép, and J. László. VI, 308 pages. 1983.

GPSR Compliance

The European Union's (EU) General Product Safety Regulation (GPSR) is a set of rules that requires consumer products to be safe and our obligations to ensure this.

If you have any concerns about our products, you can contact us on ProductSafety@springernature.com

In case Publisher is established outside the EU, the EU authorized representative is:

Springer Nature Customer Service Center GmbH
Europaplatz 3
69115 Heidelberg, Germany

Batch number: 10704763

Printed by Printforce, the Netherlands